Progress in Molecular and Subcellular Biology 3

Progress in Molecular and Subcellular Biology

3

By

A. S. Braverman · D. J. Brenner · B. P. Doctor · A. B. Edmundson
K. R. Ely · M. J. Fournier · F. E. Hahn · A. Kaji · C. A. Paoletti
G. Riou · M. Schiffer · M. K. Wood

Managing Editor

F. E. Hahn

With 58 Figures

Springer-Verlag Berlin · Heidelberg · New York 1973

ISBN 3-540-06227-0 Springer-Verlag Berlin · Heidelberg · New York
ISBN 0-387-06227-0 Springer-Verlag New York · Heidelberg · Berlin

Typesetting and printing: Carl Ritter & Co., Wiesbaden. Bookbinding: Konrad Triltsch, Würzburg.

Contents

List of Contributors

ALBERT S. BRAVERMAN, New York Medical College, Metropolitan Hospital Center, New York, New York 10029, USA

DON J. BRENNER, Walter Reed Army Institute of Research, Washington, D.C. 20012, USA

BHUPENDRA P. DOCTOR, Walter Reed Army Institute of Research, Washington, D.C. 20012, USA

ALLEN B. EDMUNDSON, Argonne National Laboratory, Argonne, Illinois 60439, USA

KATHRYN R. ELY, Argonne National Laboratory, Argonne, Illinois 60439, USA

MAURILLE J. FOURNIER, Department of Biochemistry, University of Massachusetts, Amherst, Massachusetts 01002, USA

FRED E. HAHN, Walter Reed Army Institute of Research, Washington, D.C. 20012, USA

AKIRA KAJI, Department of Microbiology, School of Medicine, University of Pennsylvania, Philadelphia, Pennsylvania 19104, USA

CLAUDE A. PAOLETTI, Institut Gustave Roussy, 94 Villejuif, France

GUY RIOU, Institut Gustave Roussy, 94 Villejuif, France

MARIANNE SCHIFFER, Argonne National Laboratory, Argonne, Illinois 60439, USA

MICAL K. WOOD, Argonne National Laboratory, Argonne, Illinois 60439, USA

Reverse Transcription and the Central Dogma

FRED E. HAHN

"Molecular biologists have a religion all of their own in which Nobel prize winner Francis Crick is the prophet and the DNA molecule is the icon. Molecular biologists have a 'trinity' of three kinds of molecules — DNA, RNA and the protein molecules — which correspond to each other on a unit-for-unit informational basis. They have a 'dogma' (and they call it a dogma) which says that 'information' — that is the molecular pattern — passes from DNA to RNA to protein but does not pass in the reverse direction."

POTTER (1964)

I. Introduction

The Central Dogma of molecular biology which postulates the unidirectional transmission of genetic specifications for protein biosynthesis was enunciated by CRICK (1958) who proposed explicitly that "once 'information' has passed into protein *it cannot get out again*. In more detail, the transfer of information from nucleic acid to nucleic acid, or from nucleic acid to protein may be possible, but transfer from protein to protein or from protein to nucleic acid is impossible. Information means here the *precise* determination of sequence either of bases in the nucleic acids or of amino acids in the protein."

At the time of that writing (1958), messenger RNA as a separate macromolecular category had been neither proposed nor discovered (indications of the formation of phage T2 messenger RNA obtained by VOLKIN and ASTRACHAN (1956) had gone largely unrecognized). The transcription of RNA from DNA, in general, was awaiting discovery and OCHOA (1958) still considered polycondensation of nucleoside diphosphates through reversal of the polynucleotide phosphorylase reaction to represent biosynthesis of RNA. Transfer RNAs, than called "soluble RNA", had not yet been shown to be the set of amino acid adaptors excogitated by CRICK (1957), and the cryptanalysis of the amino acid code was bogged down in abstract speculations on the nature of symbols comprising a putative nucleic acid alphabet and on formal reasons why an assumed alphabet of 4^3 nucleotide triplets might be intrinsically restricted to the unambiguous designation of precisely 20 different amino acids, i.e. of the standard set of constituents of proteins (CRICK, GRIFFITH and ORGEL, 1957). If one accepts one of Webster's Seventh Collegiate Dictionary's definitions of "dogma" as "a point of view or tenet put forth as authoritative without adequate grounds", the Central Dogma of 1958 certainly was a *dogma*.

However, when POTTER (1964) wrote his spirited remarks above on the Central Dogma of molecular biology, the cryptanalysis of the RNA code which determines amino acid sequence in protein biosynthesis, was nearly completed, and the two-step biochemical decipherment of structural genes of DNA through consecutive operations called "transcription" and "translation", was envisaged, at least in general outline. At that time, molecular biologists were, therefore, justified to expand the proposition of the Central Dogma to denote the unidirectional passage of "information" concerning sequential molecular pattern from DNA through RNA into protein.

II. Cryptology and the Central Dogma[1]

To molecular biologists interested in cryptology, it should have also been apparent at that time that a *plaintext*, the linear covalent amino acid sequence in protein, is *superenciphered* in the ciphertext of its determinant structural gene in chromosomal DNA. The first encipherment consists of a substitution transformation in which one set of symbols (the amino acids) is replaced by another set of symbols (the codons in messenger RNA). The superencipherment involves a second substitution transformation in which the RNA codons are replaced by their complementary triplets in the transcribable DNA strand. The biological *decipherment* requires, therefore, two separate procedures in reverse: (1) the decipherment of the second substitution: this is known as "transcription" and yields messenger RNA in *placode*, followed by (2) the decipherment of the first substitution, known as "translation" which yields the amino acid plaintext. CRICK (1970) calls this "information transfer from one polymer with a defined alphabet to another".

Since in substitution transformation the letters in the plaintext lose their *identities* but retain their *positions*, the postulation of the "sequence hypothesis" (CRICK, 1958), which assumed colinearity of amino acids in protein and of corresponding symbols in nucleic acids, was tantamount to postulating that the genetic ciphertext must be the result of a *substitution* transformation instead of a *transposition* in which the letters retain their *identities* but change their positions. One might consider the three-dimensional rearrangement of linear polypeptide chains into functional proteins which brings topographically distant amino acids into proximity to represent an encipherment by a transposition transformation; this process appears to be an inherent deterministic function of key amino acid sequences and does not require the operation of a separate *ad hoc* cryptographic machinery except in those instances in which existing covalent bonds are broken or new covalent bonds are formed in order to stabilize the eventual biologically active three-dimensional protein configuration.

Considering in cryptological terms the DNA ciphertext an encicode, there exists no *a priori* formal reason why free passage of information in both directions, that is decipherment *and* encipherment could not biologically occur. It is mechanistically apparent, however, that a transmission of biological information in both directions might require separate cryptographic machineries. This is obvious for the "translation" step for which no mutually deterministic relationship appears to exist between the symbols of the codon alphabet and those of the amino acid alphabet; it is not so obvious for the "transcription" step for which a deterministic relationship between DNA and RNA symbols *does* exist with base complementarity as the key and the only mechanistic requirement remains for polymerizing enzymes to link template-determined nucleoside triphosphates by repetitive condensations.

Out of the cryptological framework of reference, the 1964 version of the Central Dogma might have been restated to say that biological systems are only equipped (1)

[1] A glossary of cryptological terms is printed at the end of this article.

to *de*cipher DNA (by transcription) but not to *superen*cipher information (from RNA) as to the sequence of symbols in DNA, and (2) to *de*cipher the messenger RNA placode (by translation) but not to *en*cipher (from a plaintext amino acid sequence) the sequence of codons in messenger RNA. The recent discovery of reverse transcription shows that the first of these two tenets is not invariably valid.

III. Reverse Transcription: Experimental Evidence

In 1964 Lee-Huang and Cavalieri demonstrated the first instance of reverse transcription in an *in vitro* model system by showing that a DNA polymerase preparation from *E. coli* synthesized poly (dA + T) on a template of poly (U + rA); the authors discussed their results only in terms of enzymology but did not interpret them as to their possible biological significance. In the same year Temin [1964 (1)] hypothesized that the replication of the RNA of RNA-containing tumor viruses proceeds through a DNA intermediate. This would require the action of an enzyme capable of catalyzing a reversed transcription, i.e. the *biosynthesis of DNA on an RNA template*. Such enzymatic activity was discovered simultaneously by Baltimore (1970) and by Temin and Mizutani (1970) in Rauscher mouse leukemia and Rous sarcoma viruses. The enzymatic reaction was demonstrated by incubating suspensions of the purified virions with the four deoxyribonucleoside triphosphates, including tritiated thymidine triphosphate, and magnesium ions. In these experiments, using virus particles as a source of both the RNA template and the reverse transcriptase enzyme, tritium was incorporated into acid-insoluble, i.e. polymeric products which were susceptible to hydrolysis by deoxyribonuclease. The enzymatic reaction was precluded by pretreating the virus suspensions with ribonuclease, suggesting that the RNA of the virus particles was essential for the polymerization reaction.

Temin's hypothesis [1964 (1)] further predicted that a DNA, complementary to virus RNA, should appear in infected cells during the course of viral replication and should be demonstrable by molecular hybridization techniques. In fact, he presented some evidence in favor of this prediction [1964 (2)]. Spiegelman, Burny, Das, Keydar, Schlom, Travnicek and Watson [1970 (1)] proceeded to show not only the occurrence of the RNA-dependent polymerase reaction catalyzed *in vitro* by six different RNA containing tumor viruses but they also demonstrated that these viruses synthesized DNA-RNA hybrids using the single-stranded virus RNAs as templates; finally, hybridization experiments proved that the DNA strands were, indeed, complementary to the virus RNAs. The formation of hybrid DNA-RNA was soon confirmed and is species-specific for the homologous virus RNA (Rokutanda, Rokutanda, Green, Fujinaga, Ray and Gurgo, 1970; Duesberg and Canaani, 1970; Hatanaka, Huebner and Gilden, 1971).

These discoveries were rapidly extended by numerous additional examples of reverse transcriptase activities in tumor viruses (Hatanaka, Huebner and Gilden, 1970; Green, Rokutanda, Fujinaga, Ray, Rokutanda and Gurgo, 1970; Scolnick, Rands, Aaronson and Todaro, 1970). A total of 27 isolated preparations of RNA tumor viruses was shown to contain RNA-dependent DNA polymerase activity (Schlom, Harter, Burny and Spiegelman, 1971). The enzymatic activity is imbedded in the core of the virus particles (Gerwin, Todaro, Zeve, Scolnick and Aaronson, 1970), is unmasked by treatment of virus suspensions with non-ionic detergents such as Nonidet P-40 or with ether and is enhanced more strongly by Mn^{2+} than by Mg^{2+} (Green et al., 1970; Scolnick et al., 1970). The product DNAs are of relatively small molecular size, having sedimentation coefficients of 2-4 S (Hatanaka et al., 1970) or 7 S (Green et al., 1970).

The Mn^{2+} preference of the polymerases and their susceptibility to rifamycin derivatives (Gallo, Yang and Ting, 1970; Scolnick, Aaronson, Todaro and Parks, 1971; Gurgo, Ray, Thiry and Green, 1971) as well as to streptovaricins (Brockman and Carter, 1971) are reminiscent of properties of bacterial RNA-polymerase enzymes. It should be noted that Lee-Huang and Cavalieri (1964) considered their *E. coli* DNA polymerase which transcribed poly (dA + T) from poly (U + rA) to be a subunital hybrid of DNA and RNA polymerases. In contrast, mammalian DNA-dependent RNA polymerases such as that of liver nuclei (Wehrli, Nüesch, Knüsel and Staehelin, 1968) or of Ehrlich ascites cells (Mizuno, Yamazaki, Nitta and Umezawa, 1968) are not inhibited by rifamycins or streptovaricins.

Viral DNA polymerases exhibit a bewildering lack of template specificity. Originally, the enzyme was found to depend upon endogenous viral RNAs (Baltimore, 1970; Temin and Mizutani, 1970). The correct operational definition of this type of enzyme remains, therefore, that of a polymerase which synthesizes DNA on a single-stranded RNA template. In fact, Schlom, Spiegelman and Moore (1971) insist that this definition and experimental proof of the formation of a DNA-RNA hybrid be applied as stringent criteria in the evaluation of all instances of assumed reverse transcription in different life forms.

Following the original discoveries of Baltimore (1970) and of Temin and Mizutani (1970) it was found that native or denatured DNAs of different biological origins were also utilized as templates [Mizutani, Boettger and Temin, 1970; Spiegelman et al., 1970 (2); Říman and Beaudreau, 1970; McDonnell, Garapin, Levinson, Quintrell, Fanshier and Bishop, 1970; Fujinaga, Parsons, Beard, Beard and Green, 1970] as well as yeast RNA (Bosmann, 1971), or certain synthetic polynucleotides, foremost poly dA.dT, poly rA.dT, poly dA (Mizutani et al., 1970), poly dC.rG, poly rI.rC, poly dI.rC [Spiegelman et al., 1970 (3)], poly rA.dT (Scolnick et al., 1971) or poly rA.rU (Stone, Scolnick, Takemoto and Aaronson, 1971). RNA-dependent DNA polymerases show much greater activity with certain synthetic polynucleotides than they exhibit with homologous RNAs. For this reason, experiments with synthetic primers/templates are useful in the detection of such enzymes.

Duesberg, Helm and Canaani (1971) succeeded in solubilizing and purifying a DNA polymerase preparation from Rous sarcoma virus which utilized as templates native homologous viral RNA and denatured salmon DNA but had low activity with heat-dissociated homologous virus RNA or with the RNAs of influenza or tobacco mosaic viruses. Similar studies have been reported by McDonnell, Taylor, Levinson and Bishop (1971) who found that the purified enzyme did not function with poly rA.rU as a template. While the broad range of template utilization could suggest that more than one species of DNA polymerases might occur in the various biological sources studied, Duesberg et al. (1971) and McDonnell et al. (1971) observed homogeneity of enzyme activity in centrifugation analysis, which leads to the inference that activities stimulated by different nucleic acid templates may reside within biophysically homogeneous enzyme preparations. On the other hand, Mizu-

TANI, TEMIN, KODAMA and WELLS (1971) have reported that the virions of Rous sarcoma virus contain, in addition to RNA-dependent DNA polymerase, DNA ligase and exonuclease activities, i.e. "many of the enzymes usually implicated in DNA replication, recombination and repair" to give the virus "the complete machinery necessary to transfer its information from RNA to double-stranded DNA integrated in the host DNA".

IV. Reverse Transcription and Cancer

The discovery of reverse transcription exhibited *in vitro* by RNA-containing tumor viruses, caused excitement. Not only did this process offer itself as one explanation of the molecular mechanism of viral carcinogenesis but it gave rise to hopes that the RNA-dependent DNA polymerase reaction might become a tool in the diagnosis and even in the treatment of certain cancers, foremost leukemia. For example, it was thought possible to follow the sequence of remissions and relapses during the chemotherapy of leukemia by essaying the reverse transcription reaction. A further obvious idea was to design antimetabolites with specific action upon reverse transcriptase because this reaction was thought to be one long-sought biochemical difference between tumor cells and normal cells which might be exploited in terms of the design of selectively toxic antitumor drugs. GALLO et al. (1970) found, indeed, that an RNA-dependent DNA polymerase was present in lymphoblasts of leukemic patients but not of normal donors; this enzyme was inhibited by rather high concentrations of N-demethylrifampicin. Shortly thereafter, an Editorial (1970) in Nature reported that SPIEGELMAN and his associates had demonstrated the transcription of double-stranded DNA from single-stranded RNA templates not only by RNA tumor viruses but also in the leucocytes of more than forty leukemic patients and in the cells of two osteosarcomas and one chondrosarcoma; by contrast, cells of normal human blood or from patients with non-malignant blood disorders did not exhibit reverse transcriptase activity. Additionally, SCHLOM, SPIEGELMAN and MOORE (1971) discovered the presence of RNA-dependent DNA polymerase activity in particles of virus-resembling morphology isolated from human milk. These particles may be similar to type B mouse mammary tumor viruses, and their incidence in the milk of American women is statistically correlated with a familial history of breast cancer (FELLER, CHOPRA and BEPKO, 1967). SPIEGELMAN reported at an Annual Meeting of the American Society of Biological Chemists (June, 1971) that he and MOORE had confirmed and extended their work, using milk from Parsee women of Bombay; the Parsees have practiced intermarriage for 1200 years and have a high statistical incidence of breast cancer.

Two lines of findings, however, have deemphasized the idea of an exclusive role of reverse transcription in the molecular pathogenesis or pathology of RNA virus-induced cancers. (1) Mammalian RNA viruses which have not thus far been implicated in the causation of cancers, such as the visna virus which causes a slow, progressive and fatal neurological disease in sheep and primate syncytical ("foamy") viruses of no known pathogenicity, also exhibit DNA polymerase activities (LIN and THORMAR, 1970; SCOLNICK et al., 1970; SCHLOM, HARTER, BURNY and SPIEGELMAN, 1971; STONE, SCOLNICK, TAKEMOTO and AARONSON, 1971; PARKS, TODARO, SCOLNICK and AARONSON, 1971). One might conjecture that reverse transcriptase activity of mammalian RNA viruses does not mandatorily correlate with carcinogenicity. (2) Poly rA.rU-dependent DNA polymerase activity has been detected in normal mouse and human cells (Scolnick, Aaronson, Todaro and Parks, 1971); the same authors also reported that tumor cells from humans in which no known RNA-containing tumor virus has been detected contain polymerase activity and concluded that if all enzyme activities detected "are manifestations of a latent viral genome, then it would seem to be ubiquitous".

The critical evaluation of observations which argue against an exclusive role of reverse transcription in RNA virus carcinogenesis comes from two lines of reasoning. (1) As stated above, Schlom, Spiegelman and Moore (1971) insist that experiments with synthetic templates and those which do not demonstrate the formation of DNA-RNA hybrids fail to prove conclusively the presence of RNA-dependent DNA polymerase. These enzymological criteria have not been uniformly satisfied for reported enzyme activities which appear to be unrelated to RNA virus carcinogenicity. (2) Failures to detect carcinogenic activity of polymerase-containing viruses or to detect viruses in polymerase-containing mammalian cells are in the category of negative results which are difficult to prove conclusively. At the time of this writing, the argument is incapable of resolution by discussion of results published so far. It is perhaps safe to assume that reverse transcription does play a role in virus carcinogenesis but that there may exist instances of reverse transcription which are unrelated to the pathogenesis of pathology of cancers.

V. Is There Reverse Transcription in Bacteria?

One such instance is the discovery of reverse transcription of heterologous 5 S and ribosomal RNAs by a DNA polymerase from *E. coli* (Cavalieri and Carroll, 1970) The substrate and ionic requirements of this reaction are the same as for the RNA-virus reverse transcriptase, and the reaction products are DNA-RNA hybrids. The demonstration of reverse transcription by a bacterial enzyme may well be related to the observations of San-Chuin, Mang-Ming, Rui-Chu, Wai-Chu and Wen-Lin (1961, 1962) who have reported type transformation to penicillin resistance in *Bacillus subtilis* with *ribonucleic acid* from a resistant strain of this organism as the "transforming principle". Their work has received little attention among molecular biologists and was not cited in an article by Kirtikar and Duerksen [1968 (1)] who obtained increased penicillinase production in three bacteria, among them *B. subtilis*, by treatment with RNA from penicillinase-constitutive *Bacillus cereus*. The phenomenon was RNA concentration-dependent and persisted for "at least three generations of recipient cultures", implying "replication of the introduced" RNA "fraction by cellular polymerases to a limited extent in some unknown manner". The authors suggested the existence of "the active RNA component in an autonomous or cytoplasmic state" but did not consider the possibility of having accomplished type transformation with RNA. The induction of penicillinase production was caused by one defined RNA fraction and was antagonized by other RNAs [Kirtikar and Duerksen, 1968 (2)]; it is not a conclusive counterargument that Ciferri, Barlati and Lederberg (1970) failed to find penetration of several synthetic polyribonucleotides into cells of *B. subtilis* which were competent to take up type-transforming DNA. In fact, Ciferro et al. (1970) anticipate further work of "others who may have more ingenious approaches to the problem". Should the principal result of the experiments of San-Chuin et al.

(1961, 1962) and of KIRTIKAR and DUERKSEN [1968 (1, 2)] be substantiated, a plausible mechanism for the penicillinase$^+$ marker to become part of the hereditary endowment of *B. subtilis* could be: reverse transcription, integration of the product DNA into the bacterial chromosome and, from then on, conventional DNA replication. Clearly, a search for the occurrence of reverse transcription in organisms other than those involved in the pathogenesis or pathology of mammalian cancers should be undertaken in order to delineate the biological scope and significance of this process.

VI. Biological Significance of Reverse Transcription

The immediate mechanistic significance of reverse transcription (apart from its bearing on the Central Dogma) lies in the fact that it offers one additional hypothesis of DNA biosynthesis, albeit of unknown biological scope: RNA-dependent DNA polymerase must now be taken into account along with the classical DNA polymerase I. system of the Arthur Kornberg group and with DNA polymerase II, whose study (T. KORNBERG and GEFTER, 1971) originated from the isolation of DNA polymerase I$^-$ bacterial mutants (DELUCIA and CAIRNS, 1969), when it comes to unravelling the "DNA replication mystery" as it has been called in an Editorial (Nature, 1971).

While it might seem premature to speculate in teleological terms on the general biological significance or utility of reverse transcription, some such speculations have already been offered in the literature. Reference has been made in IV. to the possible role of reverse transcription in the pathogenesis and pathology of RNA virus carcinogenicity. Additionally, and even before RNA-dependent DNA polymerase activity had been discovered, TEMIN [1964 (1)] proposed that reverse transcription might be biologically useful as a mechanism for "somatic information storage", for example, in differentiation, antibody synthesis and memory. To the extent to which such a proposal in this general form would imply that "information", external to chromosomal endowments, becomes inscribed in the form of *ad hoc* synthesized RNA which then, by reverse transcription, inserts this information into chromosomal DNA to become hereditary, it would suggest a molecular mechanism for the working of Michurinian genetics. BOSMANN (1971), on the other hand, has made the interesting suggestion that reverse transcription may be a molecular device for internal "gene amplification". Considering the vast abundance of repeated sequences in the DNA of the genomes of higher organisms (BRITTEN and KOHNE, 1968), it is an ingenious thought that certain DNA substructures might be first conventionally transcribed into RNA and then, by reverse transcription, reenter DNA, giving rise to progressive abundance of such repeated sequences.

Apart from teleological speculations on the biological role or utility of reverse transcription, its discovery in mammalian viruses and cells has far-reaching consequences for theories of biochemical evolution. The classical scheme, based upon the stability and continuity of the genetic endowment, modified only by random mutations followed by selection of mutants for competitive survival capacity, must now make allowance for the insertion of entire new determinants through the machinery of reverse transcription. Indeed, premediated changes of heredity by reverse transcription of selected ribonucleic acids could potentially become a method in genetic experiments.

VII. Cryptography and the Central Dogma

The view has been introduced above that the Central Dogma is, in fact, the expression of a set of operational rules governing biological cryptography. Those who object to the terminology of biochemical genetics as being anthropomorphic (CHARGAFF, 1963) and consider its use one indication of an epistemological twilight of science (CHARGAFF, 1970) might also take exception to the application of the terminology of the secret writing of man to the biological processes of transformations and transmission of genetic specifications. However, the use of the terms "alphabet" and "words" in relating nucleic acids to protein synthesis (GAMOW, 1954), propositions of various forms "codes" (GAMOW, RICH and YCAS, 1956) and the use of the term "code" for an RNA template in protein synthesis (CRICK and WATSON, 1956) — "cipher" would have been the correct designation — indicate that the early theorists of protein synthesis were aware of the compelling formal analogy between voluntary human and involuntary genetic cryptography. Conversely, the author of an elementary text on cryptology (KAHN, 1967) has discussed the nucleic acid "code of life" in his treatment of the art of secret writing.

This article elaborates on the formal analogy between cryptographic principles followed by man and those inherent in biochemical genetics, not for the purpose of injecting teleological or anthropomorphic speculations into molecular biology, but rather for the evident reason that the task of transforming and transmitting a linear set of symbols, comprising a meaningful text, is practically accomplished according to certain common logical principles. The discovery of reverse transcription has, in fact, brought the knowledge of biological cryptography more closely in line with such principles.

It is, therefore, appropriate to review the processes of encipherment and decipherment for whose operations indications do or do not currently exist in molecular biology. Such review has also been made by CRICK (1970) without reference to cryptography.

1. Concerning an encipherment of the plaintext amino acid sequence, i.e. a specification of RNA by protein, nothing of this nature has been observed and the adherence to the Central Dogma has discouraged the search for such occurrences and for their machinery at the present state of biochemical evolution. Experiments have, however, been aimed at detecting present-day deterministic amino acid-codon relationships, that is a cryptographic key (WOESE, DUGRE, SAXINGER and DUGRE, 1966), and the suggestion has been made that a prebiotic "autocatalytic cycle" may have involved "polynucleotides of certain compositions and polyamino acids of certain compositions, the synthesis of the one being catalyzed by the other *and vice versa*" [my italics], representing a form of "primitive translation" (WOESE, 1968). This envisages explicitly a bidirectional transfer of structural specifications to have operated in a primordial state of biological or prebiological cryptography before the evolution of the translation apparatus restricted the passage of genetic specifications unidirectionally to the decipherment of the RNA placode into the plaintext amino acid sequence.

2. No hypotheses or experimental data exist concerning a direct copying mechanism for amino acid sequences from existing protein to new protein, i.e. for the transmission of sequential structural information *in cleartext*.

3. Transcription of DNA into messenger RNA by the cryptographic device of DNA-dependent RNA polymerase, i.e. the first step in the *de*cipherment of the DNA encicode, and the recently discovered reverse transcription of RNA into DNA by RNA-dependent DNA polymerase, i.e. the superencipherment of the RNA placode, show that between the two categories of nucleic acids, the passage of precise information concerning the sequence of symbols occurs biologically in both directions, using base complementarity as a key and being mediated by enzymatic machineries whose operating mechanistic principles require further study.

4. RNA itself can be copied in the case of certain bacterial RNA viruses (SPIEGELMAN and DOI, 1963) and does serve here in the dual capacity of being the viral chromosome *and* the virus messenger RNA. The genetic system of bacterial RNA viruses, hence, does not use superencipherment by a DNA polymerase and has only one deciphering step from RNA placode to protein.

5. DNA itself can be copied as postulated by WATSON and CRICK (1953) and experimentally demonstrated (MESELSON and STAHL, 1958; CAIRNS, 1963; GOULIAN, KORNBERG and SINSHEIMER, 1967), although the details of the *in vivo* copying machinery remain to be elucidated.

6. Finally, there exists preliminary evidence that denatured DNA can *in vitro* direct the polymerization of amino acids when under the influence of streptamine-containing antibiotics (MCCARTHY and HOLLAND, 1965; MASAKUWA and TANAKA, 1967). Whether this constitutes a direct and precise decipherment of DNA triplet sequences into amino acid sequences remains to be shown.

VIII. Conclusion

Returning at the end to the significance of the discovery of reverse transcription for the validity of the Central Dogma of molecular biology, CRICK's original version (1958) envisaged the possibility of a "transfer of information from nucleic acid to nucleic acid". Since nucleic acids, when complementary in sequential structure through point-counter-point base pairing "in register", are mutual determinants of each others base sequences, it is perhaps, in retrospect, not too surprising to find enzymes devoid of stringent template specificities *in vitro* which catalyze the monotonous and repetitive condensation reactions between nucleoside triphosphates when they are correctly aligned on templates. While template specificity *in vivo* may well involve the selective role of discrete initiation sites and of enzyme factors such as σ, one might say that the essential cryptographic process in the polycondensation of nucleic acid building blocks is the readout of bases against their complements in templates and that polymerizing enzymes merely "print out" the results.

It is difficult, however, to envisage mechanistically a reversal of the translation process: transfer RNAs are one-way adaptors for the sequentialization of amino acids in protein biosynthesis but they can neither react with constituent amino acids of proteins in peptidic linkage nor can they, on a unit-per-unit basis, organize nucleoside triphosphates for polycondensation into nucleic acids. One would need to postulate an entirely different and separate biochemical machinery for "reverse translation", an unlikely prospect which has led Lancet in an Editorial (1970) to ascribe to a noted molecular biologist the remark that he would "become a theologian" if reverse translation were discovered.

It appears, therefore, that the key statement of Crick's (1958) Central Dogma which holds that "once information has passed into protein *it cannot get out again*" will remain valid and can now be reiterated on safer grounds 15 years later since the mechanistic details of the translation machinery have become better understood.

IX. Glossary of Terms

In this article, when discussing the Central Dogma of molecular biology and its validity for the transformation and transmission of genetic specifications, use has been made of some of the basic terminology of cryptology. The reader might find a glossary of this terminology useful.

1. *Cryptology:* In the most general sense the science of secret writing including cryptography and cryptoanalysis.

2. *Cryptography:* The techniques of secret writing through the use of various transformations of the plaintext.

3. *Cryptanalysis:* The "breaking" or solution of a cryptic message without possessing the key; the methods by which codes or ciphers are broken.

4. *Plaintext:* The message which is put into secret form by transformation.

5. *Code:* Codes operate on plaintext groups of variable length: codegroups or codenumbers replace entire plaintext elements.

6. *Cipher:* Ciphers operate on plaintext units of regular length, in the simplest form on single letters of an alphabet. In the genetic "code" the basic unit of the plaintext is the single amino acid.

7. *Decipherment:* The procedures by which the ciphertext is converted into the plaintext in routine instances in which the key is available. This is in contrast to cryptanalysis.

8. *Encipherment:* The procedures by which the plaintext is converted into the ciphertext.

9. *Substitution Transformation:* One of two general types of encipherment in which one set of symbols is substituted for another set of symbols, the sequence remaining the same.

10. *Transposition Transformation:* One of two general types of encipherment in which the symbols of the plaintext are retained but are "transposed", i.e. changed in sequence.

11. *Placode* (from plain code): The result of encoding the plaintext by only one transformation; also the intermediate result of the partial decipherment of a superenciphered code.

12. *Superencipherment:* The result of an additional encoding of a placode by a second transformation.

13. *Encicode* (from enciphered code): The ciphertext resulting from a superencipherment.

14. *Ciphertext:* The final enciphered message transmitted.

15. *Cleartext:* The plaintext message transmitted without encipherment, i.e. in "clear" or in plain language.

References[2]

Baltimore, D.: RNA-dependent DNA polymerase in virions of RNA tumor viruses. Nature (Lond.) **226**, 1209 (1970).

Bosmann, H. B.: RNA-directed DNA synthesis: Identification in L5178Y mouse leukemic cells and distribution of the polymerase in a synchronized L5178Y cell population. FEBS Letters **13**, 121 (1971).

Britten, R. J., Kohne, D. E.: Repeated sequences in DNA. Science **161**, 529 (1968).

Brockman, W. W., Carter, W. A.: Streptovaricins inhibit RNA-dependent DNA polymerase present in an oncogenic RNA virus. Nature (Lond.) **230**, 249 (1971).

Cairns, J.: The bacterial chromosome and its manner of replication as seen by autoradiography. J. molec. Biol. **6**, 208 (1963).

Cavalieri, L. F., Carroll, E.: RNA as a template with *E. coli* DNA polymerase. Biochem. biophys. Res. Commun. **41**, 1055 (1970).

Chargaff, E.: Amphisbaena. In: Essays on nucleic acids. Amsterdam: Elsevier 1963.

Chargaff, E.: Vorwort zu einer Grammatik der Biologie. Experientia (Basel) **26**, 810 (1970).

Ciferri, O., Barlati, S., Lederberg, J.: Uptake of synthetic polynucleotides by competent cells of *Bacillus subtilis*. J. Bact. **104**, 684 (1970).

Crick, F. H. C.: Discussion. In: The structure of nucleic acids and their role in protein synthesis. Cambridge: University Press 1957.

Crick, F. H. C.: On protein synthesis. In: The biological replication of macromolecules, 138. New York: Academic Press 1958.

Crick, F.: Central dogma of molecular biology. Nature (Lond.) **227**, 561 (1970).

Crick, F. H. C., Griffith, J. S., Orgel, L. E.: Codes without commas. Proc. nat. Acad. Sci. (Wash.) **43**, 416 (1957).

Crick, F. H. C., Watson, J. D.: Virus structure: General principles. Ciba Foundation Symp. on the nature of viruses, 1956, p. 5.

DeLucia, P., Cairns, J.: Isolation of an *E. coli* strain with a mutation affecting DNA polymerase. Nature (Lond.) **224**, 1164 (1969).

Duesberg, P. H., Canaani, E.: Complementarity between Rous sarcoma virus (RSV) RNA and the *in vitro*-synthesized DNA of the virus-associated DNA polymerase. Virology **42**, 783 (1970).

Duesberg, P., Helm, K. V. D., Canaani, E.: Properties of a soluble DNA polymerase isolated from Rous sarcoma virus. Proc. nat. Acad. Sci. (Wash.) **68**, 747 (1971).

Editorial: Two ways to protein. Lancet **1970**, II, 31.

Editorial: Roundabouts and swings. Nature (Lond.) **228**, 1255 (1970).

Editorial: The DNA replication mystery. Nature (Lond.) **230**, 11 (1971).

Feller, W. F., Chopra, H., Bepko, F.: Studies on the possible viral etiology of human breast cancer. Surgery **62**, 750 (1967).

Fujinaga, K., Parsons, J. T., Beard, J. W., Beard, D., Green, M.: Mechanism of carcinogenesis by RNA tumor viruses. III. Formation of RNA-DNA complex and duplex DNA molecules by the DNA polymerase(s) of avian mycoblastosis virus. Proc. nat. Acad. Sci. (Wash.) **67**, 1432 (1970).

Gallo, R. C., Yang, S. S., Ting, R. S.: RNA-dependent DNA polymerase of human acute leukemic cells. Nature (Lond.) **228**, 927 (1970).

Gamow, G.: Possible relation between deoxyribonucleic acid and protein synthesis. Nature (Lond.) **173**, 318 (1954).

Gamow, G., Rich, A., Ycas, M.: The problem of information transfer from the nucleic acids to proteins. Advanc. biol. med. Phys. **4**, 23 (1956).

Gerwin, B. I., Todaro, G. J., Zeve, V., Scolnick, E. M., Aaronson, S. A.: Separation of RNA-dependent DNA polymerase activity from the murine leukemia virion. Nature (Lond.) **228**, 435 (1970).

Goulian, M., Kornberg, A., Sinsheimer, R. L.: Enzymatic synthesis of DNA. XXIV. Synthesis of infectious phage ϕ X174 DNA. Proc. nat. Acad. Sci. (Wash.) **58**, 2321 (1967).

[2] This covers the literature on RNA-dependent DNA polymerase from June 1970 to the end of May 1971, i.e. for the first 12 months after the discovery of the polymerase reaction.

GREEN, M., ROKUTANDA, M., FUJINAGA, K., RAY, R. K., ROKUTANDA, H., GURGO, C.: Mechanism of carcinogenesis by RNA tumor viruses. I. An RNA-dependent DNA polymerase in murine sarcoma viruses. Proc. nat. Acad. Sci. (Wash.) **67**, 385 (1970).

GURGO, C., RAY, R. K., THIRY, L., GREEN, M.: Inhibitors of the RNA and DNA dependent polymerase activities of RNA tumor viruses. Nature (Lond.) **229**, 111 (1971).

HATANAKA, M., HUEBNER, R. J., GILDEN, R. V.: DNA polymerase activity associated with RNA tumor viruses. Proc. nat. Acad. Sci. (Wash.) **67**, 143 (1970).

HATANAKA, M., HUEBNER, R. J., GILDEN, R. V.: Specificity of the DNA product of the C-type virus RNA-dependent DNA polymerase. Proc. nat. Acad. Sci. (Wash.) **68**, 10 (1971).

KAHN, D.: The Codebreakers, the story of secret writing. London: Weidenfeld and Nicolson 1967, p. 942.

KIRTIKAR, M. W., DUERKSEN, J. D.: (1) A penicillinase-specific ribonucleic acid component from *Bacillus cereus*. I. Ribonucleic acid extraction and definition of the *in vivo* test system. Biochemistry **7**, 1172 (1968).

KIRTIKAR, M. W., DUERKSEN, J. D.: (2) A penicillinase-specific ribonucleic acid component from *Bacillus cereus*. II. Partial characterization of the active component. Biochemistry **7**, 1183 (1968).

KORNBERG, T., GEFTER, M. L.: Purification and DNA synthesis in cell-free extracts: Properties of DNA polymerase II. Proc. nat. Acad. Sci. (Wash.) **68**, 761 (1971).

LEE-HUANG, S., CAVALIERI, L. F.: Isolation and properties of a nucleic acid hybrid polymerase. Proc. nat. Acad. Sci. (Wash.) **51**, 1022 (1964).

LIN, F. H., THORMAR, H.: Ribonucleic acid-dependent deoxyribonucleic acid polymerase in visna virus. J. Virol. **6**, 702 (1970).

MCCARTHY, B. J., HOLLAND, J. J.: Denatured DNA as a direct template for *in vitro* protein synthesis. Proc. nat. Acad. Sci. (Wash.) **54**, 880 (1965).

MCDONNELL, J. P., GARAPIN, A.-C., LEVINSON, W. E., QUINTRELL, N., FAUSHIER, L., BISHOP, M. O.: DNA polymerase of Rous sarcoma virus: Delineation of two reactions with actinomycin. Nature (Lond.) **228**, 433 (1970).

MCDONNELL, J. P., TAYLOR, J., LEVINSON, W., BISHOP, J. M.: Soluble DNA polymerase from Rous sarcoma virus. Fed. Proc. **30**, 1163 Abs. (1971).

MASAKUWA, H., TANAKA, N.: Stimulation by aminoglycoside antibiotics of DNA-directed protein synthesis. J. Biochem. (Tokyo) **62**, 202 (1967).

MESELSON, M., STAHL, F. W.: The replication of DNA in *Escherichia coli*. Proc. nat. Acad. Sci. (Wash.) **44**, 671 (1958).

MIZUNO, S., YAMAZAKI, H., NITTA, K., UMEZAWA, H.: Inhibition of DNA-dependent RNA polymerase reaction of *Escherichia coli* by an antimicrobial antibiotic, streptovaricin. Biochim. biophys. Acta (Amst.) **157**, 322 (1968).

MIZUTANI, S., BOETTIGER, D., TEMIN, H. M.: A DNA-dependent DNA polymerase and a DNA endonuclease in virions of Rous sarcoma virus. Nature (Lond.) **228**, 424 (1970).

MIZUTANI, S., TEMIN, H. M., KODAMA, M., WELLS, R. T.: DNA ligase and exonuclease activities in virions of Rous sarcoma virus. Nature (Lond.) **230**, 232 (1971).

OCHOA, S.: Biosynthesis or ribonucleic acid. In: Recent progress in microbiology, 122. Stockholm: Almquist & Wiksell 1958.

PARKS, W. P., TODARO, G. J., SCOLNICK, E. M., AARONSON, S. A.: RNA-dependent DNA polymerase in primate syncytium-forming (foamy) viruses. Nature (Lond.) **229**, 258 (1971).

POTTER, V. R.: Society and science. Science **146**, 1018 (1964).

ŘÍMAN, J., BEAUDREAU, G. S.: Viral DNA-dependent DNA polymerase and the properties of thymidine labelled material in virions of an oncogenic RNA virus. Nature (Lond.) **228**, 427 (1970).

ROKUTANDA, M., ROKUTANDA, H., GREEN, M., FUJINAGA, K., RAY, R. K., GURGO, C.: Formation of viral RNA-DNA hybrid molecules by the DNA polymerase of sarcoma-leukemia viruses. Nature (Lond.) **227**, 1026 (1970).

SAN-CHIUN, S., MANG-MING, H., RUI-ZHU, C., HUI-ZHU, C., WEN-LIN, Z.: Ribonucleic acid as a transforming principle in bacteria. Abstracts Vth Intern. Congr. Biochem., 409 (1961).

SAN-CHUIN, S., MANG-MING, H., RUI-CHU, C., WAI-CHU, C., WEN-LIN, C.: Ribonucleic acid as a transforming principle in bacteria. Scientia Sinica **11**, 233 (1962).
SCHLOM, J., HARTER, D. H., BURNY, A., SPIEGELMAN, S.: DNA polymerase activities in virions of visna virus, a causative agent of a "slow" neurological disease. Proc. nat. Acad. Sci. (Wash.) **68**, 182 (1971).
SCHLOM, J., SPIEGELMAN, S., MOORE, D.: RNA-dependent DNA polymerase activity in virus-like particles isolated from human milk. Nature (Lond.) **231**, 97 (1971).
SCOLNICK, E., RANDS, E., AARONSON, S. A., TODARO, G. J.: RNA-dependent DNA polymerase activity in five RNA viruses: Divalent cation requirements. Proc. nat. Acad. Sci. (Wash.) **67**, 1789 (1970).
SCOLNICK, E. M., AARONSON, S. A., TODARO, G. J., PARKS, W. P.: RNA-dependent DNA polymerase activity in mammalian cells. Nature (Lond.) **229**, 318 (1971).
SPIEGELMAN, S., DOI, R. H.: Replication and translation of RNA genomes. Cold Spr. Harb. Symp. quant. Biol. **28**, 109 (1963).
SPIEGELMAN, S., BURNY, A., DAS, M. R., KEYDAR, J., SCHLOM, J., TRAVNICEK, M., WATSON, K.: (1) Characterization of the products of RNA-directed DNA polymerases in oncogenic RNA viruses. Nature (Lond.) **227**, 563 (1970).
SPIEGELMAN, S., BURNY, A., DAS, M. R., KEYDAR, J., SCHLOM, J., TRAVNICEK, M., WATSON, K.: (2) DNA-directed DNA polymerase activity in oncogenic RNA viruses. Nature (Lond.) **227**, 1029 (1970).
SPIEGELMAN, S., BURNY, A., DAS, M. R., KEYDAR, J., SCHLOM, J., TRAVNICEK, M., WATSON, K.: (3) Synthetic DNA-RNA hybrids and RNA-RNA duplexes as templates for the polymerases of the oncogenic RNA viruses. Nature (Lond.) **228**, 430 (1970).
STONE, L. B., SCOLNICK, E., TAKEMOTO, K. K., AARONSON, S. A.: Visna virus: A slow virus with an RNA-dependent DNA polymerase. Nature (Lond.) **229**, 257 (1971).
TEMIN, H. M.: (1) Nature of the provirus of Rous sarcoma. Nat. Cancer Inst. Monogr. **17**, 557 (1964).
TEMIN, H. M.: (2) Homology between RNA from Rous sarcoma virus and DNA from Rous sarcoma virus-infected cells. Proc. nat. Acad. Sci. (Wash.) **52**, 323 (1964).
TEMIN, H. M., MIZUTANI, S.: RNA-dependent DNA polymerase in virions of Rous sarcoma virus. Nature (Lond.) **226**, 1211 (1970).
VOLKIN, E., ASTRACHAN, L.: Phosphorus incorporation in *Escherichia coli* ribonucleic acid after infection with bacteriophage T2. Virology **2**, 149 (1956).
WATSON, J. D., CRICK, F. H. C.: The structure of DNA. Cold Spr. Harb. Symp. quant. Biol. **18**, 123 (1953).
WEHRLI, W., NÜESCH, J., KNÜSEL, F., STAEHELIN, M.: Action of rifamycins on RNA polymerase. Biochim. biophys. Acta (Amst.) **157**, 215 (1968).
WOESE, C. R., DUGRE, D. H., SAXINGER, W. C., DUGRE, S. A.: The molecular basis for the genetic code. Proc. nat. Acad. Sci. (Wash.) **55**, 966 (1966).
WOESE, C. R.: The fundamental nature of the genetic code: Prebiotic interactions between polynucleotides and polyamino acids or their derivatives. Proc. nat. Acad. Sci. (Wash.) **59**, 110 (1968).

X. Addendum

After the completion of this article in July 1971, intensive research on reverse transcription has continued. One aim of this work has been to separate and distinguish reverse transcriptases of tumor viruses from other DNA polymerases (for example, ROSS, SCOLNICK, TODARO and AARONSON, 1971; GOODMAN and SPIEGELMAN, 1971). The second aim has been the study of the effects of inhibitors of the RNA-dependent DNA polymerase reaction (for example, MÜLLER, ZAHN and SEIDEL, 1971; FRIDLENDER and WEISSBACH, 1971). Both lines of investigation converge upon the importance of reverse transcription in oncology. Neither the results of the work cited nor those of numerous other investigations modify or contradict the principal ideas set forth in the main body of this article.

One new mechanistic feature of the reverse transcription reaction has been discovered by VERMA, MEUTH, BROMFELD, MANLY and BALTIMORE (1971) who have found that the reverse transcriptase from avian myeloblastosis virus initially synthesizes from the endogenous RNA template a covalently linked DNA-RNA complex which contains an oligoribonucleotidic primer entity, i.e. a molecular species which is required to *initiate* the polymerization of DNA. Analogous regions in DNA, required for the initiation of induced messenger RNA transcription, exist in the form of "promoters" which have been mapped in bacterial genetic analyses of the regulatory segment of operons. The discovery of oligoribonucleotidic "primers" of the reverse transcription reaction strengthens, therefore, the analogy between forward and reverse transcriptions.

References Cited in Addendum

FRIDLENDER, B., WEISSBACH, A.: DNA polymerases of tumor virus: specific effect of ethidium bromide on the use of different synthetic templates. Proc. nat. Acad. Sci. (Wash.) **68**, 3116 (1971).

GOODMAN, N. C., SPIEGELMAN, S.: Distinguishing reverse transcriptase of an RNA tumor virus from other known DNA polymerases. Proc. nat. Acad. Sci. (Wash.) **68**, 2203 (1971).

MÜLLER, W. E. G., ZAHN, R. K., SEIDEL, H. J.: Inhibitors acting on nucleic acid synthesis in an oncogenic RNA virus. Nature (Lond.) New Biol. **232**, 143 (1971).

ROSS, J., SCOLNICK, E. M., TODARO, G. J., AARONSON, S. A.: Separation of murine cellular and murine leukaemia virus DNA polymerases. Nature (Lond.) New Biol. **231**, 163 (1971).

VERMA, I. M., MEUTH, N. L., BROMFELD, E., MANLY, K. F., BALTIMORE, D.: Covalently linked RNA-DNA molecule as initial product of RNA tumor virus DNA polymerase. Nature (Lond.) New Biol. **233**, 131 (1971).

The Isolation of Genes: A Review of Advances in the Enrichment, Isolation, and *in vitro* Synthesis of Specific Cistrons [1, 2]

MAURILLE J. FOURNIER, Jr., DON J. BRENNER, and B. P. DOCTOR

I. Introduction

Answers to many of the outstanding questions about the biosynthesis of nucleic acids and proteins will require direct studies with isolated genes. These questions include details about: (1) the structure of the gene; (2) the mechanism and regulation of gene expression, and (3) in at least certain cases, post-transcriptional events which modify a primary gene product.

1. Gene structure. Although it is clear that a gene may contain more information than is reflected in the primary structure of its products, little else is known about its substructure. Specifically, information is lacking about the chemical and physical nature of the RNA polymerase initiation and termination sites and possibly other sites which may function in the regulation of transcription. In the case of some cistrons, the nature of the 'extra' template information which directs the synthesis of a precursor RNA larger than the final product is not known.

2. The transcription process. With preparations of intact genes, the multistep transcription process can be examined in detail. Further, such preparations provide a means of detecting transcription factors which influence the rate and extent of RNA synthesis.

Another matter that has long puzzled molecular biologists is the function of the minor nucleotides which occur in DNA and possibly RNA templates [71]. Do these modified bases play a role in biosynthesis, transcription or degradation? Are they found in genes? If so, in which elements? Preparations of specific genes should prove useful in resolving these issues.

3. Post-transcriptional events. It has recently been demonstrated that several species of cellular RNA are derived from larger precursor molecules. Thus far, precursor RNAs have been described for ribosomal RNA [119], 5s RNA [119], and transfer RNA [2, 16, 37, 38].

At the time of writing there is no information about the role of these supernumerary oligonucleotides. By transcribing isolated genes *in vitro*, it should be possible to prepare precursor species and to determine their biological significance.

In addition to nucleolytic "tailoring", there are a number of other post-transcriptional modification reactions which alter gene products. Among these reactions

[1] The literature survey pertaining to this review was concluded in March, 1972.

[2] The abbreviations used in this article follow the recommendations of the IUPAC-IUB Combined Commission of Biochemical Nomenclature (CBN).

are a myriad of specific enzyme reactions involved in the formation of minor bases in ribosomal and especially transfer RNA [71]. Although this question has received much attention in recent years, it is not yet clear what effect modification reactions have on the function of the gene product. Progress on this important matter has been severely hampered by the unavailability of unmodified precursor RNA. Purified ribosomal and tRNA cistrons could be used to generate unmodified RNA for such comparative studies. This unmodified RNA could then be used as a substrate to isolate and characterize the enzymes which convert the primary gene product to fully modified RNA.

Preparations of purified genes may prove to be important for the understanding of yet another phenomenon. It has been discovered recently that messenger RNAs (mRNA) from mammalian cells [52, 60, 89, 91] and from several animal viruses [81, 111, 118] contain stretches of poly (A) sequences. It is not clear if these sequences are part of the primary gene product or are attached to the RNA after transcription. Nor is it known what effect the presence of poly (A) sequences has on the function of these mRNAs. If the information for these sequences is not part of the corresponding DNA cistron, RNA products transcribed *in vitro* from appropriate gene preparations could be used to elucidate the modification process and to determine its biological significance.

II. Procedures for the Preparation of Specific Cistrons

There are more than thirty reports in the literature describing preparations of DNA that have been at least partially enriched for specific cistrons. The methods used to obtain these preparations include:

A. Isolation by physical and biochemical techniques
B. Enrichment by genetic manipulation
C. A combination of genetic and physico-chemical methods
D. Chemical synthesis
E. Enzymatic synthesis

Each of these approaches will be discussed first in a general way with a view toward presenting the rationale of the method and then, specific studies where the methods have been used with at least a modicum of success will be reviewed. In later sections the nature of specific cistron preparations and their suitability for *in vitro* studies of gene structure and transcription will be discussed. Finally, in a concluding section, we will summarize the properties of all the cistron preparations described in the review.

A. Fractionation and Purification of Genetic Material by Physical and Biochemical Techniques

Certain genes can be isolated or at least highly enriched with relative ease, either because their nucleotide composition or the size of the DNA molecule of which they are part renders them physically distinct from the bulk of the other genetic material in the cell, or because a physical distinction can be artificially conferred upon these genes by some manipulation — such as formation of a RNA:DNA hybrid.

Nucleic acid species which differ significantly in guanine plus cytosine (G+ C) content also differ in density and thermal stability. These species can therefore be fractionated by techniques based on these physical properties. Accordingly, cistrons that are G+ C-rich (or poor) relative to the remainder of the genome or are contiguous to such a region are logical candidates for purification. For this reason, and because of their biological importance, several investigators have sought to purify rRNA cistrons using such techniques. Thus far, sequences complementary to ribosomal RNA (rDNA) have been prepared from *Mycoplasma sp.* (Kid) [128], *Escherichia coli* [53], *Bacillus subtilis* [141], yeast [95], toad [20], and sea urchin [117].

1. *Fractionation by Density and Differential Thermal Stability*

In 1966, DAVISON demonstrated that *E. coli* DNA fragments containing rDNA can be enriched some 5 to 20 fold merely by subjecting the DNA to isopycnic centrifugation in cesium chloride. This is due to the fact that in *E. coli* rRNA, and thus rDNA, has a G+ C content some 6 to 8% higher than unfractionated DNA [100, 105].

Fig. 1 shows the distribution of denatured DNA (of undetermined molecular weight) sedimented to equilibrium in CsCl. Two *E. coli* DNA profiles are shown solid-line plots) to emphasize the fact that the distribution is heterogeneous and often skewed toward the region of higher density (right profile). When fractions from this gradient were pooled and tested for their ability to hybridize with ribosomal and transfer RNA, the profiles shown in Fig. 2 were obtained. It can be seen that pooled fractions 1 and 2 (from Fig. 1) which contain only 3 to 5% of the total DNA are clearly enriched for rDNA sequences and perhaps 4s RNA cistrons. The author indicates that the enrichment may be up to 20 fold and that as much as 8% of the DNA in the densest fractions could be rDNA [53].

Although this method is simple and effective, it is not suited for processing more than a few hundred micrograms of DNA. Inasmuch as the rRNA cistrons comprise about 0.3% of the genome in *E. coli* [137], several hundred milligrams of unfractionated DNA would have to be processed to obtain enough rDNA to permit detailed analyses of the type outlined in the Introduction. Further, the cost of cesium salts precludes the use of this technique on a large scale. Thus, purification of genes by equilibrium density-gradient centrifugation in CsCl is best suited for use at later stages of purification.

It may not have been clear at the time that the study by DAVISON was in progress, but certain other DNAs are much better suited for the isolation of rDNA by density fractionation than is *E. coli* DNA. For example, in *E. coli* the G+ C content of rRNA and total DNA differ by only 6 to 8% (58% and 50 to 52% respectively, references 100, 105). The G+ C content of rDNA is relatively high, between 55 and 65% in most organisms. Therefore, the lower the G+ C content of total DNA, the more marked the difference between it and the G+ C of rDNA. In eukaryotes total DNA G+ C is remarkably constant between 35 and 44% [132]. Among bacteria, the G+ C content of total DNA varies between 20 and 80% [113].

In *B. subtilis* [109], the rDNA contains 12 to 14% more G+ C pairs than are present in total DNA. Since the G+ C pairs have greater thermal stability [105, 108], TAKAHASHI was able to enrich the rDNA cistrons 4 to 9 fold by a technique of thermal

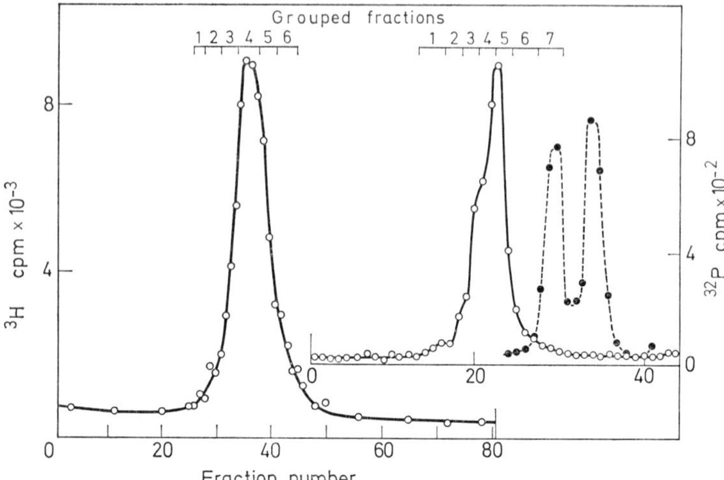

Fig. 1. Profiles of denatured DNA from *Escherichia coli* centrifuged to equilibrium in CsCl gradients. In order to minimize the centrifugation time required to reach equilibrium there were layered above and below the DNA (in 1 to 1.5 ml of CsCl of appropriate density) CsCl solutions of 10%-lower and 10%-higher density. The tubes were centrifuged for 20 h at 30,000 rev/min and then for 48 h at 25,000 rev/min (Spinco Model L, rotor SW 39). The dashed curve (right) shows the distribution of denatured DNA (^{32}P label) from *Bacillus megaterium* G phage run in a separate tube but at the same time as the DNA from *E. coli*. Thus the widths of the bands but not their relative positions are significant. (From Davison, 1966)

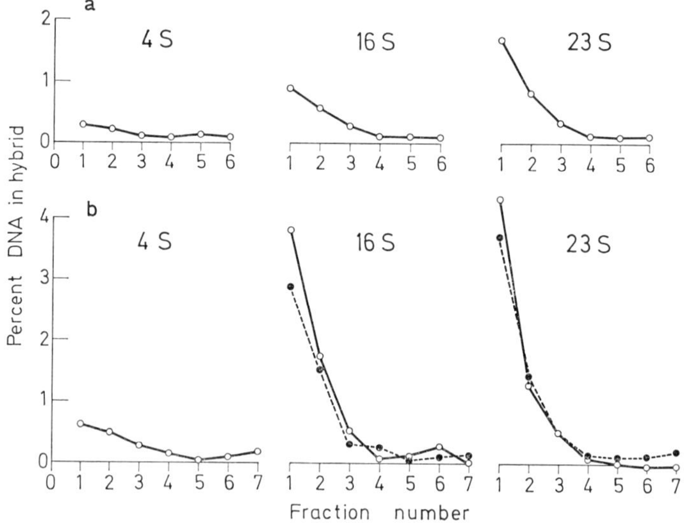

Fig. 2. The ability of the grouped samples shown in Fig. 1 as fractions 1 to 6 (a) and 1 to 7 (b), respectively, to hybridize with 4s, 16s, and 23s RNA. Between 0.2 and 5 µg DNA was adsorbed to each filter for the assays, and sufficient ^{32}P-RNA was used to saturate the DNA. Experiments with two different RNA preparations are shown in (b) by the full and dashed lines. (From Davison, 1966)

fixation [141]. In this procedure, DNA is partially denatured, quickly cooled, and immediately passed through a cellulose nitrate filter. All single-stranded DNA passes through the membrane while double-stranded DNA is trapped. When the filtrates from samples denatured at different temperatures were assayed for rDNA by hybridization with ribosomal RNA, it was observed that at 4.5° above the Tm of total DNA, about 55% of the rRNA cistrons are still at least partially double-stranded and can be trapped by the filtration technique (Table 1). (Tm is used here to designate the temperature at which 50% of the native DNA has undergone the hyperchromic transition associated with melting, i.e. unstacking of the bases prior to strand separation.)

Although simple and rapid, this method is not easily scaled up and is useful only for isolating cistrons with a G+ C content markedly higher or lower than that of

Table 1. Concentration of rRNA cistrons of *B. subtilis* by heating the DNA at a temperature above the T_m and filtration through a millipore filter

^{14}C Thymidine-labeled DNA of *B. subtilis* and ^3H-labeled rRNA prepared from the same strain were used. Ribosomal RNA cistrons in the filtrate were measured by the DNA-rRNA hybridization method at excess rRNA level. The degrees of concentration are expressed as $^3H/^{14}C$. (From Takahashi, 1969)

Temperature of treatment	Radioactivity found in filtrate (counts/min)	[^3H]rRNA hybridized (counts/min)	$^3H/^{14}C$
No treatment	430	4340	10.0
75°	300	3690	12.3
80°	96	2790	29.1
82.5°	46	2450	53.1
—	0	(120)	—

total DNA. On the other hand, genes concentrated by this technique can be recovered as double-stranded material, unlike cistrons prepared by fractionation of denatured DNA. It will be important to have native DNA for certain of the studies of gene structure and function outlined above.

Ryan and Morowitz [128] have also capitalized on a difference in thermal stability to devise a procedure for enriching *Mycoplasma* rRNA and tRNA genes some 50 and 20 fold respectively. These RNA species possess G+ C contents of 48 and 54% [128] in a genome that is only 25% G+ C overall [100].

Under conditions of low ionic strength and at temperatures well below the melting point of the nucleic acid, hydroxyapatite will bind double-stranded DNA. Raising the temperature of the column will cause the DNA to dissociate and the separated single strands will be eluted. If the temperature is increased slowly with continuous elution, it is possible to fractionate the DNA into early and late melting fractions. DNA which is low in G+ C will melt and be eluted before G+ C-rich DNA.

In the procedure developed by Ryan and Morowitz, double-stranded DNA fragments prepared by sonication are first adsorbed to a hydroxyapatite column and then eluted with a temperature gradient. The bulk of the DNA melts and elutes at

79.5 °C but a G+ C-rich fraction is retained until the temperature is raised to 89° (Fig. 3). This fraction corresponds to about 1 to 1.5% of the total DNA and is enriched some 44 fold. Hybridization assays show the material to be 12.7% rDNA and 3.2% tDNA. Although the yield of rDNA and tDNA is not reported by the investigators, it would appear that the G+ C-rich DNA contains about 28% of the tRNA genes.

If double-stranded *Mycoplasma* cistrons are desired, the thermal elution can be terminated at a point a few degrees below the temperature at which the G+C-rich fraction dissociates and the material can then be eluted with buffer of higher molarity.

For this method to be successful, the DNA must first be sheared in order that the G+C-rich sequences can be separated from contiguous sequences of low G+ C.

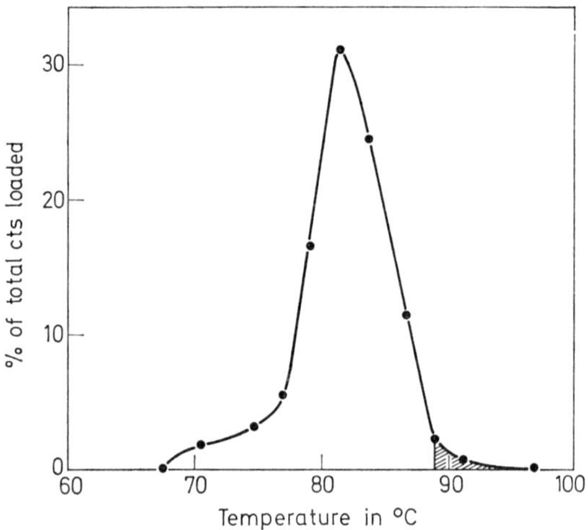

Fig. 3. Thermal elution profile of Kid DNA from hydroxyapatite using 0.17 M phosphate buffer and eluting as a function of temperature. As the sonic fragments melt, they are eluted by the buffer. The shaded area represents that DNA which was left native during an enrichment and eluted with higher molarity (0.27 M) buffer. (From RYAN and MOROWITZ, 1969)

Sonication is the method of choice for shearing as there is little denaturation of the DNA. In their study, RYAN and MOROWITZ used DNA fragments with a molecular weight of 3×10^5 daltons.

Although this technique cannot be used to obtain pure cistrons (unless a gene is discovered with a considerably higher G+ C content than all other cistrons in the genome) it can be easily scaled up to process several hundred milligrams, especially if the DNA is applied to the hydroxyapatite at a temperature where the bulk of the nucleic acid is denatured and will not be adsorbed. Further, the method could be adapted for batch operation. In addition to being a simple, effective means for enriching G+ C-rich cistrons, it allows native template material to be recovered where desired.

In 1966, BIRNSTIEL and coworkers reported that the DNA which codes for 28s ribosomal RNA in *Xenopus laevis* could be enriched some 500 fold by equilibrium density gradient centrifugation [20, 149]. This exciting discovery marked the first significant partial purification of a gene from a eukaryote.

In a later report [18], it was further determined that the cistrons for 18s rRNA cosediment with the 28s rRNA cistrons, both of which band as heavy satellite DNA in a CsCl gradient. The basis for the separation of these genes from the rest of the genome is their high G+ C content. In *Xenopus*, the G+ C content of 28s and 18s rRNA is 67% and 60%, respectively [32] while that of total RNA is about 40% [20].

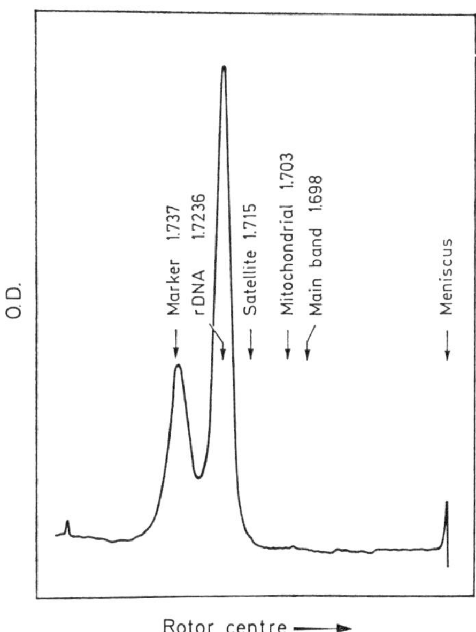

Fig. 4. Analytical centrifugation of the purified, isolated rDNA. rDNA (2 μg)was analysed in the Spinco analytical centrifuge as described in the text. Denatured *Pseudomonas aeroginosa* DNA was included as a density marker (1.737 g cm^{-3}). (From BIRNSTIEL et al., 1968)

Further, because there are several hundred copies of the ribosomal RNA genes [42, 149] and these genes are highly clustered, it is not necessary to fragment the DNA before fractionation in order to free it from contiguous G+C-poor sequences. The rDNA-containing satellite can be formed even with DNA of molecular weight 2 to 5 × 10^7 daltons [20].

The 28s rDNA fraction first isolated by the BIRNSTIEL group represents only 0.2% of the genome and approximately 30% of the 28s ribosomal RNA genes [19]. The content or yield of 18s rDNA was not determined. From these studies and others by RITOSSA and SPIEGELMAN [124] and BROWN and coworkers [34], it became clear that the rRNA cistrons in *Xenopus* are clustered on a single chromosome and associated with the nucleolar organizer region.

In a recent study [19], BIRNSTIEL et al. analyzed the purity of a fraction of rDNA satellite prepared by two successive distributions in CsCl gradients (Fig. 4). Analytical ultracentrifugation showed this material to be free of all other *Xenopus* DNA species and to be of the same density as the G+ C-rich rDNA band in unfractionated DNA [18, 20, 149].

BROWN and WEBER [34] also prepared *Xenopus* rDNA by CsCl gradient fractionation and demonstrated that the genes for 4s and 5s RNA can be separated from this material.

DAWID, BROWN and REEDER subsequently discovered that *Xenopus* rDNA can be enriched considerably by selective precipitation of bulk DNA with polylysine [54, 153]. Under their conditions over 90% of the bulk DNA can be precipitated with less than 5% loss of rDNA sequences. This technique is based on an earlier finding by LENG and FELSENFELD [90] that A+ T-rich DNA precipitates before G+ C-rich DNA in the presence of polylysine. Since the rDNA sequences in *Xenopus* are highest in G+ C content, they are the last to precipitate.

Fig. 5 shows the purification of *Xenopus* rDNA by successive precipitations with polylysine, monitored by CsCl gradient centrifugation. Samples 1, 2, and 3 are aliquots of the DNA remaining in solution after each precipitation. It can be seen that two DNA components other than rDNA ($\varrho = 1.723$) are also being concentrated. These components, however, are readily separated from rDNA by banding in CsCl. Although this method gives only partial enrichment of rDNA it does represent a major advance in the purification of *Xenopus* rDNA. When this enrichment procedure is used as a first step (before CsCl banding) it is possible to scale up the preparation of pure rDNA by several fold.

BROWN and coworkers have developed a simple and extremely effective method for purifying 5s DNA from *Xenopus laevis* that is certain to see much application [35]. Basically, the method takes advantage of the relatively high content of 5s DNA in the *Xenopus* genome combined with a magnification of the natural density difference between bulk and 5s DNA that can be obtained by differential binding of silver ions and actinomycin D. After complexing with these agents, the 5s DNA can be purified by cesium salt density gradient centrifugation.

In the *Xenopus* genome 5s DNA comprises about 0.05% of the bulk DNA, which is equivalent to 0.7% of the nuclear DNA [35]. These data and others taken from hybridization experiments indicate that the 5s DNA gene dosage is of the order of 24,000 copies per haploid complement of DNA, and that these cistrons are clustered. Because of this redundancy and a lower buoyant density in CsCl than shown by bulk DNA, BROWN's group attempted to concentrate this DNA by density fractionation. In the course of this study they learned that the apparent density difference (about 7 mg/cm^{-3}) could be magnified greatly by complexing the bulk DNA with silver ions before density fractionation. The bulk DNA binds silver ions to a greater extent than does either ribosomal or 5s DNA and the complexed DNA bands at even higher densities. Thus by successive bandings in Cs_2SO_4 it is possible to purify the 5s DNA to a high degree. After obtaining an apparently homogeneous band in Ag^+-Cs_2SO_4 the 5s DNA fraction is complexed with actinomycin D and again banded. 5s DNA binds less antibiotic than most other DNA sequences present and bands at a heavier density than the complexed material. Fig. 6 and 7 show the fractionation profiles obtained when the 5s DNA is purified in this manner.

Aliquots taken from each of the pooled 5s DNA fractions (from Fig. 7) were sedimented to equilibrium in CsCl in the absence of both Ag^+ and actinomycin D.

It is clear that after the final banding in actinomycin D-Cs_2SO_4 the 5s DNA fraction consists of DNA that appears virtually homogeneous with regard to density. These fractions are estimated to be more than 95% pure.

A comparison of the hybridization of purified 5s DNA and bulk DNA with 5s RNA indicates that the 5s RNA cistrons have been enriched about 130 fold and

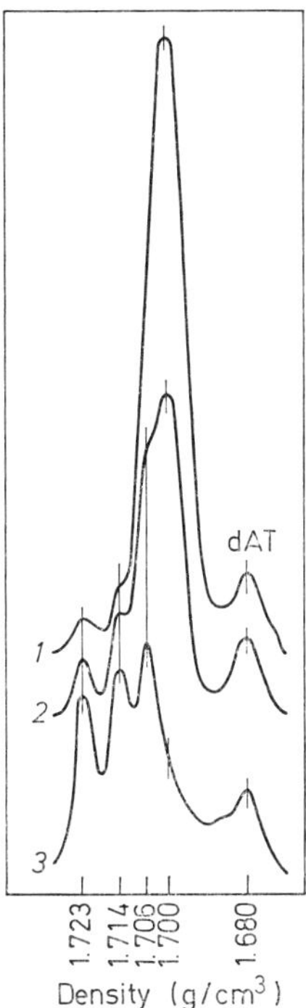

Fig. 5. Enrichment of rDNA by precipitation with poly-L-lysine. Samples of the DNA which remained in solution after each of three successive precipitations with polylysine were centrifuged to equilibrium in CsCl using dAT as a marker. In addition to rDNA (ϱ = 1.723), components with densities of 1.714 and 1.706 are enriched. The concentration of the 1.714 satellite is variable in different batches of DNA, but the satellite banding at 1.706 is always present. The percentage of the original DNA which remained in solution at each stage was 11, 2.7 and 0.5, respectively, for samples 1, 2 and 3. The rDNA component comprised 2.0, 6.4 and 28%, respectively, for samples 1, 2 and 3. Unfractionated *X. laevis* DNA contains about 0.2% of its DNA as chromosomal rDNA. (From DAWID et al., 1970)

that about one sixth of the DNA is complementary to 5s RNA. This figure must be considered a lower limit of purity since further characterization of the *Xenopus* 5s DNA suggests that the bulk of these sequences consist of so-called "spacer" DNA and that these spacer sequences are largely responsible for its low density. 5s RNA in *Xenopus* contains about 57% G+ C, yet the complementary DNA bands at a density corresponding to 33 to 35% G+ C. BROWN and his colleagues interpret thermal denaturation data from 5s DNA as evidence that the repeating 5s cistrons are separated by two regions of different density. Their estimates place the density difference of these sequences at 20% G+C or more.

Although this method appears to yield highly purified clusters of 5s DNA (molecular weight of about 1.8×10^6 daltons), the yield is rather low. The authors estimate the final yield at between 6 and 10%.

One might improve the yield by banding the DNA only twice; once in Ag^+-Cs_2SO_4 and once in actinomycin D-Cs_2SO_4. Apart from the single-stranded product, the only disadvantage of this method is that little material can be processed. Overloading

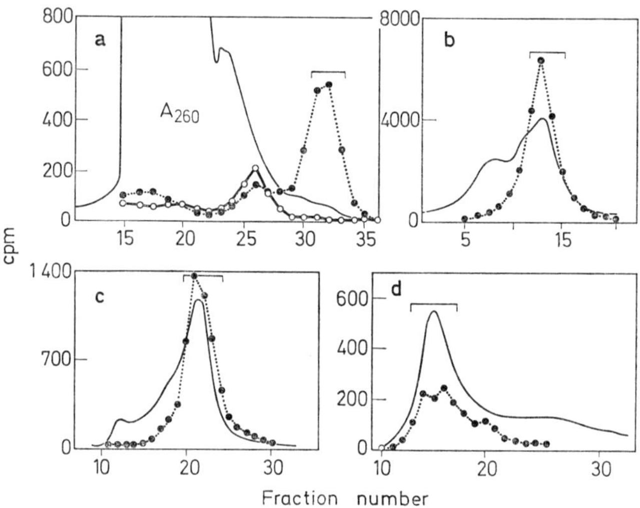

Fig. 6. Four density gradient steps for the purification of 5s DNA. a First step, Ag^+-Cs_2SO_4 gradient; b second step, Ag^+-Cs_2SO_4 gradient; c third step, CsCl gradient; d fourth step, actinomycin D-CsCl gradient. Aliquots of each fraction were hybridized with 5s (^3H) RNA (● - - - ●). In the first gradient (^{32}P)rRNA (○—○) was hybridized along with the radioactive 5s RNA; ———, absorbance at 260 nm. The absolute extent of hybridization is not comparable between the four gradients, since different aliquots were taken in each case. The *bracket* indicates the fractions containing 5s DNA that were pooled. (From Brown et al., 1968)

the gradients results in large losses due to the trapping of 5s DNA by bulk DNA. Thus, as previously stated in regard to isolation procedures based on density sedimentation, other techniques will probably have to be developed in order to obtain large-scale preparations.

In the sea urchin, *Lytechnus variegatus*, rRNA genes are contained in a G+C-rich satellite DNA which can be readily separated from main-band DNA by density gradient centrifugation [117], thermal elution chromatography on hydroxyapatite [138], or in a polyethylene glycol-dextran two-phase system [117]. In each case, resolution is possible because of substantial differences in the density or thermal stability owing to large differences in the G+C content of the two fractions. The

G+ C contents of the satellite and main band DNAs are 63% and 35%, respectively [138].

A very effective method for preparing sea urchin rDNA on a large scale has been described by PATTERSON and STAFFORD [117]. In their procedure, main-band DNA is selectively denatured by heat, quenched and then separated from double-stranded satellite DNA in a two-phase system composed of polyethylene glycol and dextran.

In this system native DNA is partitioned at 68° into the polyethylene glycol-rich top phase. By selecting a temperature at which most of the main-band DNA is

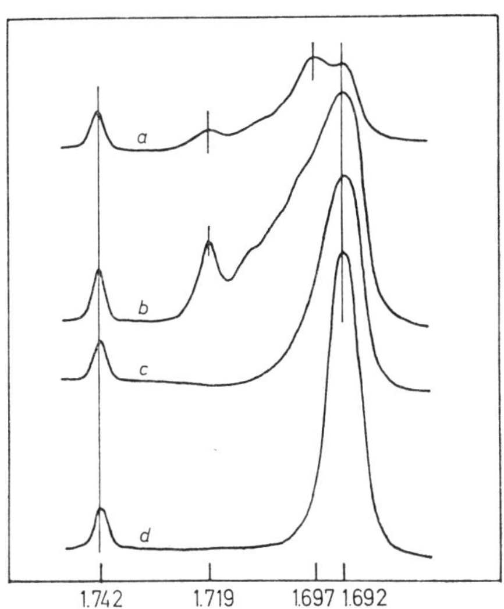

Fig. 7. Purification of 5s DNA. An aliquot of each of the pooled fractions bracketed in Fig. 6 was centrifuged to equilibrium in CsCl at 44,000 rpm for 18 h in the model E ultracentrifuge. The density marker is phage SPO 1 DNA (1.742 g/cm³). The DNA pooled at each of the four steps was 1.6, 0.43, 0.14, and 0.047% of the original starting material (120 mg of DNA in this particular experiment). From the model-E tracings, these were estimated to contain 35, 50, 83, and greater than 95% 5s DNA, respectively. The final yield is 0.047% represents about a 7% yield of the 5s DNA in the original unfractionated DNA. (From BROWN et al., 1968)

denatured with little dissociation of the satellite fraction, it is possible to effect a 700 to 1200 fold enrichment of rDNA. This material can then be further purified by CsCl gradient centrifugation.

Fig. 8 shows the CsCl density gradient profile of the distributed (polyethylene glycol phase) native, unfractionated sea urchin DNA. It is apparant that the heavy satellite ($\varrho = 1.722$ g/cc) accounts for only a small fraction of the total DNA. When bulk DNA is partially denatured at 79° or 80° and partitioned in the two-phase system, the double-stranded DNA fraction is highly enriched in satellite material (Fig. 9). Although treatment at 80° (panel b — Fig. 9) results in a greater enrichment of satellite DNA (1200 fold vs 700 fold at 79°) the total yield of satellite DNA is lower

(23% compared to 45% for the sample treated at 79°). Inasmuch as DNA prepared by treatment at either temperature can be readily resolved into satellite and main-band fractions by preparative CsCl density gradient centrifugation, PATTERSON and STAFFORD opted to use the 79° treatment in their standard procedure. Evidence that the satellite fraction contains rDNA comes from hybridization studies, the results of which are shown in Fig. 10. Here satellite enriched DNA was separated into satellite and main-band fractions by CsCl gradient centrifugation. It can be seen that essen-

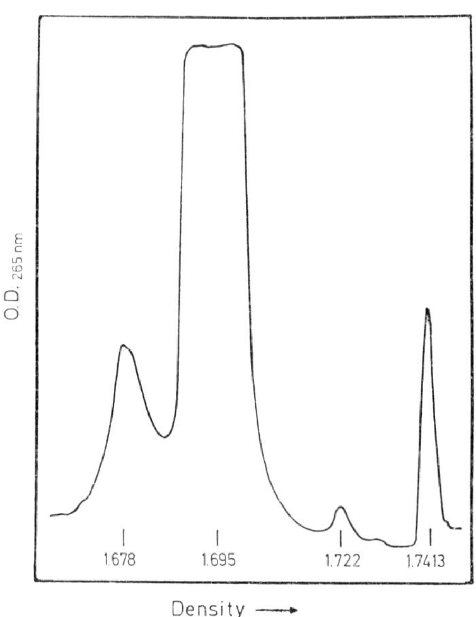

Fig. 8. Satellite in sea urchin sperm. The polyethylene glycol rich top phase (20 μl) of the two-phase system containing 100 ug of native unfractionated sea urchin DNA in 0.01 M sodium phosphate plus 170 ul of water were added to 700 ul of CsCl stock solution. Synthetic poly d(AT) (density 1.678 g/cc) and SP82 DNA (density 1.7413 g/cc) on the scale where *E. coli* DNA is 1.710 g/cc were added as reference standards. The main band (density 1.695 g/cc) was heavily overloaded. This same pattern is obtained with DNA which has not been put through the phase system. Density gradients were established by centrifugation at 44,700 rpm and 25 °C for 22 h in a Spinco Model E analytical ultracentrifuge. Samples were photographed at 265 nm and films were traced by a Joyce-Lobel microdensitometer. (From PATTERSON and STAFFORD, 1970)

tially all of the DNA complementary to sea urchin rRNA is associated with the heavy satellite band.

This fractionation method is particularly attractive because it is simple, effective, yields double-stranded template and, perhaps most important, is well suited for large-scale preparations. Unlike any other method we shall describe here, this procedure can be used to process gram quantities of DNA and is limited only by the size of the vessels available for the partition operation. Further, the procedure should be useful for the enrichment of fast-renaturing DNAs such as the circular DNAs of certain viruses, plasmids and mitochondria.

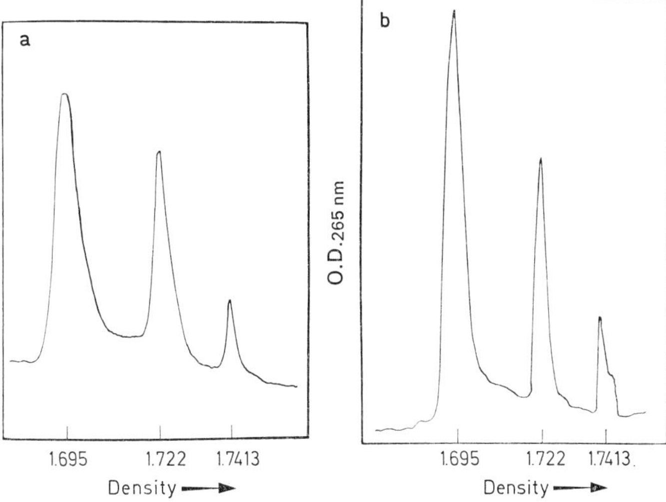

Fig. 9. Analytical CsCl density gradient banding pattern of sea urchin sperm DNA. The DNA was selectively denatured at 79 and 80 °C and separated into the polyethylene glycol rich top phase. The reference standard was SP82 DNA (density 1.7413 g/cc). Centrifugation was achieved as in Fig. 8. a DNA (2.5 μg) isolated in the polyethylene glycol rich phase after selective denaturation at 79 °C was loaded onto the CsCl in 25 μl. Satellite DNA comprised about 21 % of the total sea urchin DNA sample, an enrichment of about 700-fold; b 80° C-treated DNA. DNA (2.5 μg) was loaded onto the CsCl in 50 μl. Satellite DNA comprised about 36 % of the total sea urchin DNA sample, an enrichment of about 1200-fold. About twice as much satellite DNA is lost in the 80 °C-treated sample. Densitometer tracings of the film records are shown. (From Patterson and Stafford, 1970)

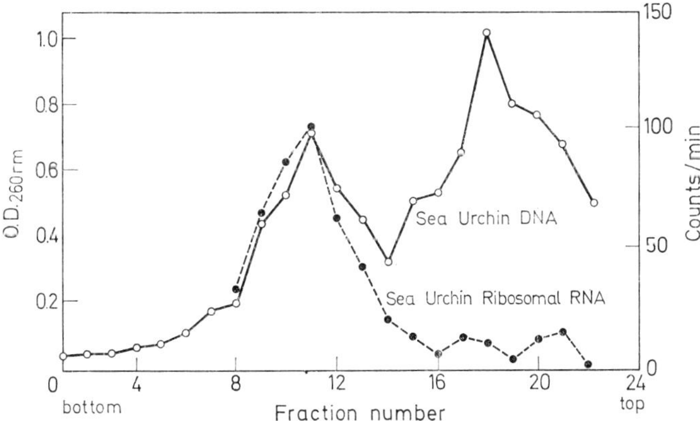

Fig. 10. Hybridization of rRNA to sea urchin satellite DNA. The DNA and RNA were preparared as described in the Materials and Methods section of the text. DNA from the 79 °C isolation in the polyethylene glycol-dextran two-phase system was centrifuged to equilibrium in parallel preparative CsCl density gradients and fractions were collected for hybridization. The counts per minute shown represent the total counts per minute minus the control counts per minute. (a) Hybridization with 7.3 μg of sea urching rRNA. The specific activity of the RNA was 348 cpm/μg. (From Patterson and Stafford, 1970)

2. *DNA Reassociation Kinetics*

When DNA is denatured and allowed to reassociate, the rate at which duplex formation occurs is dependent on the concentration of DNA. If the DNA contains only a single copy of each gene then all cistrons will reassociate at virtually the same rate. However, if there is redundancy, that is, duplicate copies of a given gene or a related family of genes, the rate at which that DNA reassociates will be faster than the DNA of a cistron which occurs only once in the genome. If there are 10 copies of the specified gene, those cistrons will reassociate ten times faster than the single-copy cistron.

By taking advantage of a difference in the kinetics of reassociation and techniques for separating single- and double-stranded DNA, genes that are highly redundant can be separated from the rest of the genome and thus concentrated.

rDNA: LUSBY and DE KLOET used this rationale to design a method of enriching the rRNA cistrons from *Saccharomyces carlsbergensis* [95]. Inasmuch as the buoyant density difference between the yeast DNA homologous to rRNA and the remainder of the DNA is small (1.698 vs 1.703 g/cc), density fractionation of rDNA is not practical. However, because the yeast haploid chromosome contains some 140 copies of the rRNA cistrons [55, 123, 130], enrichment by differential reassociation kinetics is possible. These workers found (Fig. 11) that when denatured yeast DNA is subjected to gradient elution from a hydroxyapatite column about 7% of the material elutes in the native DNA region after the main band. When the two DNA fractions were assayed for rDNA by hybridization it was learned that the native-like DNA contains 70 to 85% of the rRNA cistrons. It will be interesting to see if this fraction also includes tRNA and 5s RNA cistrons since these sequences have also been amplified to the extent of 320 to 400 copies [130]. With no further purification the method of LUSBY and DE KLOET will yield a tenfold enrichment of native rDNA. It seems, however, that the rRNA cistrons could be brought to a much higher state of purity in one or two additional steps.

If their rDNA fraction is next hybridized with rRNA and again chromatographed on hydroxyapatite, the rRNA:DNA hybrids and DNA:DNA duplexes should be separated from the minus strand of the rRNA gene [85]. Then, if the double-stranded material (RNA:DNA and DNA:DNA duplexes) is subjected to alkaline hydrolysis to degrade the RNA and again fractionated on hydroxyapatite, it should be possible to obtain single-stranded rDNA of high purity.

Eukaryotic DNA: DNA from all animals and plants, and from most of the higher protista (yeast, fungi, true algae and protozoans) examined to date contains both unique and repeated nucleotide sequences (for reviews see references 28, 31, 84; also 22, 30, 88). Unique DNA sequences are present only in a single copy per genome. Repeated DNA represents multiple copies of similar or identical sequence. A "family" of repeated DNA is a series of DNA segments sufficiently similar in base sequence to anneal with one another at arbitrarily chosen criteria for DNA reassociation. It is assumed that families arose from the multiple replication of small DNA segments, the subsequent integration of this DNA into the genome, and its dissemination throughout the species [28].

In species thus far examined [28] repeated DNA constitutes 20 to 80% of the total nuclear DNA, with from fifty to two million related sequences per family. Repeated

DNA is distributed throughout the length of the genome. All degrees of thermal stability (and hence degrees of base sequence complementarity) are seen in various families, and from this the age of families is estimated from very recent (high degree of thermal stability indicating little sequence divergence) to several hundred million years (markedly decreased thermal stability indicating extensive base sequence divergence). The repeated DNA is always expressed, and different DNA "families" are expressed in different tissues and during various stages of development.

It is possible to separate repeated DNA from unique DNA by means of rate kinetics of DNA reassociation [28, 31, 84]. Since DNA reassociation is a collision-dependent reaction that follows second-order kinetics, the speed with which a DNA

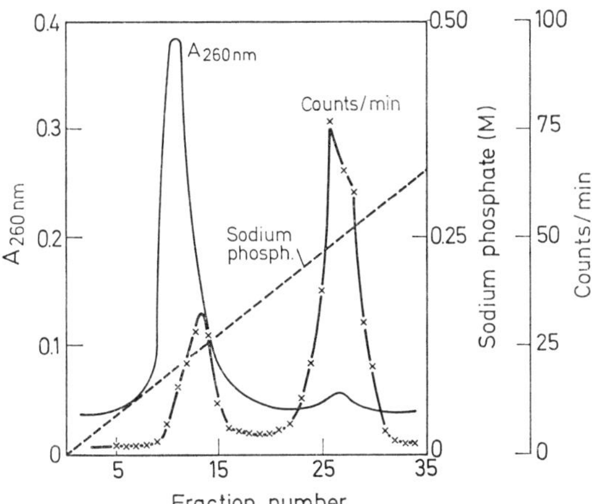

Fig. 11. Hybridization of yeast rRNA with yeast DNA fractionated on hydroxyapatite. Denatured unlabeled yeast DNA (75 μg) was fractionated on hydroxyapatite. Fractions were heat denatured in the presence of 10 μg ^{32}P-labeled rRNA (specific activity, 900 counts/min per μg), cooled to and incubated at 60 °C in 0.3 M NaCl—0.03 M trisodium citrate (pH 7.3) for 2 h. (From Lusby and De Kloet, 1970)

fraction anneals is directly proportional to its relative concentration in the incubation mixture. Once isolated, families of repeated DNA may be further fractionated by number of copies. In addition, old and new families are separable either by utilizing differences in thermal stability or by employing stringent reassociation criteria (raising incubation temperature and/or decreasing salt concentration) so that only thermally stable (new) families are able to form stable duplexes. In all these reactions reassociated DNA is separated from unreacted DNA by passage through hydroxyapatite.

Satellite DNAs have been found and studied in many, although not all eukaryotes tested [28, 31, 84]. A representative list of satellite DNAs is given in reference 47. Satellites are distinguished from the remaining DNA by their different G+C content.

Several satellite DNA fractions are present as highly repeated DNA. These are among the most readily and highly enriched by differential reassociation kinetics.

Mouse satellite [64, 148, 150] and crab satellite [134, 135, 139] are probably the best characterized of these DNAs. The function, if any, of satellite DNA is not known.

Except for rRNA and tRNA, which are repeated as many as 1000 times, the function of repeated DNA is unknown. Various hypotheses have been advanced. It may well be important in evolution, as evidenced by its wide distribution and high concentration in eukaryotes. Repeated DNA may have originated as a gene pool from which new structural genes were gleaned in rare events [28]. Another hypothesis envisions a current function for repeated DNA in a structural or regulatory role [29]. Other possible functions for the large amounts of repeated DNA in animals are in specifying antibodies and in learning.

Repeated DNA may hold the key to the next revolution in molecular biology. A detailed review of progress in determining its role must await the results of intensive studies now under way in several laboratories. In any event, the methodology for greatly enriching various portions of repeated DNA is available.

3. RNA:DNA Hybridization

Given the assignment of isolating a cistron whose RNA product can be prepared in large quantities and in purified form, an obvious approach is to form RNA:DNA hybrids and separate this material from non-hybridized DNA. This approach has been used successfully in a number of laboratories for the enrichment of ribosomal [41, 44, 45, 83, 131, 144] transfer [26, 27, 97, 99] and 5s RNA cistrons [33, 59], and also for the rII region and gene 21 region of DNA from phage T4 [11, 12, 78, 103].

Once formed an RNA:DNA hybrid can be separated from the remaining DNA by density gradient centrifugation or, if the non-hybridized DNA has not been allowed to reassociate, by any method that can distinguish between single- and double-stranded structures.

To minimize the content of contaminating DNA in the final gene preparation, the DNA moiety of the hybrid must be reduced. This can be effected before or after hybridization. When the DNA piece size is reduced before hybridization, fragmentation is usually accomplished by sonication or passage through a high pressure cell. If sheared carefully, the DNA can be reduced to fragments which are approximately the same size or only slightly larger than the gene to be isolated. On the other hand, some workers have elected to reduce the DNA fragment after hybridization by treating the RNA:DNA duplex with a DNase specific for single strands. This treatment should yield a duplex devoid of unhybridized ends because the nuclease will digest the single-stranded DNA contiguous to the hybridized cistron. This approach may not be applicable for the isolation of polycistronic units as the intercistronic sequences may not be protected from the nuclease.

rRNA:DNA hybrids: Ribosomal RNA:DNA hybrids have been isolated thus far from *E. coli* [83], *Proteus mirabilis* [83], *B. subtilis* [44, 45, 131], *Salmonella typhimurium* [144] and *Neurospora crassa* [41].

Originally devised to purify the ribosomal RNA cistrons from *E. coli* and *P. mirabilis*, the gene isolation procedure developed by KOHNE [83] can be used to purify any cistron for which sufficient quantities of gene product can be obtained.

KOHNE used hydroxyapatite column chromatography to separate rRNA:DNA hybrids from non-hybridized single-stranded DNA. The hybrids are formed in solution from sheared ^{32}P-labeled DNA (single-stranded fragment size $= 1.50 \times 10^5$

daltons) and highly purified ribosomal RNA. The hybridization reaction mixture is then applied to a hydroxyapatite column equilibrated at 60° at an ionic strength which prohibits single-stranded DNA from being adsorbed while all duplex material is retained. The bound DNA is then eluted by either raising the temperature of the column to 100 °C to dissociate the duplex material or increasing the ionic strength of the eluting buffer. Since some DNA:DNA duplexes are formed during hybridization and are retained with the hybrids on hydroxyapatite, it is necessary to repeat the hybridization and fractionation procedures. With each succeeding cycle of hybridization and chromatography the amount of non-rDNA is decreased, until after two or three cycles the amount of DNA bound by the column is constant and includes only fragments which are complementary to rRNA.

The hydroxyapatite column does not completely distinguish between RNA:DNA hybrids and DNA:DNA duplexes, and the hybridization conditions are critical to the success of the method. Conditions must be carefully controlled to ensure that hybridization of the rRNA cistrons is complete, yet only a small portion of non-rDNA is allowed to reassociate. Since nucleic acid reassociation in solution is essentially a simple, second-order process, the extent of DNA:DNA duplex formation can be governed by manipulating the DNA concentration or time of incubation. KOHNE uses the product of these parameters in describing reaction conditions and for convenience uses the term Cot (concentration × time — ref. 30) to express this value. In a RNA: DNA hybridization reaction, because of the smaller informational length of the RNA as compared to bulk DNA, the RNA Cot governs the formation of RNA: DNA hybrids while the DNA Cot determines the extent of DNA:DNA reassociation.

In the first cycle of purification it is not technically practical to use a low DNA Cot value which would virtually preclude DNA:DNA reassociation, since to do so would require the processing of large volumes of very dilute DNA. In later cycles, after most DNA has been eluted as single-stranded fragments, the DNA:DNA interaction is less favored and the content of non-rDNA falls with each cycle.

Table 2 [83] gives the hybridization data obtained when *E. coli* rDNA is purified by KOHNE's method. The results in part A of the table show that after 3 cycles of hybridization and chromatography the amount of DNA bound to the column is constant and represents about 0.27% of the initial input. This value is in good agreement with the rDNA content of the *E. coli* genome determined by others [137]. Part B of Table 2 shows the hybridization results obtained when rRNA is not included in the reaction mixture (rRNA Cot = 0). This control experiment shows that the isolation procedure is indeed dependent on the inclusion of RNA and also reveals that the content of non-rDNA fragments in the final preparation is probably no greater than 3% (0.01/0.27). That the formation of hybrid is specific is clear from Part C (Table 2) where chicken rRNA is substituted for *E. coli* rRNA. It can be seen that the level of DNA recovered after 3 cycles is almost identical to that obtained in the absence of added RNA. The denatured DNA fragments used by KOHNE are smaller (average molecular weight — 150,000 daltons) than either 16s or 23s RNA (or their precursors) which have molecular weights of 0.55×10^6 and 1.1×10^6 daltons, respectively. For this reason these cistrons cannot be complete units of transcription. Nonetheless, this work represents the first virtually complete purification of a specific functional sequence of DNA.

For nucleic acid hybridization studies the size difference is unimportant and may indeed be preferable inasmuch as the purity of rDNA will be greater. For transcription studies this shortcoming can be remedied by using DNA of higher molecular weight.

The isolation procedure developed by KOHNE can be scaled up to process a few hundred mg of DNA on a single large column of hydroxyapatite. However, if a batch procedure is used for the first cycle, then the amount of material that can be processed is limited only by the amount of hydroxyapatite one cares to contend with. Because the sorbent settles out quite rapidly, the supernatant containing the non-adsorbed single-stranded DNA can simply be decanted [57].

To avoid thermal damage to the DNA, it seems advisable to elute the hybrid material with buffer of high ionic strength or perhaps a denaturing agent such as

Table 2

Cycle	^{32}P-DNA C_0t	rRNA C_0t	Per cent of original input ^{32}P-DNA adsorbed
A 1	0.08	0.25 (*E. coli* RNA)	2.37
A 2	0.0008	0.25 (*E. coli* RNA)	0.307
A 3	1.5×10^{-5}	0.25 (*E. coli* RNA)	0.276
A 4	1.5×10^{-5}	0.25 (*E. coli* RNA)	0.265
B 1	0.08	0	2.1
B 2	0.0008	0	0.029
B 3	7×10^{-6}	0	0.01
C 1	0.033	0.3 (chicken RNA)	1.1
C 2	0.0002	0.3 (chicken RNA)	0.038
C 3	9×10^{-6}	0.3 (chicken RNA)	0.017

Data showing the fraction of original input ^{32}P-*E. coli* DNA adsorbing to hydroxyapatite when reacted (A) with *E. coli* rRNA, (B) in the absence of any rRNA, (C) with chicken rRNA. (From KOHNE, 1968).

formamide. The only outstanding disadvantage of the procedure is that it yields single-stranded product. However, it should be possible to form double-stranded template by hybridization of the hydroxyapatite cistrons with unfractionated DNA, but, for this approach to be fruitful, the complementary strands of the bulk DNA must be separated before hybridization. This should reduce the extent of undesirable DNA:DNA reassociation and make it possible to maintain the purity of the cistron preparation.

A better approach to the problem of forming double-stranded cistrons would seem to be via biosynthetic enzymes. The single-stranded cistron could be used as a template for the synthesis of its own complement by DNA polymerase I if a primer could be attached to the 3′ end of the template. This would be accomplished by hybridizing a deoxyoligo-nucleotide homopolymer to a suitable complementary homopolymer previously added to the 3′ terminus of the cistron by the enzyme DNA terminal nucleotidyl transferase [21].

Ribosomal RNA:DNA hybrids have also been isolated from *Neurospora crassa* [41] by a modification of KOHNE's hydroxyapatite procedure [83]. In this study, CHATTOPADHYAY, KOHNE and DUTTA [41] were able to enrich *N. crassa* rDNA some 100-fold to better than 90% purity.

As in the KOHNE procedure, rRNA:^{32}P-DNA hybrids formed in solution are separated from non-hybridized, single-stranded DNA by chromatography on hydroxyapatite. Under the conditions of the experiment, single-stranded DNA is not adsorbed to the gel while hybrids and reassociated DNA duplexes are retained. The

Table 3. Isolation of rRNA cistrons by DNA:RNA hybridization

^{32}P DNA C_0t	rRNA C_0t	Percent adsorbed	Percent of original ^{32}P input adsorbed	Corresponding experiment without any rRNA added: ^{32}P-DNA C_0t	Corresponding experiment without any rRNA added: Percent ^{32}P input adsorbed
1. 0.062	0.43	1.95[a]	1.95	1. 0.013	1.04
68 % apparent rDNA recovered after RNase treatment			1.238	25 % DNA recovered after RNase treatment	0.125
2. 0.00034	0.43	86.5	1.071	2. 0.000125	0.014
3. 0.00026	0.43	95.5	1.023	3. 0.000001	0.004
Base hydrolysis with NaOH					
4. 0.00014	0.43	96.1	0.983		
5. 0.00010	0.42	95.4	0.938		

Summary of results obtained in the modified isolation procedure of rRNA cistrons. Same ^{32}P labeled *N. crassa* DNA (190,000 cpm/μg DNA) is used in both experiments, (a) control experiment without any rRNA and (b) for making DNA:RNA hybrids. Percent hybridization is measured by calculating the fraction of original input ^{32}P- *N. crassa* DNA adsorbing to hydroxyapatite. (From CHATTOPADHYAY et al., 1972).

[a] 1.95 % recovered after first cycle, as a mixture of ^{32}P DNA:RNA and ^{32}P DNA:^{32}P DNA hybrids, is treated with RNase in 0.01 M phosphate buffer (PB). After RNase treatment, molarity is raised to 0.14 M PB and passed through hydroxyapatite. 68 % of this total ^{32}P-DNA (mostly r-RNA cistrons) is eluted in single stranded form.

double-stranded material is eluted by high salt or temperature, and the material is treated with pancreatic RNase at low ionic strength. This modification of the method described by KOHNE results in the degradation of the RNA moiety of the hybrid, leaving a mixture of single stranded rDNA and DNA:DNA duplexes. The sample is then reapplied to a hydroxyapatite column to separate these two species. The early-eluting single-stranded DNA is then subjected to two or more cycles of hybridization and hydroxyapatite chromatography to remove all DNA fragments not complementary to rRNA.

Table 3 summarizes the purification results [41]. After a single cycle of hybridization and column purification, the DNA is approximately 50% pure with regard to its ability to hybridize rRNA. Treatment with RNase and rechromatography increases the purity to about 80%. Finally, after two additional cycles of purification by hybridization and chromatography, the amount of DNA retained by the column is constant at 1% of the initial input DNA, and agrees with the rRNA cistron content of *N. crassa* DNA [41, 153].

Although rDNA prepared in this way is well suited for nucleic acid hybridization studies (concerned with rRNA gene conservation), the material has limited usefulness

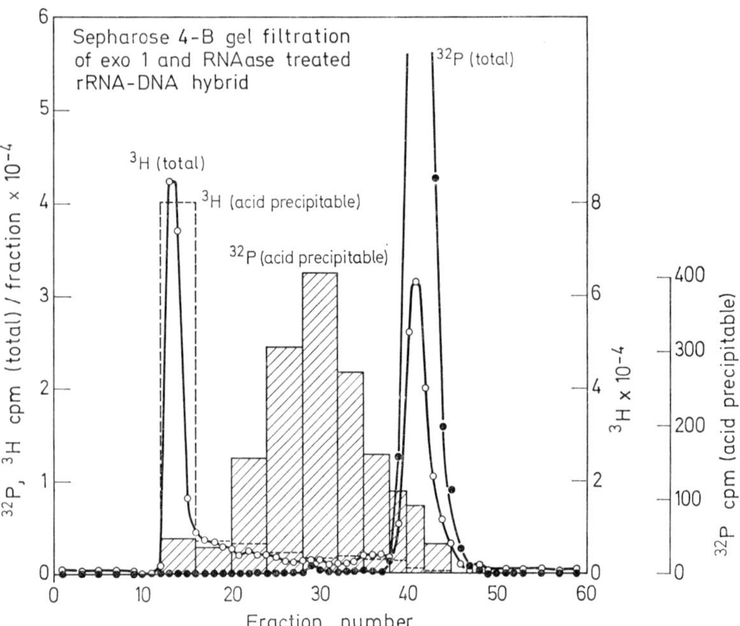

Fig. 12. Gel filtration of exonuclease I- and RNase-treated hybrid. A column of Sepharose 4 B (Pharmacia, Sweden), 70 cm high and 1.5 cm in diameter, was loaded with about 4 ml of enzymatically treated hybrid. Elution was accomplished at 4 °C with 0.3 M KCl, 0.01 M Tris, pH 7.3; and 2.5 ml fractions were collected. Aliquots of 0.1 ml were spotted on paper filters for total radioactivity determination in a Packard scintillation counter. For the analysis of acid precipitable counts, 0.5 ml aliquots were precipitated with 2 N HCl and filtered through Millipore. The excluded peak is undigested DNA and the two last coincident peaks are the hydrolysis products of RNA and DNA. In the middle is the acid-precipitable ^{32}P-labeled RNA, represented according to the enlarged scale shown on the right. (From Sgaramella et al., 1968)

as a template for transcription studies. Inasmuch as the DNA was initially sheared to single-stranded fragments of molecular weight about 125,000 daltons, the isolated rDNA sequences cannot be intact cistrons.

Sgaramella and co-workers [131] are believed to be the first investigators to combine hybridization and nuclease treatment to purify a specific DNA sequence. After hybridization of *B. subtilis* rRNA and DNA in solution, these workers add *E. coli* exonuclease I [82] to the reaction mixture to hydrolyze unhybridized, unrenatured DNA. Following this treatment a mixture of pancreatic and T_1 RNases is added to digest unhybridized RNA. The mixture is then fractionated on a column of Sepharose 4B, giving rise to the elution profile shown in Fig. 12.

The excluded material (fractions 12 to 16) is presumed to contain undigested DNA and some RNA (possibly in hybrid form). The peak of material eluting between fractions 38 to 46 contains the acid-soluble hydrolysis products. The hybrid elutes as a broad band mainly in the region between the excluded and small molecular weight products.

The hybrid fractions are pooled and then applied to a cellulose nitrate column [82]. Single-stranded structures and duplexes with single-stranded "tails" are bound while double stranded DNA and nuclease-treated RNA:DNA hybrids are not adsorbed.

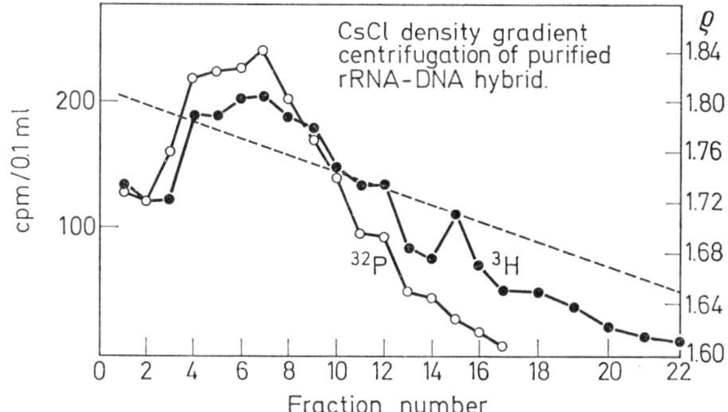

Fig. 13. CsCl density gradient centrifugation of partially purified hybrid. After a previous fractionation through a similar gradient, the fractions with a density higher than 1.73 g/cm^3 were placed in 12 ml of a solution of CsCl with average density of 1.72, and run in rotor ≠. 40 of a Spinco-L ultracentrifuge, at 33,000 rpm for 72 h, at 20 °C. 12 drop fractions were collected from the bottom and 0.25 ml aliquots of each fraction were assayed for acid-precipitable radioactivity (a quantitative recovery of the input material was obtained). No correction has been made on the specific activity of ^{3}H-DNA for the different base composition of the hybridized sequences in comparison to total DNA. The radioactivity of ^{32}P-rRNA has been normalized to that of ^{3}H-DNA. (From SGARAMELLA et al., 1968)

The hybrid is then separated from the remaining DNA contaminants by equilibrium centrifugation in a CsCl density gradient. Fig. 13 shows the CsCl fractionation profile obtained. The single broad hybrid band occupies a region of the gradient which is much less dense than RNA but more dense than either native or denatured DNA.

Table 4 summarizes the results of the purification scheme at each step. It can be seen the rDNA has been purified at least 250 fold which corresponds to a product that is approximately 75% pure if hybrid formation is complete (our calculation). The authors fail to report the molecular weight of the final product.

Although this procedure is simple and effective, the CsCl fractionation step represents a technical "bottleneck". As pointed out earlier, even ignoring the time and financial cost involved, only microgram quantities of nucleic acid can be fractionated by this method as we know it today. Further, the single greatest loss of hybrid material occurs at this stage due largely to overlapping of the broad bands — a situation that will not improve with overloading.

As the authors point out in their communication pure exonuclease I is not ideally suited for trimming away single stranded ends of a hybrid structure. This enzyme attacks from the 3′-OH terminus only and thus will not hydrolyze a single strand tail at the 5′ end. Neither is this enzyme capable of completely digesting denatured DNA which contains renatured areas. For their work, SGARAMELLA et al. used a less pure preparation of exonuclease that contained some endonuclease I. Together, these nucleases should hydrolyze all single stranded DNA present in the hybridization mixture. It is not clear, however, if the DNases digest all the single stranded DNA in a hybrid or if some short "tail" remains at each end. This point is critical from the viewpoint of knowing or predicting how much of the gene can be isolated. If the RNA polymerase recognition and regulatory sites are not protected by the RNA:DNA hybrid structure then those regions of the DNA which code for a larger precursor RNA may not escape digestion by nuclease treatment.

Table 4. Summary of the purification. (From SGARMELLA et al., 1968)

Step	DNA/RNA	Yield, %
Annealing of rRNA to total DNA	250	100
Gel filtration	20	70
Nitrocellulose filtration	10	62
CsCl centrifugation	~1	12

In characterizing the arrangement of ribosomal RNA genes in *Drosophila melanogaster* DNA, QUAGLIAROTTI and RITOSSA were able to demonstrate that rRNA:DNA hybrids could be separated from the bulk of the cellular DNA by nitrocellulose column chromatography [120]. Although these workers were not attempting to isolate rDNA in their study this work clearly demonstrates that nitrocellulose columns can be used to enrich rRNA:DNA hybrids. Under their conditions 40 to 55% of the rRNA:DNA hybrids formed in solution can be eluted from the column while the bulk of non-hybridized, single stranded DNA is retained. Because purification of rDNA was not the motive for their fractionation work, no data are available on the actual enrichment obtained. However, the elution profiles in Fig. 14 indicate that while 40% of the DNA complementary to rRNA is eluted at low ionic strength it appears that much less than 10% of the total DNA adsorbed is released by this treatment. Even if this procedure results in an enrichment of only a few fold, if it can be scaled up or modified for batch operation, the method may well prove to be useful for preparative work.

Ribosomal RNA genes have also been prepared from *B. subtilis* DNA by COLLI and OISHI [43, 44, 45]. Their approach is rather unique and involves: 1) separating the complementary strands of DNA which has previously been sheared; 2) forming RNA:DNA hybrids in solution; 3) separating the hybrids from renatured DNA by chromatography on a hydroxyapatite column and; 4) separating the hybrids from single stranded DNA by equilibrium centrifugation in a Cs_2SO_4-$HgCl_2$ density gradient. Their final preparation of rDNA has been enriched some 80 fold and is at least 30% pure [45].

Because earlier hybridization experiments by COLLI and OISHI [44] indicated that the 16s and 23s ribosomal RNA cistrons are linked on single stranded DNA fragments of molecular weight approximately 2×10^6 daltons it was decided to shear the DNA to about 3.5×10^6 daltons for hybrid formation. This size was selected so that on the average each fragment of rDNA will carry one set of 16 and 23s rRNA cistrons.

The rationale for separating the complementary DNA strands was to minimize DNA renaturation and thus, the problem of separating RNA:DNA hydrids and

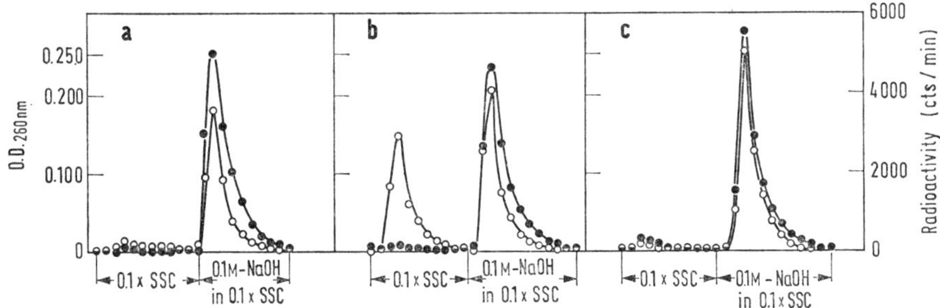

Fig. 14. Elution profiles of rRNA-DNA hybrids from nitrocellulose columns. a 35 μg DNA sheared to an average mol. wt of 3.7×10^6 daltons (before denaturation) were hybridized with 3.5 μg of a mixture of 28s and 18s rRNA labelled with ^{32}P (167,000 cts/min/μg). At the end of the incubation, 0.207% of the DNA was hybridized with rRNA. Only 7% of the hybrid, however, was eluted with 0.1 × SSC from a nitrocellulose column. b 35 μg of the same DNA preparation used in (a) were incubated under standard conditions with 12 ug of rRNA labelled with ^{32}P (167,000 cts/min/μg). The DNA in this case was hybridized to saturation with rRNA (0.337%). In this case 40% of the hybrid was eluted with 0.1 × SSC from a nitrocellulose column. c 30 μg sheared DNA (2.8×10^6 daltons before denaturation) were hybridized under standard conditions with saturating amounts of 28s rRNA (9 μg in 2 ml) which was labelled with ^3H (170,000 cts/min/μg). In spite of the fact that all DNA sequences complementary to 28s rRNA were in hybrid form (28s rRNA-DNA was 0.228%), only 4% of the hybrid could be eluted from nitrocellulose columns with 0.1 × SSC. —●—●—, O. D. 260 mu; —○—○—, radioactivity. (From QUAGLIAROTTI and RITOSSA, 1968)

DNA:DNA duplexes at a later stage. Further, by removing DNA strands which can compete with rRNA during hybridization, that is, the minus strand rDNA, the yield of RNA:DNA hybrids should be improved. As it turns out, the complementary strands of rDNA can be well separated by chromatography on methylated albumin kieselguhr [44].

Fig. 15 shows the strand separation obtained when sheared, denatured DNA is eluted from methylated albumin kieselguhr with a shallow, linear salt gradient. In agreement with the results of RUDNER and coworkers [126] COLLI and OISHI find that the DNA complementary to 16 and 23s RNA elutes with H-strand (heavy) DNA, but in the later fractions of that peak. They further determined by competition hybridization that the DNA complementary to the H strand of the rDNA elutes early in the L-strand (light) region and is well separated from H-rDNA.

Evidence that the complementary strands of the rest of the genome are also separated for the most part comes from an experiment in which the fractions of H-DNA which hybridize with rRNA were allowed to reassociate and then examined for DNA:DNA duplex formation by hydroxyapatite column chromatography.

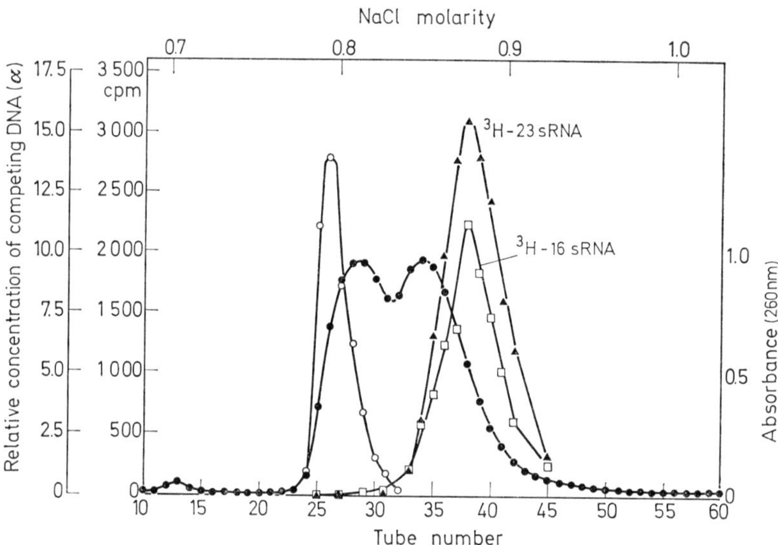

Fig. 15. Methylated albumin kieselguhr column chromatography of denatured DNA. 4 mg of denatured DNA were applied to the column (3.2 × 10.0 cm) and eluted by a linear gradient of NaCl (0.6 to 1.2 M, 500 ml total) in the presence of 0.05 M sodium phosphate buffer, pH 6.7, and 5.5 ml fractions were collected. After measuring absorbance at 260 mμ (●), 0.1 ml of each fraction was taken for hybridization with ^{3}H-16s RNA (□) and ^{3}H-23s RNA (▲). Saturating amounts of labeled ribosomal RNA (0.2 μg/ml) were used during hybridization. The relative concentration of the DNA which is competitive with ribosomal RNA during hybridization was assayed by incubating (68 °C for 2 h) 0.2 ml of each fraction of the first peak in tubes containing 2.3 μg of pooled DNA (tubes 37 to 43), 0.2 μg of ^{3}H-23s RNA and 0.3 M NaCl in a final volume of 1 ml. DNA from the first peak was omitted in the control tube. The assay of hybridized radioactivity was performed as described in Materials and Methods. The values (α, o) were obtained, after normalizing the relative inhibitory activity (β), by using the following equation $\alpha = \beta\ (1 - \beta)$ ($\beta = 1$ when inhibition is 100%). (From COLLI and OISHI, 1969)

Fig. 16 shows the salt gradient elution profile obtained from hydroxyapatite in such an experiment [45]. Under their conditions, single stranded DNA and the rRNA:DNA hybrids elute early (fractions 50 to 60) while renatured and partially renatured DNA are not released until a higher ionic strength. It can be seen that labeled RNA is also present in the renatured DNA fractions for reasons that are not clear to us and not explained by the investigators.

Colli and Oishi then used Cs_2SO_4 density gradient centrifugation to separate the RNA:DNA hybrids from single stranded DNA. In this fractionation mercuric ions were included to effect better separation of single and double stranded DNA. This is possible because more mercuric ions are bound to single stranded DNA than to native DNA and the difference in their respective densities may be magnified. They reasoned that RNA:DNA hybrids, which are partially double stranded, would

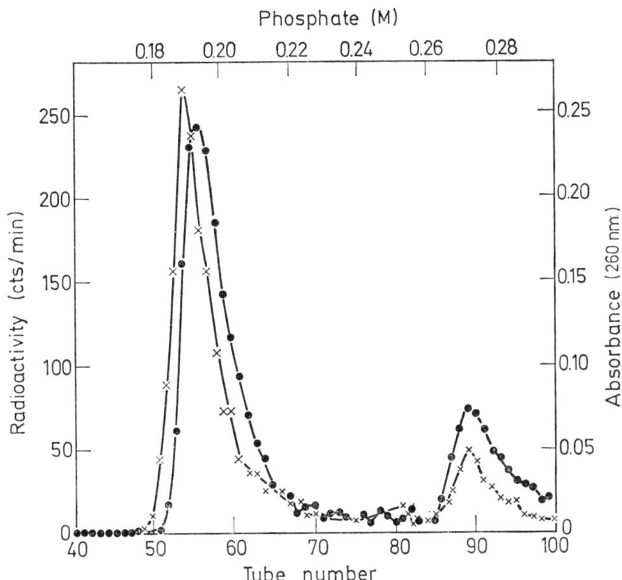

Fig. 16. Hydroxyapatite column chromatography of the DNA-RNA hybrid. The hybrid DNA-RNA obtained was applied (total input = 239,000 cts/min) to the hydroxyapatite column (1.7 cm × 4.3 cm) and washed 5 times with 5-ml portions of 0.07 M sodium phosphate buffer (pH 7.0). The elution was performed at room temperature by a linear gradient of sodium phosphate (0.07 M to 0.4 M, 240 ml total) and fractions of 2 ml were collected. The absorbance (—●—●—) of each fraction was measured at 260 mμ. The peak to the left is single stranded DNA and the peak to the right is renatured DNA. 25-μl portions were taken from each fraction, mixed with 5 ml of 0.5 M KCl and filtered through membrane filters; (— × — × —), radioactivity. (From Colli and Oishi, 1970)

bind mercuric ions to a much lesser extent than would single stranded DNA and thus band at a lower density in the gradient. Fig. 17 shows this to be the case.

Under their conditions, non-hybridized denatured DNA is more dense than either the RNA:DNA hybrid (middle peak) or native DNA, but is not sufficiently dense to permit complete separation from the hybrid material. Nevertheless, if one is willing to sacrifice yield by pooling only those fractions which are virtually free of single stranded DNA a hybrid preparation of high purity can be recovered. When the shaded fractions (Fig. 17) were assayed, the final preparation of rDNA was enriched 80 fold and was at least 30% pure. The yield of rRNA cistrons was about 3%. The per cent purity refers to the relative amount of DNA in the preparation which is complementary to rRNA and should not be confused with the purity of a preparation of hybrids. In this latter case, the purity refers to the proportion of DNA fragments which are hybridized. It should be clear that it is possible to have a preparation of DNA that is 100% pure with regard to its ability to form a hybrid — that is, every fragment contains a rRNA cistron, and yet be 30% pure when the ratio of RNA to DNA in the hybrid is considered. As COLLI and OISHI point out, the value of 30%

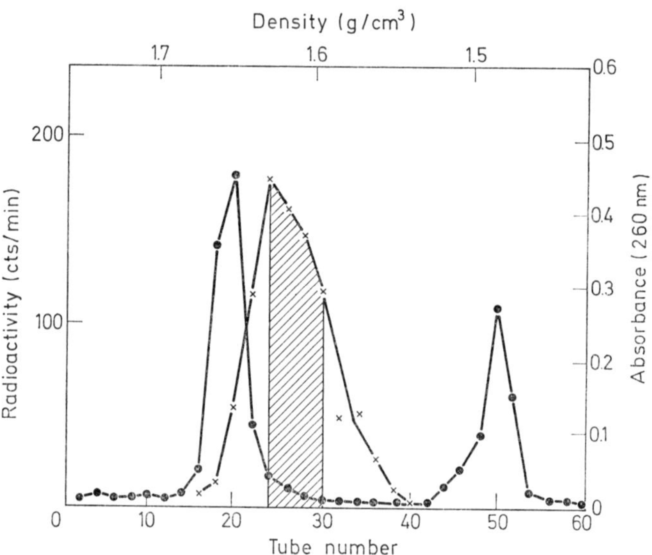

Fig. 17. Cs_2SO_4-Hg (II) density-gradient centrifugation of the DNA-RNA hybrid. 50 μg of the DNA-RNA hybrid (total radioactivity = 40,000 cts/min) as measured by the DNA content, were mixed with 25 μg of native DNA (marker), 3.6 g of solid Cs_2SO_4, 0.015 M sodium borate buffer (pH 9.0) and 1.12×10^{-5} M $HgCl_2$. The mixture was centrifuged at 34,000 rev./min for 43 h at 4°C in the Spinco SW 50.1 rotor. After centrifugation, 5-drop fractions were collected. Every other fraction was then mixed with 0.3 ml of 0.015 M NaCl—0.001 M EDTA and the absorbance (—●—●—) was measured at 260 mμ. The peak to the left is single stranded DNA and the peak to the right is the double stranded DNA marker. 20 μl portions were taken from the same fractions and radioactivity (— × — × —) was assayed. The shaded area under the curve represents the fractions which were pooled for the final assay. (From COLLI and OISHI, 1970)

for the purity of their rDNA preparation is a minimum estimate and does not intimate that the remaining 70% is not part of the rRNA transcriptional unit(s). If the anatomy of the *B. subtilis* transcriptional units is basically like those in *E. coli* and higher organisms then certain elements of these units may not hybridize with native rRNA. These include sequences of nucleotides that code for the "extra" segments in larger precursors to 16s and 23s RNA, and also "spacer" DNA which is located between rRNA cistrons and is currently without known function [6].

Table 5 summarizes the quantitative results obtained by COLLI and OISHI at each step of this purification [45]. As we noted above, the overall yield of 3% can be

Table 5. Purification of the ribosomal RNA genes. (From COLLI and OISHI, 1970)

Purification step	DNA μg(μmole)[a]	23s RNA genes			16s RNA genes			Spec. act.$_1$/Spec. act.$_2$	Mean mol. wt of DNA (daltons × 10^{-6})
		Total activity (units)[b]	Recovery (%)	Spec. act.$_1$[c]	Total activity (units)[b]	Recovery (%)	Spec. act.$_2$[c]		
Sheared and denatured DNA	13,000 (41.0)	90.2	100.0	2.2 (1)	57.4		1.4 (1)	1.57	3.6
MAK[d] column chromatography	1763 (5.6)	49.5	55.0	8.9 (4)	33.9	59.0	6.1 (4)	1.46	3.5
Concentration and dialysis	1391 (4.4)	39.1	43.0	8.9 (4)	26.6	46.0	6.1 (4)	1.46	3.5
Hydroxyapatite chromatography	352 (1.1)	9.2	10.0	8.3 (4)	7.0	12.1	6.3 (4)	1.32	2.9
Cs_2SO_4-Hg (II) centrifugation	5.6 (0.018)	3.3	3.6	182.0 (83)	1.9	3.4	106.0 (76)	1.72	2.4

[a] Nucleotide equivalent.
[b] A unit of activity is defined a 1 mμmole of RNA hydridized with DNA at 68 °C for 2 h in 0.3 M-NaCl—0.01 M-sodium phosphate at pH 7.0.
[c] Spec. act., specific activity, mμmoles of RNA hybridized per μmole of DNA (nucleotide equivalent).
[d] Methylated albumin kieselguhr.

increased somewhat at the expense of purification. It should be noted that the ratio of 16s and 23s rDNA is fairly constant throughout each operation indicating that preferential losses do not occur. This method can be applied generally for the preparation of any cistron from DNA which can be separated into complementary strands and for which sufficient gene product can be obtained. However, the procedure is limited in application. As with other methods which depend heavily on density fractionation in a cesium salt gradient, this procedure cannot be scaled up to yield milligram quantities of cistrons. The use of methylated albumin kieselguhr chromatography is also a limiting step inasmuch as the capacity of this material for DNA is not high [126, 144].

Udvardy and Venetianer [144] have recently developed a new column procedure which will fractionate rRNA:DNA hybrids from rRNA and single stranded DNA. Purification of the hybrids is achieved by repeated cycles of hybridization and chromatography on a column of deoxycholate-treated benzoylated-DEAE cellulose. With this method, the rRNA cistrons from *Salmonella typhimurium* were enriched some 190 fold with a yield of about 6%. The purity of their final preparation was approximately 64%. The purification procedure adopted by Udvardy and Venetianer consists of: (1) strand separation of sheared DNA by methylated albumin kieselguhr chromatography; (2) hybridization of the H-strand DNA fragments with rRNA; (3) fractionation of the RNA:DNA hybrids from non-hybridized RNA and single-stranded DNA by chromatography on deoxycholate-treated benzoylated DEAE cellulose and (4) a second cycle or hybridization and chromatography.

These workers opted to use only plus strand rDNA in their hybridization step for two reasons. First, this measure will improve the yield of rRNA:DNA hybrids inasmuch as the non-transcribing strand of the rRNA cistrons will not be present to compete with rRNA for duplex formation. Second, their modified benzoylated DEAE cellulose column cannot resolve RNA:DNA hybrids from reassociated DNA. Thus, the purity of the final preparation of RNA:DNA hybrids is tied directly to the extent of DNA:DNA duplex formation.

Complementary strands of DNA were separated using discontinuous salt elution from methylated albumin kieselguhr. Margulies et al. [96] previously demonstrated that this technique can be used to separate the strands of DNA from several bacteria, including *S. typhimurium*.

Udvardy and Venetianer found it necessary to rechromatograph the light and heavy fractions from methylated albumin kieselguhr in order to eliminate cross contamination. When they tested each fraction for the ability to hybridize with rRNA, it was discovered that twice as much rRNA is bound to the H-strand preparation as to unfractionated DNA and that L-strand DNA contains less than 5% of the total rRNA binding sites determined for H-DNA. Thus, it appears that virtually all of the DNA complementary to rRNA is present in the H-strand preparation. However, the authors did not demonstrate that minus-strand rDNA is not also present in their H-strand DNA. This should be tested by a competition hybridization experiment such as that used by Colli and Oishi [44] and described above (see Fig. 15).

^{32}P-labeled H-strand DNA was then hybridized with unlabeled rRNA and chromatographed on deoxycholate-treated benzoylated DEAE cellulose using a linear NaCl-acetone gradient. Fig. 18 shows the elution profile obtained. Under these conditions, free rRNA is eluted at about 0.38 to 0.45 M NaCl, DNA:DNA duplexes and RNA:DNA hybrids at 0.50 to 0.55 NaCl and non-hybridized single stranded DNA above 0.65 M NaCl. In Fig. 18, the bulk of the label elutes as a single broad band mainly in the single stranded DNA region. There is, however, a small shoulder of activity eluting before the main peak (fractions 20 to 35). When fractions were pooled and tested for the ability to rehybridize with labeled rRNA (following removal of the hybridized non-labeled RNA), it was learned that these early fractions were enriched for rDNA. Pooled fractions A and B were able to hybridize with rRNA to the extent of about 7 to 8%. When fractions A to C were rehybridized and rechromatographed on deoxycholate-treated benzoylated DEAE the rDNA was further enriched

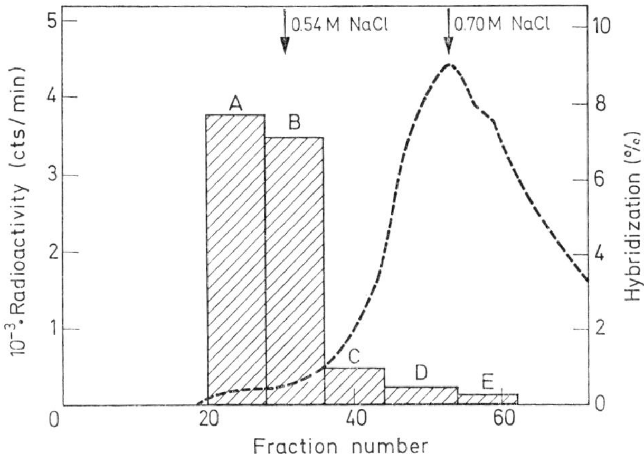

Fig. 18. Chromatography of the DNA-rRNA hybrid on deoxycholate-treated benzoylated-DEAE-cellulose. 2.7 mg (^{32}P) DNA "heavy" strand, hybridized with unlabelled rRNA was put on a 2 × 12 cm deoxycholate-treated benzoylated-DEAE-cellulose column and eluted with a linear acetone-NaCl gradient as described in Methods. Fractions indicated on the figure were pooled and after removal of the unlabelled rRNA were hybridized with (^{3}H) rRNA. The columns represent the percentage hybridization values of these fractions. (From Udvardy and Venetianer, 1971)

as shown in Fig. 19. Now, the most enriched fraction, labeled pooled fractions A, is at least 64% pure when rehybridized with labeled rRNA.

Table 6 summarizes the purification data. It is clear that rDNA of high purity can be obtained if one is willing to compromise on the yield. A final purification of nearly 200 fold (fraction A — second column) can be obtained, but with a recovery of only 6%. The molecular weight of the rDNA fragments so obtained is about 10^6 daltons and, of course, the cistrons are single stranded DNA.

The most serious obstacle to scaling up this procedure is the methylated albumin kieselguhr column strand separation. As noted above, the capacity of these columns is limited [126, 144] and, in our experience, the column size can be increased only a few fold [65].

It seems possible to us that the benzoylated DEAE cellulose column procedure of Udvardy and Venetianer can be employed successfully, even with DNA that has not been separated into complementary strands. As KOHNE [83] and we [26] have demonstrated, it is possible to regulate the rate and extent of RNA:DNA and DNA:DNA duplex formation in such a manner as to obtain rRNA:DNA [83] or

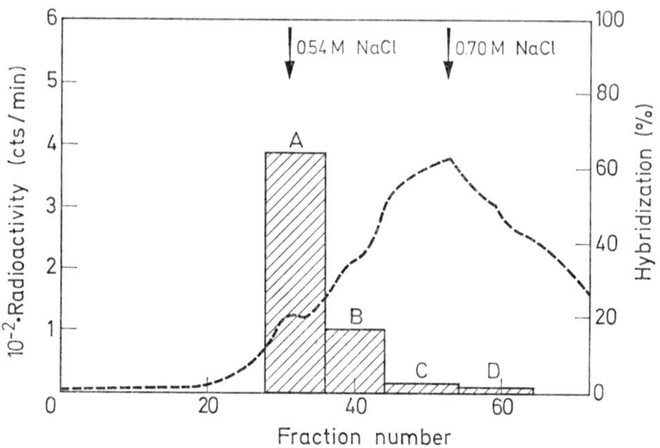

Fig. 19. Second chromatography of the DNA-rRNA hybrid on deoxycholate-treated benzoylated-DEAE-cellulose. 0.3 mg (^{32}P) DNA corresponding to fractions A—C of Fig. 18 was hybridized again with rRNA, put on a 2 × 12 cm deoxycholate-treated benzoylated-DEAE-cellulose column and eluted with a linear acetone-NaCl gradient. Fractions indicated on the figure were pooled and after removal of the unlabelled rRNA were hybridized with (^{3}H) rRNA. The columns represent the percentage hybridization values of these fractions. (From UDVARDY and VENETIANER, 1971)

Table 6. Summary of the purification procedure. (From UDVARDY and VENETIANER, 1971)

Purification step	Amount of DNA µg	Hybridization with rRNA %	Purification -fold	Recovery of material hybridizable with rRNA %
Sonicated total DNA	28000	0.34	1	100
Strand separation ("heavy" strand)	5100	0.66	2	35
First chromatography	312	2.96	9	10
Second chromatography				
fraction A	9	64.0	189	6
fraction B	25	17.0	50	4.5
fraction A + B	34	30.0	91	10.5

tRNA:DNA hybrids [26] with only a small amount of DNA reassociation. By carefully controlling the concentrations of nucleic acids and time of incubation and by increasing the number of cycles of hybridization and chromatography it should be possible to use the benzoylated DEAE cellulose column method to purify any cistron for which adequate amount of product RNA can be obtained.

The success of this procedure may well depend on the size of the DNA fragments. If the ratio of DNA to RNA in a hybrid is too high, the hybrid may behave more like single stranded DNA resulting in poor separation from that material. In the UDVARDY and VENETIANER study the ratio of DNA:RNA in the rRNA:DNA hybrids probably did not exceed 2 (for one 16s RNA molecule per DNA fragment of molecular weight 10^6 daltons the value would be 2, while one 23s RNA molecule per DNA fragment would yield a DNA:RNA ratio close to 1).

tRNA:DNA hybrids: In a preliminary report [104], MCFARLAND and FRASER attempted to purify the tRNA genes from mouse Ehrlich ascites tumor cell DNA. The DNA was hybridized with ^{32}P-labeled tRNA, treated with *E. coli* phosphodiesterase to degrade nonhybridized DNA, and then fractionated on a column of methylated albumin kieselguhr. A minor peak of radioactive material eluted in a region of the gradient where hybrids have been observed by others suggests the presence of tRNA:DNA hybrids. However, because no attempt was made to characterize this material, its identification must be considered only tentative.

Bacterial tRNA:DNA hybrids have been isolated in three laboratories using two different general procedures. These procedures differ basically in the means by which contaminating DNA contiguous to the tRNA:DNA hybrids is removed. In one case, the hybrids are treated with an endonuclease specific for single strands [97, 99] and in the second instance, the DNA is reduced in size before hybridization to fragments that are only about 4 times larger than the cistrons to be isolated [26].

MARKS and SPENCER used the nuclease procedure to purify the *E. coli* DNA complementary to unfractionated tRNA [98, 99]. Their isolation procedure consists of the following steps: (1) formation of tRNA:DNA hybrids in solution, (2) separation of non-hybridized tRNA from the other nucleic acid species by chromatography on a column of Sepharose 2B or Sephadex G-100, (3) digestion of non-hybridized, single stranded DNA with *Neurospora crassa* endonuclease, (4) fractionation of the tRNA:DNA hybrids, single and double stranded DNA by methylated albumin kieselguhr chromatography, (5) equilibrium density gradient centrifugation in Cs_2SO_4 and (6) two cycles of chromatography on Sephadex G-100 to yield a tRNA:DNA hybrid of molecular weight 50,000 with a DNA:RNA ratio of one. The overall yield is about 2% and the enrichment some 6000 fold.

The rationale for utilizing the *N. crassa* endonuclease [92, 121] held that the DNA regions adjacent to the hybridized tDNA would most likely be single stranded and thus susceptible to enzymatic digestion. Indeed, this hypothesis was borne out by their results. Although the nuclease possesses both single and double stranded activities, under the conditions employed the ratio of these activities was about 200 to one favoring the hydrolysis of denatured DNA.

Fig. 20 shows the fractionation of a ^{32}P-tRNA:DNA hybridization mixture on Sephadex G-100 and Sepharose 2B. It is clear that each column separates non-hybridized tRNA from the other components in the reaction mixture. In each case, the free tRNA is included by the gel and elutes late while the tRNA:DNA hybrids, DNA:DNA duplexes and non-hybridized single stranded DNA are excluded and appear at the void volume.

The tRNA:DNA hybrid fraction is then incubated with *N. crassa* nuclease, dialyzed and chromatographed on methylated albumin kieselguhr. The elution profile is shown in Fig. 21. It can be seen that the ^{32}P-tRNA:DNA hybrid fraction elutes before the peak of ultraviolet absorbing material and is fairly well resolved from that band. This second peak is presumed to consist primarily of renatured DNA. The ratio of DNA to RNA in the methylated albumin kieselguhr hybrid fraction is 84 (see Table 7).

The methylated albumin kieselguhr enriched tRNA:DNA hybrid fraction is then further purified by two successive bandings in Cs_2SO_4 density gradients. Although

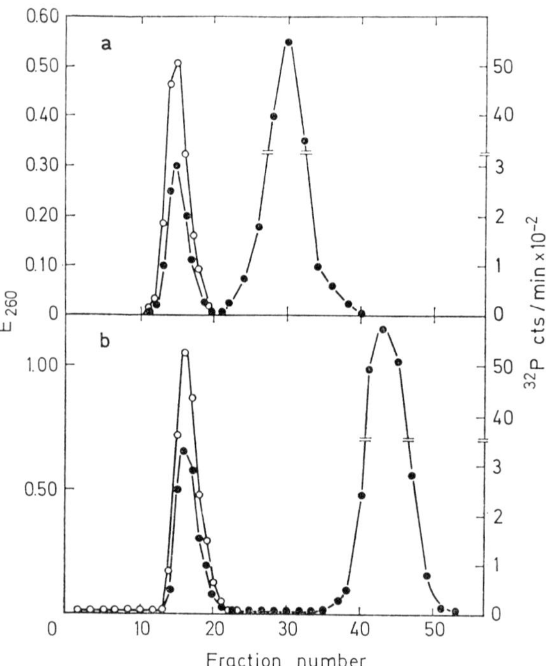

Fig. 20. Chromatography of an *E. coli* DNA-tRNA hybridization mixture on a Sephadex G-100 and b Sepharose 2B. Columns 2.5 cm × 38 cm, 10 ml of sample in 2 × SSC was applied to the column and eluted with 2 × SSC, flow rate 5 ml/h, 5 ml fractions collected. The absorbance of the collected fractions was measured with a Gilford 220 spectrophotometer. Radioactivity was measured directly by Cerenkov radiation. (—○—○—) E_{260}; (—●—●—) ^{32}P radioactivity (tRNA). (From MARKS and SPENCER, 1970)

the resulting fractionation profiles are not shown, non-hybridized DNA is removed with each banding (see Table 7). The ratio of DNA:RNA in the fractions selected drops from 84 to 14 in the first cesium gradient, and to 4 after a second banding.

Finally, the hybrids are twice chromatographed on a Sephadex G-100 column to yield a tRNA:DNA peak in which the "best" fractions exhibit a DNA:RNA ratio of one. The elution pattern of the final Sephadex G-100 fractionation is shown in Fig. 22. It is evident that the ratio of DNA to RNA is not constant across the ^{32}P-tRNA peak and that the ^{3}H-labeled DNA peak is still heterogeneous. This indicates that some non-hybridized DNA is still present.

The purification data are summarized for each step in Table 7. Although the tDNA so derived would appear to be virtually pure, a high price is exacted. Starting with approximately 2 mg of DNA, the final yield of tDNA obtained by MARKS and SPENCER is about 4 ng or about 2% of the total tRNA cistrons initially present. Even using ^{32}P-labeled DNA with a specific activity in excess of 10^6 cpm per μg — which, in our experience, is at the practical upper limit, the amount of tDNA recovered is woefully inadequate for most analyses with the possible exception of labeled tDNA: DNA hybridization studies. Because of the limitations inherent with methylated

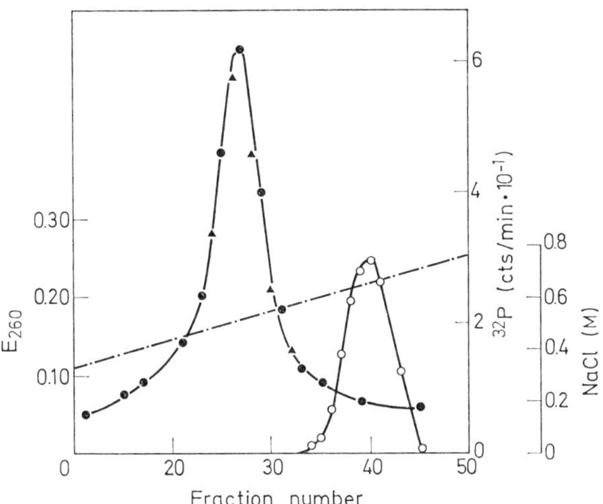

Fig. 21. Chromatography of *E. coli* DNA-tRNA hybrid on MAK following digestion with *N. crassa* endonuclease. The hybrid was prepared as described except for no addition of marker ^3H-labeled tRNA after *N. crassa* endonuclease digestion. MAK column 2.4 × 10 cm, elution with a linear gradient of NaCl in 0.05 M phosphate buffer, pH 6.7, 5-ml fractions collected. The absorbance of the collected fractions was measured with a Gilford 220 spectrophotometer. Alternate fractions, 24 to 32 inclusive, were treated with pancreatic RNase (20 μg RNase/ml for 10 min at 20 °C) before assay. Radioactivity was measured by acid-precipitable cts/min. (—○—○—) E_{260}; (- - - -) NaCl gradient; (●—●—●) ^{32}P cts/min before ribonuclease; (—▲—▲—) ^{32}P cts/min after ribonuclease. (From MARKS and SPENCER, 1970)

albumin kieselguhr and Cs_2SO_4 fractionation it seems unlikely that the procedure can be scaled up more than a few fold.

Perhaps more serious for some is the question of whether the tDNA isolated in this way is a complete cistron or unit of transcription. As stated in an earlier portion of the review, if the nuclease indeed digests all the single stranded DNA adjacent to the hybrid, those portions of the gene not protected by the RNA will not escape hydrolysis. If the natural gene product is a larger precursor RNA, then that segment of the gene which codes for the extra nucleotides will be missing from the final cistron preparation. Similarly, other portions of the gene important for the transcription may also be lacking.

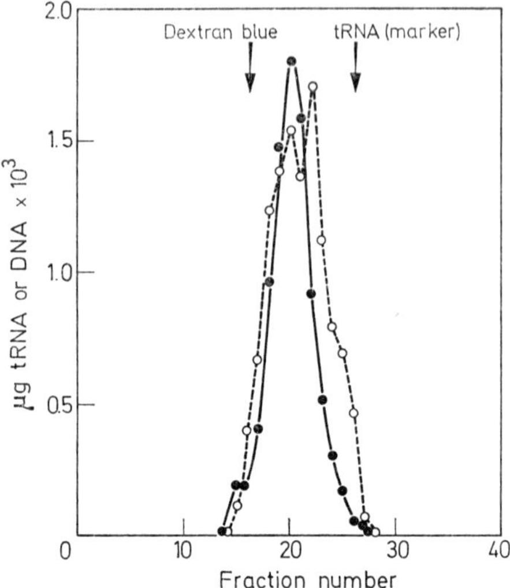

Fig. 22. Chromatography of purified *E. coli* tRNA-gene hybrid on Sephadex G-100. The hybrid preparation from Cs_2SO_4 density centrifugation was chromatographed on Sephadex G-100 and the main fractions rechromatographed on Sephadex G-100. Conditions as in Fig. 20. The position of elution of dextran blue indicates the exclusion volume of the column. The amounts of DNA and tRNA were calculated from the acid-precipitable cts/min in each fraction. (—●—●—) ^{32}P-labeled tRNA; (- - ○ - - ○ - -)^{3} H-labeled DNA. (From Marks and Spencer, 1970)

Table 7. Purification of tRNA-gene hybrid. (From Marks and Spencer, 1970)

Purification step	Procedure	Recovery (%) Step	Recovery (%) Cumulative	Fold purification Step	Fold purification Cumulative	Purity of hybrid ratio DNA/tRNA
	Hybridization					
	Sepharose 2B					
	N. crassa endonuclease					
1	MAK[a]	65		70		84
	Flash evaporation and dialysis	90	59	—	—	
2	Cs_2SO_4	38	22	6	420	14
3	Cs_2SO_4	43	10	3.5	1470	4
4	Sephadex G-100	31	3	3	4410	1.4
5	Sephadex G-100	60	1.8	1.4	6174	1

[a] Methylated albumin kieselguhr.

Miller, in our laboratory, has also prepared *E. coli* tRNA cistrons by nuclease digestion of tRNA:DNA hybrids [58, 107]. Although the method used is basically like that of Marks and Spencer described above [99] the final yield of tDNA is higher. This result appears to be due to a difference in the extent of tRNA:DNA hybridization. MARKS and SPENCER report a saturation hybridization value of 0.018% for the ratio of tRNA bound to DNA while MILLER finds a value of 0.036%. Inasmuch as 0.018% of the total DNA is only enough to code for about 20 tRNAs of molecular weight approximately 27,000 it seems likely that some tDNA may not be hybridized under their incubation conditions. Consistent with this view is the finding that, MILLER's yield of tRNA:DNA hybrids is twice that obtained by MARKS and SPENCER.

The tDNA isolation procedure developed by MARKS and SPENCER for the total complement of *E. coli* tRNA cistrons has also been used with some shortcuts to

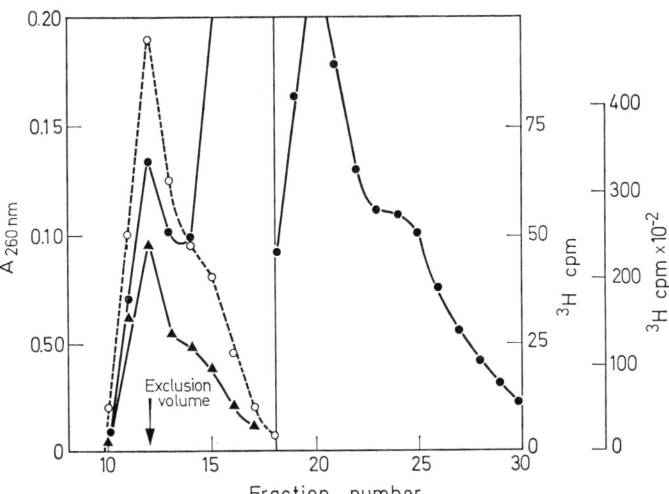

Fig. 23. Separation of a ø80 psu$^+_{\text{III}}$ DNA-tyrosine-tRNA hybrid from unhybridized tyrosine tRNA by gel filtration. Denatured ø80 psu$^+_{\text{III}}$ DNA, 250 µg, was hybridized with 1 µg of ^{3}H-labelled tyrosine-tRNA (400,000 cpm/µg) in 5 ml of 2 × SSC (0.3 M NaCl, 0.03 M sodium citrate, pH 7.0) for 2 h at 72 °C. Following slow cooling the mixture was chromatographed on a Sephadex G-100 column, 2.5 cm × 37 cm, equilibrated with 2 × SSC, at room temperature. Fractions of 5 ml were collected every 30 min and DNA in each fraction measured by absorbance at 260 nm (○ - - - ○). Aliquots of 2 ml were precipitated with 5% trichloroacetic acid (●—●), or tested for DNA-tRNA hybrid formation by filtering directly through a nitrocellulose filter (Millipore HA, 0.45 µm), washing with 2 × SSC, and digestion of the filter with pancreatic RNase, 20 µg/ml, in 2 × SSC for 60 min, at room temperature (▲—▲). (From MARKS et al., 1971)

enrich the DNA complementary to tyrosine tRNA from the transducing phage ø80 psu$^+_3$ [97]. In this preparation, MARKS and co-workers eliminated all but one of the purification steps following digestion of the hybrid material with *N. crassa* endonuclease.

In brief, the steps adopted for the enrichment of tDNA^{tyr} include: 1) hybridization of phage DNA with tRNA, 2) removal of non-hybridized tRNA by Sephadex G-100 chromatography, 3) nuclease digestion and 4) fractionation by Sephadex G-100 chromatography or Cs_2SO_4 gradient centrifugation. Due to the absence of any purification data in this preliminary report, a thorough analysis of the isolation method is not possible. It will be useful however, to review the fractionation patterns obtained by these investigators.

Fig. 23 shows the separation of non-hybridized ^{3}H-tRNA from ^{3}H-tRNA:DNA hybrid material as effected by filtration through Sephadex G-100. For reasons that are not clear to us, the separation obtained here is not as satisfactory as that realized by MARKS and SPENCER for the *E. coli* tDNA purification (compare with Fig. 20 above). Following enzymatic digestion, the reaction mixture was either rechromatographed on Sephadex G-100 or fractionated by Cs_2SO_4 gradient centrifugation.

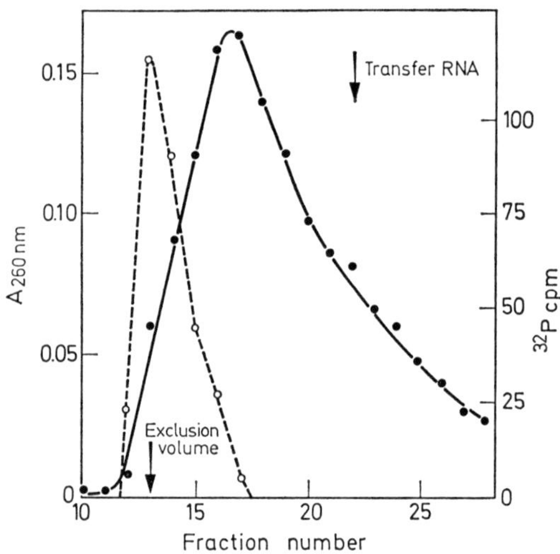

Fig. 24. Chromatography of a ⌀80 psu$^+_{III}$ DNA-tyrosine-tRNA hybrid on Sephadex G-100 following digestion with *N. crassa* endonuclease. A ⌀80 psu$^+_{III}$ DNA-^{32}P-tyrosine-tRNA (100,000 cpm/μg) hybrid (fractions 11 to 14 from Fig. 23) was digested in 5 ml of 0.1 M NaCl, 0.01 M $MgCl_2$, 0.1 M Tris-HCl buffer, pH 7.5 with 50 units *N. crassa* endonuclease at 30 °C for 8 h. The digest was chromatographed on a Sephadex G-100 column, 2.5 cm × 40 cm, as described in Fig. 23. Fractions were analyzed for presence of DNA by absorbance at 260 nm (○ - - - - ○) and tyrosine-tRNA by acid-precipitable cpm (●—●). The arrow represents the elution position of tRNA alone in a parallel chromatography. (From MARKS et al., 1971)

A post-digestion gel-filtration profile is shown in Fig. 24. In this experiment, the ^{32}P-tRNA:DNA hybrid elutes as a very broad band covering the region from the void volume to a point well into the elution position of free tRNA. Although the general position of the hybrid is consistent with a species possessing a molecular weight between 100,000 (G-100 exclusion limit) and 27,000 daltons (MW of tRNA) the width and asymmetry of the peak suggest a high degree of heterogeneity — probably in the form of hybrids with DNA tails of varying length. Because the phage selected for the study carries two genes for tyrosine tRNA which may be nearly tandem [127] one can imagine the occurrence of a variety of tRNA:DNA hybrid species of different size and DNA:RNA ratios.

The heterogeneity is also evident if the nuclease-treated hybrid material is fractionated in a cesium density gradient in lieu of gel-filtration. Fig. 25 shows the fractionation pattern obtained and the position at which marker tRNA bands. Again, the hybrid peak is broad and skewed indicating that a family of hybrids exists which vary in density owing to differences in the ratio of RNA to DNA. Although certainly not pure, this tDNA may be useful for transcription studies, especially since the sequence of the suppressor tyrosine tRNA is known [68].

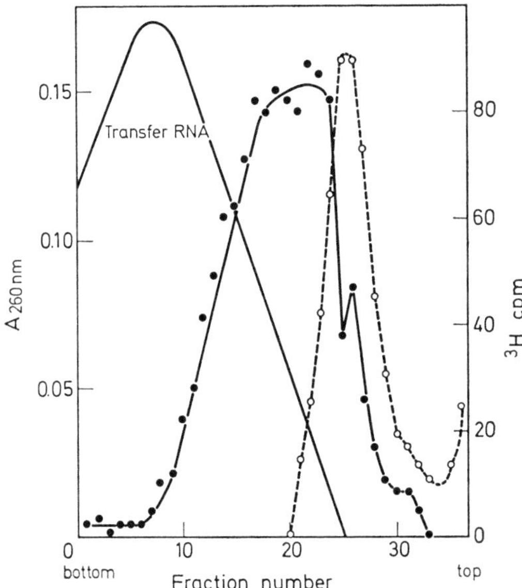

Fig. 25. Centrifugation of a ⌀80 psu$^{+}_{III}$ DNA-tyrosine-tRNA hybrid in a cesium sulphate density gradient following digestion with *N. crassa* endonuclease. A ⌀80 psu$^{+}_{III}$ DNA-^{3}H-tyrosine-tRNA (400,000 cpm/μg) hybrid (fractions 11 to 14 from Fig. 23) was digested with *N. crassa* endonuclease as described in Fig. 24. The digest was adjusted to a final density of 1.540 by addition of cesium sulphate and centrifuged for 72 h at 33,000 rpm in a Spinco SW 39 rotor at 20 °C. Ten-drop fractions were collected from the bottom of the tube. Each fraction was diluted with 1 ml H_2O and analyzed for the presence of DNA by absorbance at 260 nm (○ - - - - ○) and tyrosine-tRNA by acid-precipitable cpm (● ●). The solid line represents the sedimentation of tRNA alone in a parallel gradient. (From Marks et al., 1971)

We have also isolated *E. coli* tRNA cistrons, but without the use of nuclease [26]. Our procedure is basically similar to that developed by Kohne for the isolation of rDNA [83]. Transfer RNA cistrons have been routinely isolated in our laboratory by repeated hybridization of DNA with tRNA and fractionation of tRNA:DNA hybrids and non-tRNA cistrons by chromatography on hydroxyapatite columns. Several cycles of hybridization and chromatography are required to remove all DNA noncomplementary to tRNA. The purification procedure is schematized and described briefly in Fig. 26.

^{32}P-DNA sheared to an average single stranded fragment of about 125,000 daltons is hybridized with an excess of unlabeled, unfractionated tRNA. The nucleic acid components of the reaction mixture are depicted in Fig. 26 (Panel A).

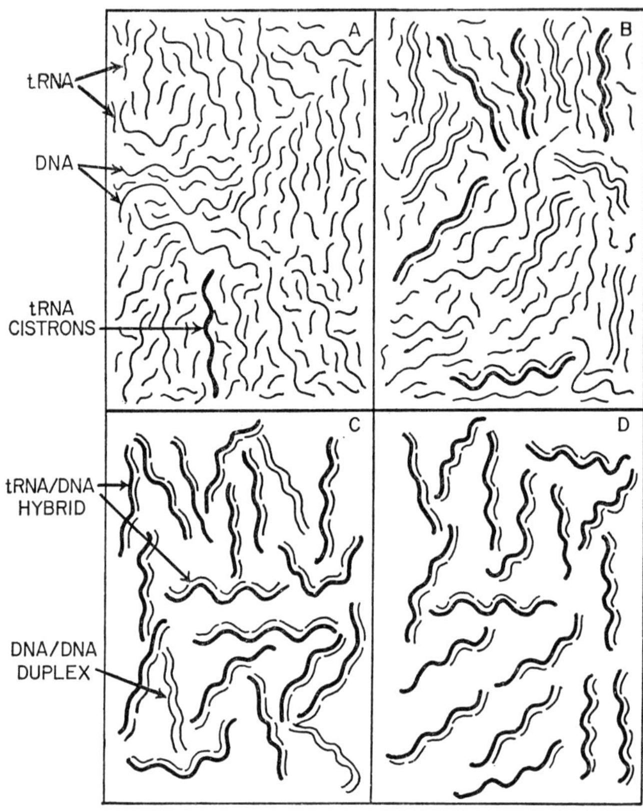

Fig. 26. Procedure for the isolation of the tRNA cistrons. A. Representation of the composition of the hybridization reaction mixture before hybrid formation is allowed to occur. Sheared, $^{32}PO_4$-labelled DNA is denatured and mixed with an excess of tRNA. Those single strands of DNA that are complementary to tRNA are then permitted to form DNA-tRNA duplexes as shown in B. It is evident that some DNA-DNA duplex formation also occurs. The reaction mixture is than applied to a hydroxyapatite column in conditions which prohibit the adsorption of single stranded DNA. Both tRNA-DNA hybrids and DNA-DNA duplexes are adsorbed and subsequently eluted. The nature of the adsorbed material is shown in C. After three to five cycles of hybridization and chromatography the only DNA remaining is that which hybridizes with tRNA (D). (From BRENNER et al., 1970)

The hybridization reaction conditions have been designed to permit all of the tRNA cistrons to react completely with tRNA while only a small percentage of the bulk DNA is allowed to reassociate. Panel B in Fig. 26 illustrates the composition of the reaction mixture after hybridization. In addition to tRNA:DNA hybrid formation some DNA:DNA reassociation also occurs. After incubation, the mixture is applied to a hydroxyapatite column equilibrated at 60° and at an ionic strength that prohibits single stranded DNA from being adsorbed while tRNA:DNA hybrids and DNA:DNA duplexes are retained. The bound material, shown in panel C of Fig. 26 is then eluted by increasing the ionic strength of the buffer or by raising the temperature of the column to a point where dissociation occurs.

After about 3 subsequent cycles of hybridization and fractionation, the only DNA fragments to be adsorbed to the hydroxyapatite are complementary to tRNA (see Panel D of Fig. 26).

Table 8 summarizes the results of our purification method. When *E. coli* DNA is incubated with *E. coli* tRNA (part A), the amount of DNA that adsorbs to hydroxy-

Table 8

	Cycle	tRNA C_0t	DNA C_0t	Percentage of original input DNA adsorbed to hydroxyapatite
(A)	1	0.36	0.08	1.12
	2	0.36	3×10^{-4}	0.091
	3	0.36	1×10^{-5}	0.070
	4	0.36	6×10^{-6}	0.060
	5	0.36	3×10^{-6}	0.057
(B)	1	0	0.05	2.96
	2	0	4×10^{-4}	0.054
	3	0	1×10^{-6}	0.013
	4	0	5×10^{-7}	0.001
(C)	1	0.36	1×10^{-5}	0.24
	2	0.36	1×10^{-7}	0.004

The fraction of original input $^{32}PO_4$-labeled *E. coli* B DNA adsorbing to the hydroxyapatite after incubation in the presence or absence of tRNA at the indicated C_0t. (A) Bulk DNA incubated with *E. coli* tRNA; (B) bulk DNA incubated in the absence of tRNA; (C) DNA that did not react with tRNA during the first cycle of incubation was reincubated with tRNA. (From Brenner et al., 1970).

apatite after five cycles of hybridization and isolation is 0.057% of the initial input of DNA. In different experiments this value was between 0.04 and 0.06% in good agreement with results from other laboratories which indicate that tRNA cistrons comprise 0.04 to 0.06% of the *E. coli* genome [109, 154].

Table 8 B shows the result obtained when DNA is incubated in the absence of tRNA and then subjected to chromatography on hydroxyapatite. Only 0.001% of the DNA is adsorbed after only four cycles of purification, indicating that the DNA-tRNA hybridization reaction is specific and, further, that the amount of DNA in the fifth cycle preparation not specifically hybridized to tRNA is probably less than 2% of the total (0.001/0.05).

Evidence that essentially all the tRNA cistrons are hybridized under our conditions comes from results shown in Table 8 C. Here, DNA that failed to react with tRNA in the first hybridization cycle was again incubated with tRNA. After only two cycles of hybridization and purification, 0.004% of this DNA is adsorbed to hydroxyapatite. Since these values have been corrected for a 5 to 7% loss experienced with each cycle, the actual yield of tDNA after four cycles is about 80%. In a typical preparation, 50 mg of unfractionated DNA will yield approximately 15 to 20 μg of tDNA.

These results raise the question of how the tRNA cistrons are arranged in the genome. The data indicate that virtually all of the tDNA is recovered and the yield agrees with the tRNA information content of the genome. This implies there is little

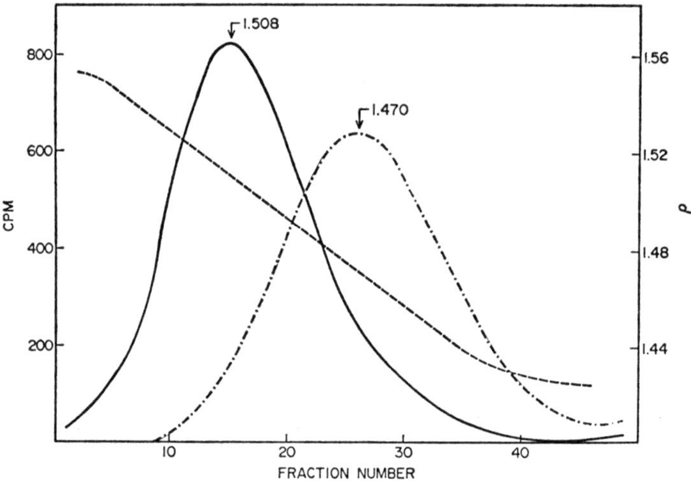

Fig. 27. Cs_2SO_4 density gradient centrifugation of tDNA and tDNA-tRNA hybrid. Both the samples were adjusted to input density of 1.485 with Cs_2SO_4 in 0.5 × SSC buffer and centrifuged in 40 rotor (Spinco Model L) at 33,000 rpm for 72 h. The fractions were obtained by tapping the bottom of the tubes and radioactivity in each fraction was measured. (From DOCTOR et al., 1972)

non-tDNA in the preparation. Because the tDNA fragments are about 125,000 daltons or some 4 to 5 times longer than a native tRNA molecule, and because the yield of DNA is not 4 to 5 times the amount of DNA complementary to tRNA, it would appear that at least some of the tRNA cistrons may be clustered to a high degree; perhaps even contiguous. Further study seems to support this hypothesis.

Analysis of purified tRNA: ^{32}P-DNA hybrids and non-hybridized tDNA by Cs_2SO_4 density gradient fractionation shows the material to be free of non-hybridizable DNA and reassociated DNA, respectively. Fig. 27 shows the patterns obtained. The tRNA:DNA bands at a density of 1.508 g cm^{-3} similar to that reported for other RNA:DNA hybrids [39] and the tDNA bands at a density of 1.470 g cm^{-3}, corresponding to single stranded DNA [140]. Under these conditions native DNA has a density of 1.424 g cm^{-3} [140].

Inasmuch as our hybrid bands at a density which corresponds to hybrids with a RNA:DNA ratio close to one it appears that the bulk of our hybridized DNA fragments are covered with tRNA. In some experiments the hybrid band is skewed toward lower densities suggesting that, while most of the DNA fragments bind several tRNAs, some fragments are only partially hybridized. These results are consistent with data from genetic mapping which indicate that many, but not all, of the tRNA genes are clustered in *E. coli* [49, 142].

Further evidence that tRNA cistrons are clustered in *E. coli* comes from a more accurate determination of the amount of tRNA bound per fragment of tDNA. In this experiment, ^{35}S-tRNA was hybridized to a preparation of ^{32}P-tDNA judged to be 75 to 80% pure with regard to its ability to form hybrids. The non-hybridized

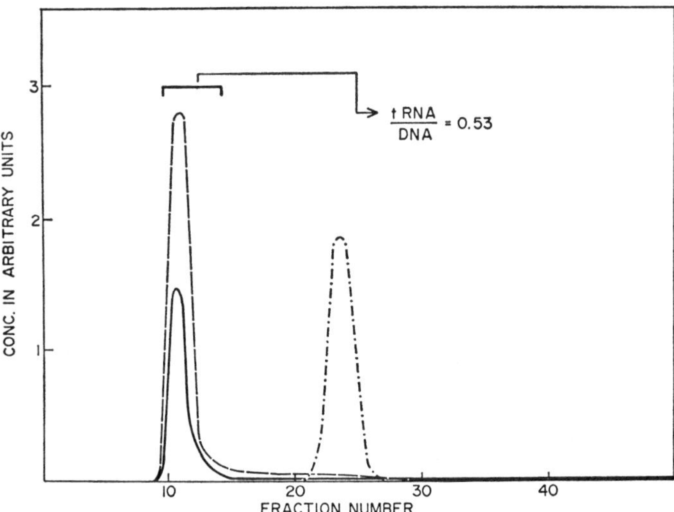

Fig. 28. Rechromatography of tDNA-tRNA hybrid on Sephadex G-100 column. A portion of the tDNA-tRNA (shown by horizontal broken line) was rechromatographed on Sephadex G-100 column. ——— ^{32}P DNA; — — — ^{35}S tRNA; —·—·— tRNA marker showing the position of free tRNA under this condition. The results of ^{32}P and ^{35}S counts are normalized to represent weight basis and expressed in arbitrary units. (From DOCTOR et al., 1972)

tRNA is separated from the other components by Sephadex G-100 chromatography. The material from the hybrid region was rechromatographed as shown in the profile in Fig. 28. The observed ratio of tRNA to DNA on a wt/wt basis is 0.53. When the value is corrected for the content of non-hybridizable DNA fragments, it is estimated that 2/3 to 3/4 of the nucleotides of the tDNA fragments are hybridized with tRNA. This amounts to, on the average, 3 to 4 tRNA cistrons per fragment of DNA. We are currently repeating this double label study with larger tDNA fragments in an effort to determine the extent of clustering.

Our data do not permit us to comment on the purity of tDNA derived by our method. While we can say that better than 95% of the DNA fragments isolated in this way can hybridize with tRNA and that, on the average, each fragment can bind 3 to 4 tRNAs, the nature of the non hybridized regions is not known. Those segments could be promoter regions, intercistronic gaps with no function, precursor tRNA information or simply other cistrons.

We have successfully scaled up our procedure ten fold by using larger columns of hydroxyapatite and can routinely generate 20 μg of tDNA from 60 to 75 mg of DNA. A batch procedure [57] can be substituted for the hydroxyapatite column for the first and second cycles of purification. After hybridization the raction mixture is merely added to a large vessel of hydroxyapatite equilibrated at 60°. After stirring briefly, the sorbent is allowed to settle and the supernatant removed by decanting. Because settling occurs in only a few minutes, reassociation can still be controlled to a high degree and thus, the extent of DNA renaturation held to a minimum. After two cycles of hybridization and batch fractionation, the hybrid fraction contains approximately 0.5% of the starting DNA and the tDNA is enriched some 200 fold.

Table 9. Concomitant purification of 5s DNA and tDNA from Escherichia coli

Cycle	Hybridization reaction[a]	% of original DNA bound to HA[b]
1	Total E. coli DNA + 5s RNA + tRNA	6.1
2	Back peak DNA from Cycle 1 + 5s RNA + tRNA	1.3
3	Back peak DNA from Cycle 2 + 5s RNA + tRNA	0.75
Unincubated background elimination	Back peak DNA from cycle 3	0.56
4	Front peak DNA from background elimination + 5s RNA + tRNA	0.17
5	Back peak DNA from cycle 4 + 5s RNA	0.07 Back peak 0.10 Front peak

[a] The hybridization reaction was carried out initially with 3 mg of labeled DNA at 50 μg/ml + 25 μg/ml each of 5s RNA and tRNA. The nucleic acids were denatured at 100 °C for 4 min, incubated at 60 °C for 15 min and then passed through HA as described in the text. After each cycle the fractions eluted from HA to 0.3 M PB[c] were diluted to 0.1 M PB and hydrolysed with 0.2 M NaOH for 2 h at 60 °C to degrade RNA. After the solution was neutralized by the addition of HCl the ribonucleotides were removed by dialysis, the solution was then concentrated and again dialysed against 0.1 M PB before the next hybridization cycle was carried out. DNA concentration decreased in each cycle, whereas the concentration of RNA, incubation temperature and duration were identical in all cycles. (From Doctor and Brenner, 1972).

[b] HA = hydroxyapatite.

[c] PB = phosphate buffer.

We are also exploring the possibility of fractionating non-hybridized single stranded DNA from tRNA:DNA hybrids by countercurrent distribution. Preliminary results indicate that an ammonium sulfate: ethyoxyethanol two phase solvent system can effect this separation with only a small number of transfers [57].

5s RNA cistrons: Our standard procedure for isolating *E. coli* tDNA has also been modified to permit the concomitant isolation of 5s RNA and tRNA cistrons [59]. After four cycles of purification of DNA hybridized with both 5s RNA and tRNA a cycle of hybridization and chromatography is performed with only 5s RNA. As expected the non-hybridized tDNA is not adsorbed to the column and is thus separated from the 5s RNA:DNA hybrids. The adsorbed and non-adsorbed fractions are then purified separately for 5s DNA and tDNA respectively. The purification data are given in Tables 9 and 10. Table 9 shows the hybridization data for the cycles in which both cistrons are jointly enriched. The yield of 5s DNA is about 0.008% of the starting DNA and like our tDNA, better than 90% of these fragments can form hybrids with 5s RNA. As was also the case with our tDNA preparations, the yield of 5s DNA agrees well with the content of 5s RNA cistrons in the genome [116]

Table 10. Isolation of 5s DNA and tDNA from E. coli

	Hybridization reaction[a]	% of original DNA bound to HA[b]
RNA Cycle	—	(0.10)
5	Front peak DNA from cycle 5 hybridization with 5s RNA + tRNA	0.061
6	Back peak DNA from cycle 5 + tRNA	0.055
5s RNA Cycle	—	(0.070)
6	Back peak DNA from cycle 5 + 5s RNA	0.037
Elimination of unincubated background	Back peak DNA from cycle 6	0.018
7	Front peak DNA from background elimination + 5s RNA	0.0081
8	Back peak DNA from cycle 7 + 5s RNA	0.0077

[a] Hybridization reactions were carried out as described in Table 9. (From Doctor and Brenner, 1972).
[b] HA = hydroxyapatite.

4. Circular DNAs — Viral, Plasmid, Mitochondrial

Circular DNAs of known function occur naturally in both prokaryotic and eukaryotic cells (for review, see ref. 73). To date, cyclic duplex forms of DNA have been described for the chromosomes of some bacteria, the genomes of bacterial and animal viruses, the DNA of bacterial plasmids and for mitochondria of higher organisms. Likewise, circular DNAs of unknown function have been found in both prokaryotic and eukaryotic systems [73].

For many investigations dealing with gene replication and expression, small intact DNA molecules may be quite suitable if the function of the gene under study can be measured in the presence of other genetic information or gene products. Indeed, there is an obvious advantage in selecting for such studies a gene contained in a circular DNA molecule. Except for circular bacterial chromosomes, the cyclic DNAs listed above are much smaller than other DNA configurations which carry the bulk of the genetic information of the cell. Fractionation of such small circular DNAs from the remainder of the cellular DNA thus represents a large enrichment for the genes carried by these molecules. The final enrichment will depend on the molecular weight of the DNA, the size of the gene, the gene dosage, and, of course, the degree to which the circular DNA is resolved from other cellular DNA. In such studies the strain must be chosen carefully, as many strains contain several natural circular episomes.

It is beyond the intended scope of this review to describe individually the large number of preparations of circular DNA reported in the literature. Thus, we will limit our discussion to the general approaches used to purify the circular DNAs listed above and refer the reader to the excellent review by HELINSKI and CLEWELL for specific papers [73].

Because of their unique size and structure a number of physical chemical techniques can be used to fractionate circular DNA from large amounts of undesirable chromosomal DNA. Techniques which have been used successfully to purify circular DNAs include: (1) chromatography on methylated albumin kieselguhr [56, 63] or benzoylated, naphthoylated DEAE cellulose [86]; (2) DNA reassociation, and (3) differential sedimentation or buoyant density centrifugation. Fractionation by centrifugation can be effected where the base composition of the circular DNA differs from that of the chromosome or where density differences can be introduced artificially by use of the intercalating dye ethidium bromide [122]. Covalently closed circular DNA binds less ethidium bromide than non-circular or nicked circular DNA. This differential binding of dye results in bouyant densities distinct for the two DNA forms.

In certain cases where plasmid DNA has a buoyant density very similar to that of host DNA, purification can be made relatively simple by transferring the plasmid to a recipient cell with DNA of different density. This method can be used now for some bacterial episomes, colicinogenic factors and drug resistance factors. In particular, the bacterial sex factor F has been transmitted at high frequency between strains of *Escherichia*, *Salmonella* and *Shigella* [77]. F-lac and other episome elements have been transferred between *Salmonella* and *Serratia marcescens* which possess DNAs which are 50 and 58% GC, respectively [62]. Similarly *E. coli* F-factors can be transferred from *E. coli* into several species of *Proteus* with 38 to 50% GC, and also to species of *Pasteurella* and *Vibrio* [9, 101].

In the next section, dealing with enrichment of specific genes by genetic manipulation, we describe work in which this approach was used to purify F-lac, P-lac [63, 152] and F-rDNA [17].

B. Enrichment by Genetic Manipulation

A cistron in the chromosome of a bacterial cell can usually be enriched at least 100-fold by preparing F′ factors or specialized transducing phage which carry the specified host marker. Although these particles usually incorporate host DNA from regions of the chromosome adjacent to their appropriate attachment sites, the specificity of insertion is not absolute. On rare occasion genetic information located far from the normal incorporation site becomes attached to the phage or episome DNA. Once purified, this DNA should be well suited for certain *in vitro* studies of gene activity which require double stranded DNA in native conformation.

1. *F-Merogenotes*

Although it has long been known that *E. coli* sex factors can incorporate host chromosomal markers, only recently has it become possible to isolate F′ factors for virtually any *E. coli* gene [94]. Once a strain containing appropriate F′ has been created it should be possible to purify the episomal DNA using the techniques listed above for the isolation of circular DNAs.

Classically, the isolation of F′ factors involved selection for early transfer of very late Hfr markers in a Hfr × F^- cross interrupted after 30 to 60 min of mating. Because interruption occurs before normal Hfr cells are able to transfer late markers as part of the continuum of the chromosome, those hybrids expressing a late marker most likely inherited it from an appropriate F′ factor [77]. Since the technique cannot be used to isolate F′ factors which carry early Hfr markers the method is limited by the Hfr strains available.

However, Low has recently made an important discovery which makes the isolation of an F′ factor for any gene in *E. coli* relatively simple [94]. In this procedure a recombination-deficient strain of *E. coli* K-12 (rec A) is used as the recipient in a Hfr × F^- cross. The Hfr strains selected contain rec A as a distal marker. Therefore, the formation of normal recombinants for early Hfr markers is virtually eliminated and the absence of normal recombinational events means that inherited DNA must be present in F′ factors.

A second method of stabilizing a F′ merogenote in the host cell is to introduce the episome into a non-homologous host in an interspecies cross [114]. This reduces the possibility of recombination and can facilitate purification of the episomal DNA.

F′-lac DNA: Although geneticists have constructed a large number of partial diploids in *E. coli* and closely related bacteria, only a few episomal DNAs carrying bacterial markers have been isolated and characterized by physical and biochemical methods. To our knowledge Falkow and co-workers [63] were the first to concentrate a specific cistron by purification of an appropriate episomal DNA. These workers were able to partially purify *E. coli* episomal DNA containing genes from the lac operon. Because *E. coli* episomal and chromosomal DNAs have the same G + C content (50 to 51%, ref. 63) *E. coli* F′-lac was transferred into *Proteus mirabilis* to facilitate its purification. Inasmuch as *Proteus* main band DNA is 39% G + C [63] and structurally distinct from episomal DNA it is possible to separate resident host DNA and lactose episome DNA by density gradient centrifugation [63], methylated albumin kieselguhr chromatography [63] or filtration through cellulose nitrate membranes following partial denaturation [152]. When the DNA from F′-lac containing *Proteus mirabilis* is fractionated on a methylated albumin kieselguhr column the material elutes in two peaks. Fig. 29 shows the elution profile obtained. CsCl density gradient analyses of some of the early fractions identifies the material in the first column peak to be highly enriched episomal DNA. Fig. 30 shows sedimentation patterns obtained from column fractions 24, 28 and 42. Native episomal DNA bands at density 1.710 g cm^{-3} while main band DNA is found at density 1.698 g cm^{-3}. Although purification data are not given in this report it is clear that the two species are fairly well resolved and can be purified to a high degree.

P-lac DNA: In another report from the same laboratory episomal DNA from a *Proteus* strain with a P-lac episome was purified to about 90% by two cycles of preferential denaturation and filtration through Millipore filters [152]. Under the conditions of the experiment, chromosomal DNA is preferentially denatured and trapped by the membrane while episomal DNA remains native and is not retained by the filter.

In the bacterial strains used in this study episomal DNA makes up about 4% of the total cellular DNA. It is possible to increase the number of episomes per cell from a few to several hundred by selectively inhibiting the synthesis of chromosomal DNA, or by choosing a recipient in which the episome replicates independently from the chromosome. Other extra-chromosomal DNA elements, namely the resistance factors (R) and colicinogenic (Col) factors have been amplified in this way [80].

F'-rDNA: BIRNBAUM and KAPLAN have been able to enrich *E. coli* rDNA from a partial diploid strain of *P. mirabilis* which contains an *E. coli* episome [17]. For this work an F-merogenote that carries *E. coli* rDNA (F'-rDNA) was transferred into a F⁻

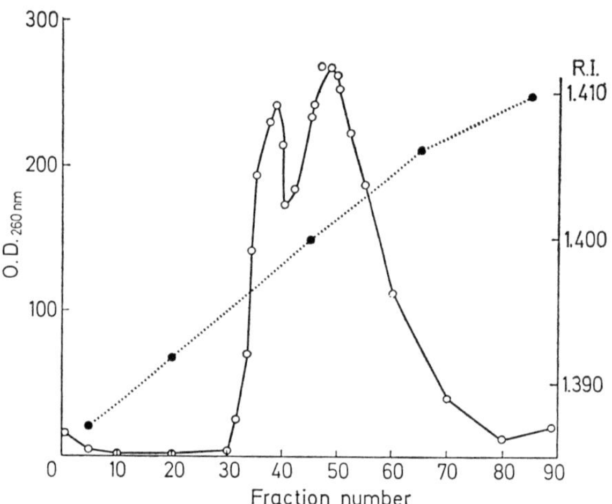

Fig. 29. MAK column chromatography of PM-1 F'-13 DNA. Approximately 1.5 mg of PM-1 F'-13 DNA was diluted in 0.5 M NaCl (buffered to pH 6.8 with 0.05 M phosphate) to 20 μg/ml and loaded on a MAK column. The DNA was eluted with a NaCl gradient of 0.5 to 0.8 M NaCl and 5-ml fractions were collected. R. I. refers to refractive index measurements used to monitor the gradient (dotted line). (From FALKOW et al., 1964)

Proteus recipient and the episomal rDNA enriched by chromatography on benzoylated-DEAE cellulose.

Their fractionation scheme is based on a 5 °C difference in the Tm between *E. coli* and *P. mirabilis* bulk DNA (50% G+C vs. 39% G+C, ref. 100) and the ability of benzoylated-DEAE to distinguish between single and double stranded DNA. Conditions were found under which 80% of the *Proteus* DNA is denatured while the *E. coli* episomal DNA remains double stranded. Since single stranded DNA has a greater affinity for benzoylated DEAE cellulose than has native DNA it was possible to partially resolve these species and enrich the F'-rDNA about 3-fold.

rRNA:DNA hybridization studies show the rDNA content of the F⁻ *Proteus*, partial diploid, and benzoylated-DEAE cellulose enriched episome DNA to be 0.35%, 0.55% and 0.99% respectively. It should be possible to purify the F'-rDNA even further using the techniques applied by FALKOW and coworkers for F'-lac DNA [63, 152] and also by reassociation kinetics.

BARON, GEMSKI, JOHNSON and WOHLHIETER [10] have been able to form partial diploids between *E. coli* Hfr strains and strains of either *S. typhosa* and *P. mirabilis*. The diploid DNA can originate at varied regions of the chromosome, and in the case of *E. coli* and *P. mirabilis* the diploid region can be isolated by density gradient centrifugation. While most of the diploid regions are large (7 to 45% of the chromosome) smaller diploids can be selected using the same methodology. The mapping data on the extent of the diploid region has been confirmed by DNA:DNA hybridization experiments [24]. *E. coli* and *P. mirabilis* share only some 1% of their DNA at stringent hybridization criteria [24], and differ greatly in G + C content. Therefore,

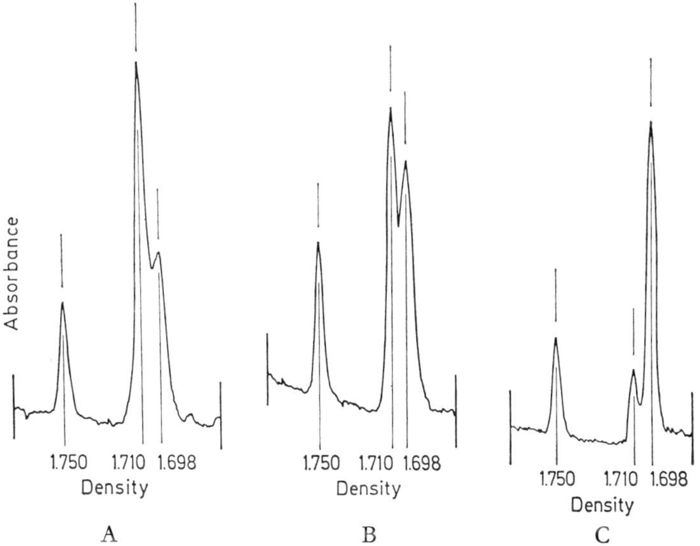

Fig. 30. Fractionation of PM-1 F′-13 DNA. Portions of column fractions shown in Fig. 29 were centrifuged in CsCl at 44,770 rev/min for 20 h. A: microdensitometric tracing of DNA from fraction 34. B: fraction 38. C: fraction 42. The band of buoyant density 1.750 g/cm^3 represents the density standard. (From FALKOW et al., 1964)

E. coli diploid DNA in *P. mirabilis* can be isolated either by CsCl density sedimentation or by hybridization of the diploid strains with *E. coli* DNA. These techniques have the potential to enable enrichment or isolation of small or large DNA segments from well-defined regions at many points on the genetic map.

2. *Specialized Transducing Phage*

Much of our current knowledge about gene structure and function has been derived from genetic and biochemical studies of bacteriophage. The nucleic acids of these particles are well suited for *in vitro* studies for a number of reasons. First, the amount of genetic information is small when compared to that of a bacterium or other microorganism. For example, a typical phage genome is only about 1% the size of the bacterial chromosome and accordingly contains a relatively small number of genes. Second, certain of the phage gene products are readily identified as virus specific and can be analyzed at the molecular level. Third, bacteriophage are excellent subjects for genetic manipulation and analysis and finally, because phage are easily grown and isolated, large quantities of template material can be prepared with relative ease.

Until recently it was believed that only those host genes near the phage attachment site could become incorporated into the virus genome and thus only a limited number of bacterial cistrons could be enriched in this manner. However, it now seems possible to prepare specialized transducing phage which carry any specified bacterial cistron [69]. Simply stated, the genetic manipulations include: (1) isolation of an F′ factor carrying the bacterial gene [94]; (2) insertion of this episome into the bacterial chromosome near a particular phage attachment site (such as ⌀80 or λ); and (3) isolation of transducing phage which have picked up the bacterial gene. Thus far, this general method has been used to incorporate the lactose [13] arabinose [69] galactose [110] and tryptophan [102] operons into the genomes of transducing phage. Although these techniques were developed primarily for the analysis of gene regulatory systems in *E. coli*, it is clear that the DNA from such transducing phage can also be useful for *in vitro* investigations of the replication and transcription mechanisms and studies of the gene product. It should be possible to obtain preparations of the bacterial gene product by hybridization of RNA transcribed from a transducing phage with DNA from the parent phage lacking the bacterial determinants. The purity of these preparations will depend on the amount of the bacterial chromosome incorporated into the phage DNA.

With our current knowledge, enrichment of phage DNA containing bacterial genes is far simpler and more fruitful than preparing episomal DNA containing the same genes. First, because the number of phage per infected cell is usually 10 to 100 times (or more) the number of F′ merogenotes per cell, it is easier to prepare quantities of phage DNA. Second, it is simpler to purify phage DNA than episomal DNA. Whereas episomal DNA must be separated from chromosomal DNA after extraction of total cell DNA, phage DNA is usually prepared from purified phage particles. Phage can be readily purified by differential centrifugation, equilibrium banding in CsCl gradients or chromatography on hydroxyapatite [15]. Therefore, pure phage DNA is easily obtained.

In addition to a large number of *in vitro* studies of phage gene replication and transcription, phage DNAs have also been used to study the expression of certain bacterial genes. For example, DNA from specialized transducing phage has been used to study the transcription of the lac operon [4, 145] and the synthesis of tRNA [51, 76, 155].

C. Gene Enrichment by a Combination of Genetic and Physico-Chemical Methods

Deletions: In any case where both a wild type strain and a well mapped deletion mutant are available, either the mRNA or the DNA from the deleted region can be highly enriched or isolated in pure form. Bautz and Hall first used a deletion technique to enrich mRNA from the rII region of phage T4 [11]. They adsorb denatured DNA from T4 to a phosphocellulose column and show that T4-specific mRNA, but not *E. coli* mRNA hybridizes to the DNA in the column. They next prepare DNA-cellulose columns using DNA from a T4 mutant in which the entire rII region, some 1% of the T4 genome, is deleted. Messenger RNA from wild-type T4 is hybridized to the rII deleted DNA. The mRNA that does not hybridize, presumably including rII-specific mRNA, is repeatedly reincubated with the deletion DNA. After three cycles of hybridization 2.8% of the initial bulk mRNA does not adsorb to the cellulose-bound DNA. Therefore the resulting mRNA, which retains its capacity to hybridize with wild-type T4 DNA, contains approximately 1/3 of mRNA specific for the rII region; about a 50-fold enrichment.

This method is restricted because phosphocellulose only binds glucosylated DNAs. In a later report [12] Bautz and Reilly utilize nitrocellulose [67, 115] to bind single stranded DNA and to carry out hybridizations with mRNA. Wild-type, denatured T4 DNA is immobilized on a nitrocellulose powder slurry and placed in a column. A small amount of nitrocellulose is placed on top and above that is placed nitrocellulose containing DNA from a rII deletion mutant. Wild-type or mutant T4 mRNA is very slowly passed through the column at 60 °C. The mRNA mainly binds to the mutant DNA, except for the rII message which is hybridized to the wild-type DNA below. The enriched rII mRNA is further fractionated by hybridization against mutants deleted in only the A or the B cistron of the rII region.

This method was later modified [103] to obtain r II DNA. Both wild-type and rII-enriched mRNA from the very small rII deletion mutant r1519 are isolated by the deletion method as above. The mRNA from each column layer is eluted and then hybridized with sheared wild-type DNA. The hybrids are then centrifuged to equilibrium in CsCl. The 1519 mRNA should hybridize to those r^+DNA fragments capable of transforming the r1519 region, but not to another deletion, r386, physically distinct from 1519. The hybrid fraction of the gradient therefore should show a high ratio of r1519/r386 transformants. Alternatively, when the experiments are carried out with mRNA enriched for the portion of rII including r386, but not r1519, the hybrid fraction should show a high ratio of r386/r1519 transformants. In this way they obtain as much as a 10-fold enrichment for the selected marker.

The work of Bautz and his coworkers is elegant within the framework of the existing technology. The major difficulties encountered by these investigators were the inability to hybridize quantitatively and the large (2×10^6 daltons) DNA fragments. Newer techniques, such as outlined below, include DNA strand separation, single strand specific nucleases and better means of quantitatively distinguishing between single stranded and double stranded DNA. Their utilization should greatly increase the sensitivity of the deletion method of gene isolation.

Jayaraman and Goldberg, also working with phage 4T [78], combine the DNA strand separation techniques of Szybalski and coworkers [70] and digestion of DNA using an endonuclease specific for single stranded DNA [5] to obtain specific mRNA: DNA hybrids. T4 DNA is separated into H and L strands [70] and shown to be pure by self-annealing experiments on hydroxyapatite and banding in CsCl. Radio-labeled mRNA is prepared from wild-type T4 as well as from mutants deleted in the rII or gene 21 region. Messenger RNAs are then incubated with both H and L strand DNA. The incubation mixtures are passed through nitrocellulose filters where the hybrids bind to the filter. Endonuclease treatment removes non-hybridized DNA, and RNA is removed by thermal denaturation. The resultant DNA is used for transformation in a bacterial spheroplast system. They show that a marker is protected from endonuclease digestion only when hybridized with mRNA in which the marker is not deleted. They further demonstrate the time course of transcription of markers by using mRNA made at various times after infection. If the marker is transcribed, the mRNA hybridized to and protected the DNA from endonuclease digestion.

With slight alterations their method should be very effective in isolating specific genes. One must either first isolate specific mRNA (or DNA) by repeatedly hybridizing wild-type sheared DNA (or mRNA) with DNAs from various deletion mutants and selecting for the nucleic acid that does not hybridize.

An example of this approach is the enrichment of lac operon mRNA from *E. coli* [25]. Pulse-labeled RNA is isolated from *E. coli* under conditions favoring β-galactosidase synthesis. The RNA is repeatedly reacted with DNA from an *E. coli* strain containing a deletion of the entire lac operon. The percentage of nonhybridizable mRNA levels at about 3% of the input. This mRNA hybridizes well with wild-type *E. coli* DNA and with DNA from other lac$^+$ enteric bacteria. The lac operon is only part of the deletion in the strain employed. The deletion is some 2% as determined by interrupted mating experiments. These experiments could be repeated with DNA and with any set of strains where a wild-type and a well-defined deletion are available.

The deletion method of gene isolation has not been utilized to anywhere near its potential. Methods outlined below may be more useful in obtaining genes from systems where transducing phage genetics are quite sophisticated. The deletion technique, however, can be used in any bacterial or other system, where deletions can be obtained regardless of the existence or absence of host or viral mating systems.

Lac operon DNA: Although a number of physical and genetic manipulations can be used to enrich or purify selected genes, most of the methods developed thus far do not yield pure preparations of individual genes which are intact. For many experiments dealing with transcription, it will be important to have a gene preparation that is free of DNA from other operons, and is completely double stranded. One means of obtaining such a preparation of bacterial genes has been developed by BECKWITH and colleagues for the isolation of lac operon DNA [133] and has also been used by LITTAUER and coworkers to prepare double stranded tDNA [50]. The general method can be extended to other bacterial genes.

The method involves annealing strands of DNA from two specialized transducing phages which carry the specified gene in opposite orientations. If appropriate mutant phage are constructed, only the portion of the DNA strand which corresponds to the object gene will be able to anneal. The non-homologous single stranded DNA is then removed by nuclease treatment. The key to successful purification lies with the selection of suitable transducing phage. Fig. 31 summarizes the general procedure developed by BECKWITH's group for purification of a portion of the *E. coli* lac operon [133]. The 'sense' strand of the lac operon (markers i, p, o, z, y, a) is in the light strand of phage A and in the heavy strand of phage B. When the seperated H-strands of the two phage are combined and allowed to reassociate only the lac DNA is able to anneal. The single stranded non-lac DNA is then 'trimmed' away with single strand-specific nuclease from *N. crassa* to yield pure, double stranded lac DNA.

In their study, BECKWITH et al. used $\varnothing$80 and λ transducing phage which were constructed such that the only chromosomal homology between the two was the DNA of the integrated lac operons. To ensure this, all non-lac chromosomal DNA was eliminated from one strain via deletion mutations. Each phage contains intact lac operator and promotor regions and the structural gene z which codes for β-galactosidase. After separating the complementary DNA strands of the phage by equilibrium CsCl gradients in the presence of poly UG [74] the lac 'sense' strands were identified by hybridization with *in vitro* labeled lac mRNA. Consistent with the origin

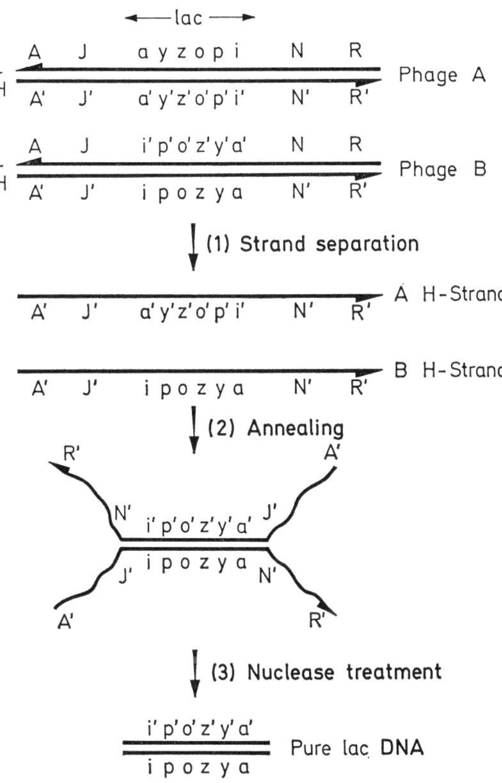

Fig. 31. Purification of lac operon DNA. We have drawn the genomes of both lac transducing phages as vegetative maps of λ derivatives. A, J, N and R are markers on the λ chromosome. The lac markers are the repressor structural gene (i), the promoter (p), the operator (o) the β galactosidase structural gene (z), the lac permease structural gene (y), and the galactoside transacetylase structural gene (a). An arrowhead is located at the 5' terminus of each DNA chain. The primes indicate complementary sequences. (From Shapiro et al., 1969)

of these phages the lac mRNA hybridizes to the L-strand of the λ lac DNA and to the H-strand of the ⌀80 lac DNA. The CsCl gradient fractionation patterns are shown in Fig. 32.

After recovering separated strands, the H-strands from the two phages were mixed and allowed to hybridize. Electronmicroscopic examination show that heteroduplexes are indeed formed. Further, the linear length of the duplex structure closely approximates the predicted length of lac DNA calculated from their genetic data.

To show that double stranded lac DNA is present in the H-strand duplexes, labeled lac mRNA was hybridized to annealed and denatured H-strand DNA and the binding results compared. It was found that at least 5 times more lac mRNA is able to hybridize with denatured H-strand DNA than with the annealed H-strand preparation. The obvious conclusion is that the duplex region of the heteroduplex contains lac DNA, which is not available for hybridization with lac mRNA.

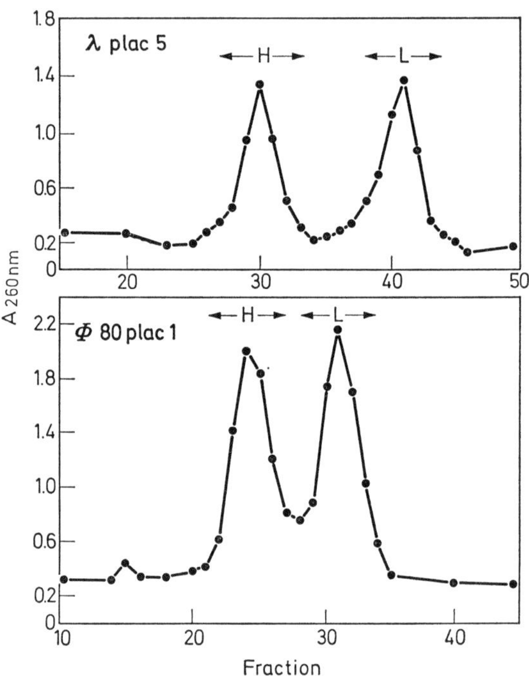

Fig. 32. Separation of the DNA strands of λ plac 5 and ⌀ 80 plac 1. The experimental details are given in the text. Absorbance at 260 nm was read on 150 μl samples in a Zeiss spectrophotometer. The fractions included under the double-headed arrows were pooled for further experiments. The λ plac 5 gradient contained 75 fractions and the ⌀80 plac 1 gradient 68 fractions. (From SHAPIRO et al., 1969)

Fig. 33, panel a, shows the length distribution of the heteroduplex lac DNA as visualized in the electron microscope. The rather narrow symmetrical distribution is consistent with the formation of a single class of hybrid species.

The next step in the purification is the removal of the four single strand tails from the heteroduplexes. This is accomplished with the single strand specific *N. crassa* nuclease described by LINN and LEHMAN [93].

The resulting duplex structures when examined by electron microscopy appear to be free of single stranded segments. The length distribution of the nuclease treated duplex lac DNA is depicted in panel b of Fig. 33. It is evident that nuclease digestion did not change the apparent length of the duplex segment by more than a few per cent (1.4 to 1.5 μm for untreated versus 1.3 to 1.5 μm for nuclease treated).

An RNA:DNA hybridization experiment with lac mRNA and native and denatured nuclease resistant heteroduplex DNA revealed that virtually all the lac

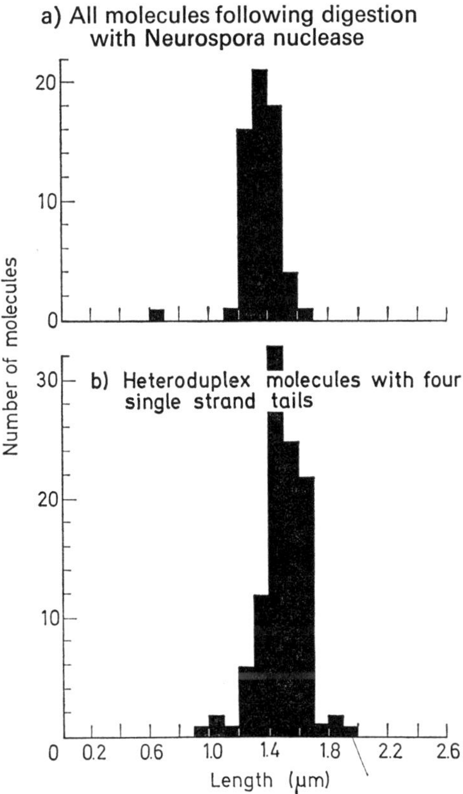

Fig. 33. Length distribution of duplex DNA visualized in the electron microscope. The length determinations were made by enlarging electron micrographs to a final magnification of 50,000 to 100,000 times on a translucent screen, tracing the duplex DNA, and measuring the contour lengths of the tracings with a curvimeter. The results from micrographs of at least two completely independent DNA preparations were pooled to give each histogram. 106 four-ended heteroduplex molecules and 62 short duplex molecules were measured. (From Shapiro et al., 1969)

sequences present are in native DNA; that is, only when it was denatured could the DNA hybridize with lac mRNA.

From the genetic data on the two mutant phage Beckwith and colleagues are able to state that the lac DNA duplexes are free of other chromosomal DNA. Further, it also seems reasonable to conclude that all of the duplex material is lac DNA as no homology between the H-strands of ⌀80 and λ phage DNA has yet been detected [133].

The purity of a preparation of genes isolated in this way depends on the homology of the phage DNAs and the amount of undigested single stranded DNA left at the ends of the heteroduplex after nuclease treatment. Since these 'tails' may cause interference in studies of RNA polymerase binding and chain initiation [147], the size and influence of these segments must be investigated thoroughly.

This method of preparing pure, double stranded genes may well have the greatest potential for studies which require intact transcriptional units free from contaminating DNA.

With our current knowledge, this approach can be applied to the purification of any bacterial gene [133]. As stated above, it is possible to incorporate virtually any gene into the genome of a specialized transducing phage [69, 119] and in each of the two possible orientations [14, 69].

tDNA: The DNA which codes for tyrosine tRNA in the transducing phage ⌀80 psu_3^+ has been isolated by Daniel et al. [50] using the general technique devised by Beckwith and co-workers for lac DNA.

After verifying that the tRNA genes of two strains of ⌀80 psu_3^+ were carried in opposite orientations [51, 106] by hybridization of tRNA to separated strands of phage DNA the H-strands from the two phage were annealed. Electron micrographs of the products formed when H-strands from the two phage are incubated together and separately reveal that duplex structures are formed in each case. Because the extent of self-annealing is low, it was presumed that most of the duplexes formed when the two H-strands are annealed are heteroduplexes.

Following hybridization, the heteroduplexes were digested with *Neurospora* endonuclease [92] to remove single stranded segments of DNA. This treatment resulted in an apparent decrease in molecular weight from about 1.7×10^6 daltons to about 1.5×10^6 daltons (based on mean length determinations from electronmicrographs) or 1.3×10^6 daltons vs 1.08×10^6 daltons if the mode values are used.

The size of the fragment suggests that much of it is not tDNA. Even if the ⌀80 su_3^+ tyrosine tRNA precursor described by Altman and Smith (40,000 to 45,000 daltons [ref. 3]) is smaller than the primary gene product it seems unlikely that the tDNA content will exceed 10%.

The contaminating DNA may be *E. coli* chromosomal segments that are adjacent to the tRNA gene. In ⌀80 su_3^+ mutants it has been determined that the tyrosine tRNA cistron comprises about 4% of the bacterial DNA incorporated into the phage genome, and that this *E. coli* DNA fragment is larger than 5% of the bacterial genome carried by the phage [129]. Nevertheless this procedure enriches the tRNA gene some 30 fold from the phage or some 2000 fold compared to its content in *E. coli.*

Although tDNA prepared in this way is probably not pure, it does function as a template for the *in vitro* synthesis of tRNA-like RNA. With their preparation, Littauer and his associates find that 25 to 80% of the RNA synthesized can compete with tRNA in hybridization assays.

D. In Vitro Chemical Synthesis: Yeast Alanine tDNA

Khorana and his associates have recently completed the monumental task of chemically synthesizing a gene *de novo* [1]. They chose the gene that specifies $tRNA^{ala}$ in yeast. The complete nucleotide sequence of yeast $tRNA^{ala}$ is known. It is also known that the plus strand of DNA, the strand that is transcribed to synthesize RNA, has a sequence complementary both to the sequence of RNA and to the sequence of the minus strand of DNA.

Using this knowledge KHORANA employed a three stage experimental approach to achieve the synthesis of yeast tDNA[ala]. First, deoxyoligonucleotides are prepared to correspond to the entire sequence of both DNA strands. These fragments are

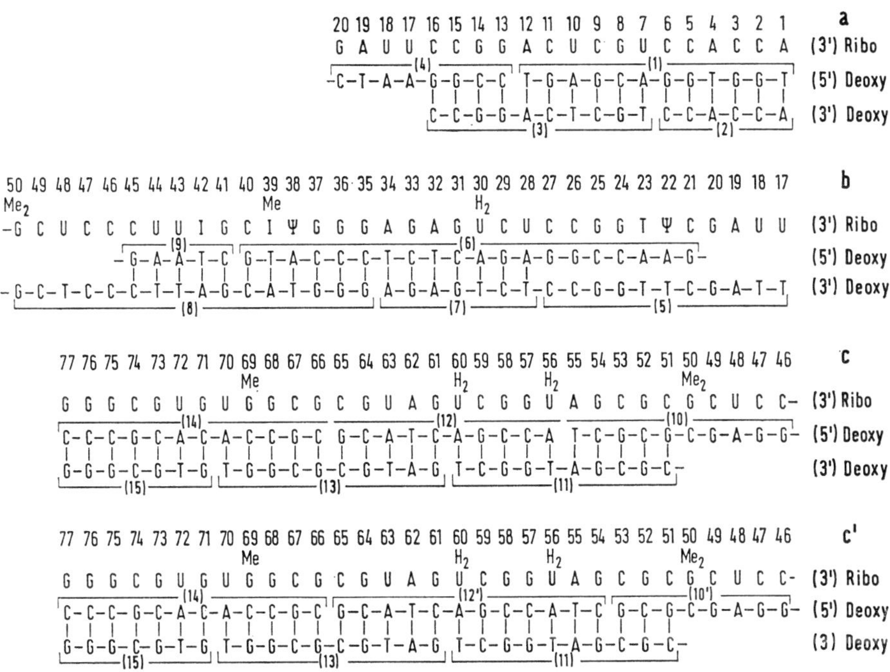

Fig. 34. Total plan for the synthesis of a yeast alanine tRNA gene. The chemically synthesized segments are in brackets, the serial number of the segment being shown within the brackets. A total of seventeen segments (including 10′ and 12′) varying in chain length from penta to icosanucleotides were synthesized. In the synthesis as well as in derivation of the gene sequence the assumption was made that the rare bases present in the tRNA arise by subsequent modification of the four standard bases used by the transcribing enzyme. Thus inosine is formed by deamination of adenosine and so comes from an A-T base pair in DNA. (From AGARWAL et al., 1970)

prepared in lengths of 8 to 12 nucleotides with free hydroxyl groups at both the 3′ and 5′ ends. The complementary fragments contain overlap regions of some 4 to 5 nucleotides. In the second stage, adenosine triphosphate, suitably labeled in the γ-phosphoryl group, in the presence of T4 polynucleotide kinase, is used to phosphorylate the 5′ hydroxyl groups of the deoxyoligonucleotides. Finally, the appropriate segments are joined head to tail to form bihelical molecules. This step is catalyzed by T4 polynucleotide ligase.

The plan for total synthesis of yeast tDNA[ala] is shown in Fig. 34. The gene is divided into three portions shown as A, B, and C (or C′). Each portion consists of several segments (as indicated by numbered brackets). These segments are synthesized and then portions A and B are joined. The joining of the A-B portion to portion C results in synthesis of total yeast $tDNA^{ala}$.

This simplistic account does not convey the formidable technical problems that KHORANA and his colleagues have overcome during a period of several years. Among these considerations are the composition of segments with respect to their purine content, the construction of segments with regard to accuracy of specific nucleotide sequence and yield, complementarity and conformation of segments for head to tail joining.

KHORANA's group is now engaged in the total synthesis of the gene specifying suppressor $tRNA^{tyr}$ from *E. coli*. Their plan is to synthesize both sequences complementary to $tRNA^{tyr}$ and sequences complementary to the larger precursor moiety of $tRNA^{tyr}$ whose sequence has been determined by ALTMAN et al. [3].

Synthesis of these tRNA genes is a superb achievement. Even more important, however, is the development of the essential methodology to a point where one should be able to synthesize any gene whose structure is known (usually from its RNA transcription product).

Results of attempts to use $tDNA^{ala}$ as a primer for transcription are eargerly awaited. Failure of transcription attempts would imply that the precursor molecule, including initiation regions of the gene are necessary for transcription. These regions can "easily" be added to the basic gene sequence as their sequences are elucidated. Alternatively, one might be able to construct one or more synthetic initiator sequences that would stimulate the transcription of any one of a series of genes.

E. In Vitro Enzymatic Synthesis of Cistrons: Globin DNA

With the recent discovery of viral RNA-dependent DNA polymerases [7, 143] a new approach to the problem of preparing DNA complementary to some mRNAs became available. Characterization of these enzymes has made it clear that a variety of natural RNAs (and DNAs) can serve as templates for the synthesis of complementary DNA and that a primer molecule is apparently required [8]. With this information and the knowledge that globin mRNA contains a poly (A) sequence at the 3′ terminus to which an oligo (dT) primer might be hybridized [36, 91], it seems only natural that an attempt should be made to synthesize a single stranded globin cistron *in vitro*. Indeed, three laboratories appear to have successfully accomplished this feat almost simultaneously [79, 125, 146]. The basic experimental approach adopted by the three is schematized in Fig. 35.

Globin mRNA is first hybridized with oligo (dT) to form a primer at the 3′ terminus. Then, DNA complementary to the RNA template is synthesized with reverse transcriptase from avian myoblastosis virus. After transcription, the reaction product is treated with alkali (or RNAse) to release the RNA moiety of the RNA:DNA hybrids, presumably leaving at least part of the 'sense' strand of the genes for globin. Since the three groups used different methods to purify the mRNA and to characterize the resulting DNA product, each report will be discussed separately.

Ross and colleagues [125] purified rabbit globin mRNA by phenol extraction of purified reticulocyte polyribosomes, chromatography on an oligo (dT)-cellulose column, and sucrose gradient centrifugation. The RNA obtained sediments at 9s, migrates as a single band on polyacrylamide gel electrophoresis and directs the synthesis of globin in a cell-free system.

The product of the transcription reaction (before treatment with alkali) has a density of about 1.5 g cm^{-3} in Cs_2SO_4 equilibrium centrifugation. This is the density expected for a RNA:DNA hybrid composed of approximately equal amounts of RNA and DNA. Fig. 36 shows the distribution profile they obtained.

When this material is treated with alkali to hydrolyze the RNA, and the DNA is examined by alkaline sucrose gradient centrifugation, two DNA peaks are observed. The sedimentation constants for these species are 5.8s and 6.3s which correspond to 127,000 and 155,000 daltons, respectively. The larger species is estimated to contain about 500 bases which is 50 nucleotides more than should be required to code for

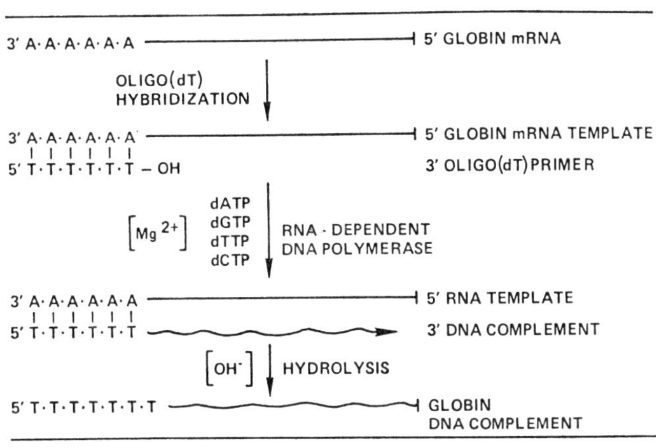

Fig. 35. Synthesis of the DNA complement of rabbit globin mRNA. (From Ross et al., 1972)

rabbit globin but may be 150 bases less than the number of bases in globin mRNA [66]. The 5.8s DNA on the other hand, will be about 410 nucleotides in length and thus may not contain all of the bases of the globin cistron. This peak appears to be heterogeneous and may also contain smaller fragments.

Hybridization experiments on hydroxyapatite with *in vitro* synthesized "globin" DNA show that the product of the reverse transcriptase reaction can hybridize with globin mRNA but not with reticulocyte ribosomal RNA. However, as the authors themselves point out, vigorous proof that the DNA synthesized in this system is globin DNA will come only from a demonstration of its ability to direct the synthesis of globin mRNA and globin in a cell-free system. The work of VERMA et al. [146] and KACIAN et al. [79] also lack this proof of identity. In a note added in proof, Ross and co-workers report that human globin mRNA can also be transcribed with the viral polymerase.

In the report from BALTIMORE's laboratory [146] 'globin' DNA was synthesized from 10s rabbit reticulocyte RNA purified by successive sucrose density gradient centrifugation and then hybridized with oligo (dT). To prevent the formation of double stranded DNA by the virus polymerase, actinomycin D was included in the reaction mixture (in the presence of the antibiotic, synthesis is decreased by about one-third). Most of the resulting single stranded DNA sediments at 8s (Fig. 37) and hybridizes efficiently with the 10s RNA which served as template. In this report too, for reasons that are not yet clear, the DNA product appears to be smaller than its

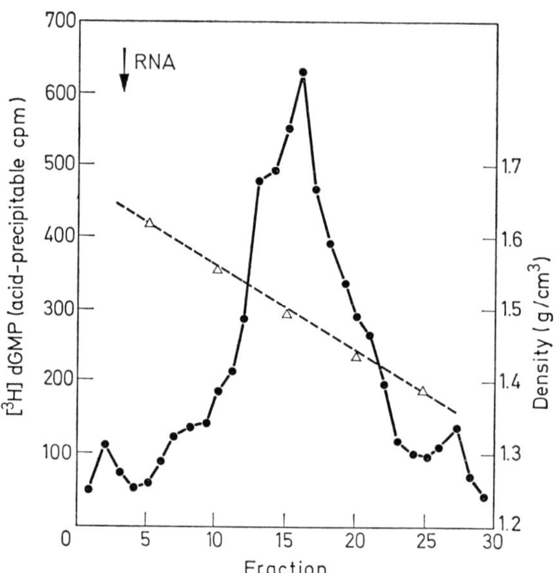

Fig. 36. Equilibrium centrifugation in Cs_2SO_4 of the reaction product. A reaction mixture, incubated for 120 min was made 1.0% in sodium dodecyl sulfate and extracted with a mixture of 10% (v/v) m-cresol in phenol. The product was precipitated from the aqueous phase by the addition of 2 volumes of ethanol, and was then dissolved in 4.8 ml of 0.01 M Tris-HCl (pH 7.2)-1 mM EDTA. To this, an equal volume of a solution of saturated Cs_2SO_4 in the above buffer was added. The sample, in polyallomer tubes, was centrifuged in the Spinco 65 rotor at 38,000 rpm for 70 h at 10 °C. About 0.32 ml fractions were collected from the bottom of each tube. The density of every fifth fraction was measured, and the radioactivity of each fraction was determined after precipitation by the addition of 10% Cl_3CCOOH. (From Ross et al., 1972)

RNA template. This could be due to incomplete synthesis or a difference in the secondary structures of the DNA and RNA complements.

The third report, by KACIAN et al. [79] demonstrates the synthesis of DNA from human and rabbit globin mRNA preparations. The human DNA product has a sedimentation value of 8.3s when compared to 10s human or mouse mRNA and hybridizes efficiently with both human and rabbit globin mRNA.

Fig. 38 shows the results of these hybridization experiments. After incubation with rabbit 18s ribosomal RNA or Qβ viral RNA the human DNA bands at a density of 1.45 g cm^{-3} in a Cs_2SO_4 gradient (panels A and B). This corresponds to the density of single stranded DNA. However, annealing of the human DNA with human or rabbit mRNA results in the formation of hybrids with densities intermediate to those for RNA or denatured DNA. The human DNA:human RNA hybrid has a density of about 1.55 g cm^{-3} (panel C) while the rabbit:human hybrid bands at a lighter density (panel D) suggesting that the heterologous sequences hybridize less well because of a lower extent of homology.

Although it has not been demonstrated conclusively by any of these groups that the RNA which acts as template for the synthesis of complementary DNA is indeed

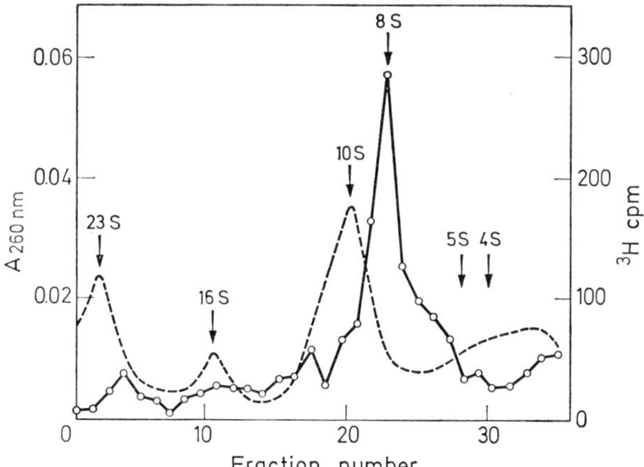

Fig. 37. Exponential sucrose gradients from 15 to 30% containing 0.1 M NaCl, 0.01 M Tris-HCl (pH 7.4), 1 mM EDTA and 0.5% SDS were prepared. The rabbit reticulocyte 10s RNA complementary DNA product was prepared using the complete reaction mixture. The samples were treated with ribonuclease reagent to hydrolyse any RNA bound to the DNA product. The gradients were centrifuged in the SW 27.1 rotor at 26,000 rpm for 31 h at 22 °C. As the gradient fractions were collected, A_{260} was monitored using a Gilford spectrophotometer and the labelled material was acid-precipitated and collected. The figure represents a combined pattern of separate but comparable gradients, one containing the ribonuclease-treated DNA product and the other containing rabbit reticulocyte 10s RNA along with ^{14}C-uridine-labelled 23s, 16s, 5s and 4s *E. coli* RNA markers. The radioactivity of the *E. coli* marker RNAs is shown only by arrows in the figure. (×—×), ^{3}H-DNA (- - -) A_{260}. (From VERMA et al., 1972)

globin RNA or that a complete copy of the template has been synthesized, it is clear that this method can be used successfully to prepare DNA complementary to some RNAs. Reverse transcriptase can use an RNA primer to synthesize a RNA:DNA hybrid, and can also use the hybrid as primer for the production of double stranded DNA [136]. This method, therefore should allow isolation of either just the plus strand or both strands of a gene. While mRNA's which naturally terminate with poly (A) sequences can probably be transcribed, other RNA's may have to be modified in order to accept a primer molecule and highly structured RNA's like tRNA or 5s RNA may not be transcribable by the RNA-dependent DNA polymerases. Even if transcription of a RNA is not complete the product may still be useful for determining gene dosage and mRNA production.

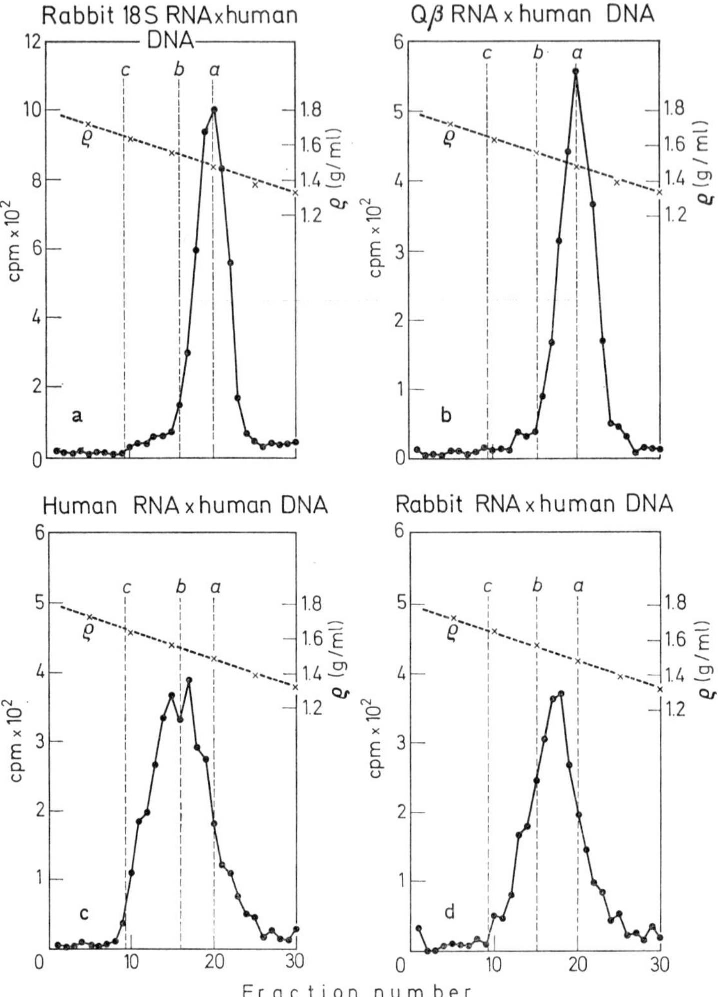

Fig. 38. Hybridization between human globin DNA and various RNAs. Hybridizations were done in 50% formamide. Hybridizations were done in a volume of 10 μl using 0.3 μg of each RNA and 10^{-4} μg of DNA. After incubation for 20 h at 37 °C, the hybridization reaction was mixed with 60 μg of calf thymus DNA and 0.003 M EDTA was added to 5.4 ml. An equal volume of saturated Cs_2SO_4 was added and centrifugation was performed in the Spinco 50.1 rotor at 44,000 rpm at 20 °C for 60 h. The vertical lines labelled a ("DNA"), b ("hybrid"), and c ("RNA") indicate the positions of material banding at densities of 1.47, 1.55 and 1.65 g/ml, respectively. (From KACIAN et al., 1972)

III. Conclusion

A. Discussion — Current and Future Approaches

We have tried to summarize studies in which specific genes, genetic regions or mRNAs have been enriched or isolated in pure form. The perceptive reader will immediately note two areas of omission. We have not discussed animal virus studies, nor have we considered in depth eukaryotic DNA found in such diverse structures as mitochondria, chloroplasts or kinetoplasts. A tremendous amount of data is available for genome transcription and fractionation in animal virus systems, and for that matter in certain plant viruses. These topics merit separate reviews.

It seems apparent that one can now isolate rDNA, 5s DNA and tDNA from any species where these RNA species can be highly purified in milligram quantities and where DNA can be labeled at a high specific activity. Advances in *in vitro* labeling techniques [46] should satisfy this second requirement in organisms where *in vivo* labeling of DNA is difficult.

Future gene isolation studies well might follow three principle lines of development: (1) scaling up existing gene isolation methods, (2) finding new methods of selecting for functional genes and fully utilizing existing gene selection techniques, and (3) developing effective transcriptional systems for isolated genes.

Some of the existing methods for gene isolation can probably be scaled up without significant modification. These include fractionation on hydroxyapatite or nitrocellulose, and polyethylene glycol-dextran phase separation. Unfortunately, neither methylated albumin kieselguhr nor density gradient sedimentation appear to be compatible to separation of large quantities of nucleic acid. Perhaps these methods can be modified to retain sensitivity in the presence of increased nucleic acid input. Alternatively, they may be employed only in the later stages of purification, after most contaminating nucleic acid has been removed by other methods.

A possible alternative approach is to use a small amount of pure gene product as template for a DNA polymerase. Theoretically one could use either a single strand or a double strand product as template. In practice the extent of reaction is limited and exact conditions must be worked out for each system. The potential of this approach is so great that, undoubtedly, one can expect significant advances in the not too distant future.

E. coli contains sufficient DNA to specify some 3,000 average size genes, whereas vertebrates contain enough DNA for at least 2×10^6 genes [84]. It seems a fair assumption that any gene that can be selected for, can be isolated in highly enriched or essentially pure form. The problem is how to select!

How then can one attempt to select for individual genes? In bacteria the outlook seems bright, at least in strains where genetic transfer systems are known or where the genetics are well worked out. The use of well defined deletion mutants is certainly applicable to gene isolation studies. The potential of this method is virtually untapped. Small, well defined deletion mutants are available for many regions on the genetic maps of *E. coli* and *S. typhimurium*. The number of deletions can be extended in these species and in a vast array of other organisms.

The use of episomes to enrich for specific genes should be a fruitful approach, especially among the gram negative bacteria. Episomes have the advantage of being promiscuous. They can, therefore, be transferred from one species to another almost at will. One must, however, deal with the F or other transfer factor to which genes are attached. Hybridization of episome to purified transfer factors should remove the transfer factor portion of the episome and leave the desired gene(s) as unhybridized DNA.

As previously stated the isolation of genes by using phage mutants that carry the bacterial gene in opposite orientation may be the method of choice in organisms where these phage can be isolated. The problem here is constructing the proper strains and building up sufficient genetic data. The method is at present restricted to some of the enterobacteria. The next most likely candidates seem to be species of *Bacillus*.

Purified RNA is just as effective as DNA in selecting for a specific gene. In these cases the problem is how to obtain pure mRNA for a given gene. Several available approaches should significantly enrich for specific mRNAs. In phage and viruses early and late genes are expressed at different times and are under different control mechanisms. Therefore RNA pulse labeled at different times during replication results in differing fractions of labeled mRNA. In bacteria one can utilize synchronous cultures to enrich for early genes [48], and mutants that "overproduce" a given gene product or mutants with temperature sensitive biochemical blocks to enrich for specific mRNA species.

In the bulk of microbes molecular biology and genetics are as yet virtually undeveloped. In many cases this is due to difficulties inherent in cell cultivation. In the case of many microorganisms, significant advances await the utilization of molecular and genetic techniques by workers in the field and the realization by molecular biologists that many organisms other than *E. coli*, *S. typhimurium* and *B. subtilis* offer exciting research possibilities.

In the eukaryotes, with very few exceptions (notably certain yeasts, fungi and *Drosophila*) gene isolation is a greater problem due to fewer means of genetic manipulation and larger genome sizes. One can utilize differential DNA reassociation to fractionate repeated from unique DNA. If the desired genes belong to a family of repeated DNA, one can further fractionate to a point where all members of the family and other families of similar gene dosage are isolated.

In certain cases enrichement of specific mRNA has already been accomplished. Isolation of mRNAs for histones [23, 112], hemoglobin [40, 61] and immunoglobulins [87, 151] has been reported.

Significant advances in selection are necessary before gene isolation in animals becomes generally feasible. A fruitful approach may be to exploit animal viruses. If animal viruses containing specific portions of animal genomes can be isolated they can be utilized in the same manner as phage in bacterial systems.

Once isolated, genes can be used to determine gene dosage in a given organism by reassociation rate kinetics. Isolated genes may also be used to determine the rate of evolutionary divergence in specific genetic regions as opposed to divergence in the genome as a whole. This is accomplished in interspecies hybridization reactions.

The main reason for isolating genes, however, is to ask questions about their structure and function. In order to do this one must be able to transcribe the genes. Furthermore, transcription must yield an RNA product of high fidelity so that its role can be assayed in terms of translation, structural function or control function.

In vitro transcription of the lac operon carried by phage ⌀80 p lac has been reported [75], as has transcription of tyrosine tDNA [129, 155], and rDNA [72]. Both quality and quantity of transcription product must be improved, however, the tyrosine tRNA obtained in at least one case [155] is biologically active. This tremendously fertile area of investigation should blossom into full maturity in this decade.

Table 11. Preparation of specific cistrons and mRNAs

Basis of preparation	Organism	Degree of purity	Structure[a]	Reference
	Ribosomal RNA cistrons			
buoyant density	*E. coli*	low	s. s. DNA	53
buoyant density	*X. laevis*	high	d. s. DNA	19, 20
thermal stability	*B. subtilis*	low	d. s. DNA	141
thermal stability	*Mycoplasma*	low	d. s. DNA	128
reassociation rate	*S. carlsbergensis*	low	d. s. DNA	95
liquid phase partition	*L. variegatus*	high	d. s. DNA	117
RNA:DNA hybridization	*E. coli*	high	s. s. DNA	83
RNA:DNA hybridization	*P. mirabilis*	high	s. s. DNA	83
RNA:DNA hybridization	*N. crassa*	high	s. s. DNA	41
RNA:DNA hybridization	*B. subtilis*	high	s. s. DNA	44, 45, 131
RNA:DNA hybridization	*S. typhimurium*	high	s. s. DNA	144
Differential precipitation	*X. laevis*	low	d. s. DNA	54
F-merogenote isolation	*E. coli*	low	d. s. DNA	17
	Transfer RNA cistrons			
thermal stability	*Mycoplasma*	low	d. s. DNA	128
RNA:DNA hybridization	*E. coli*	high	s. s. DNA	26, 27, 99, 107
RNA:DNA hybridization	*E. coli* (∅80 su_3)	high	s. s. DNA	97
DNA:DNA hybridization	*E. coli* (∅80 su_3)	high	d. s. DNA	99
in vitro chemical synthesis	yeast	high	d. s. DNA	1
	5s RNA cistrons			
RNA:DNA hybridization	*E. coli*	high	s. s. DNA	59
Buoyant density	*X. laevis*	high	s. s. DNA	35
	Lactose operon			
DNA:DNA hybridization	*E. coli*	high	d. s. DNA	133
F-merogenote isolation	*E. coli*	low	d. s. DNA	63, 152
F-merogenote isolation	*P. mirabilis*	low	d. s. DNA	63, 152
transducing phage	*E. coli*	low	d. s. DNA	13
deletion mutant	*E. coli*	low	mRNA	25
	Arabinose operon			
transducing phage	*E. coli*	low	d. s. DNA	69
	Galactose operon			
transducing phage	*E. coli*	low	d. s. DNA	110
	Tryptophan operon			
transducing phage	*E. coli*	low	d. s. DNA	102
	rII cistrons			
deletion mutant	T4	low	s. s. DNA	103, 78
deletion mutant	T4	low	mRNA	11, 12
	Gene 21			
deletion mutant	T4	low	s. s. DNA	78

Table 11. (continued)

Basis of preparation	Organism	Degree of puriry	Structure[a]	Reference
	Globin			
reverse transcriptase	vertebrate	high	s. s. DNA	79, 125, 146
physico-chemical	vertebrate	high	mRNA	40, 61
	Histones			
physico-chemical	vertebrate	high	mRNA	23, 112
	Immunoglobulins			
physico-chemical	vertebrate	high	mRNA	87, 151

[a] s. s. = single stranded; d. s. = double stranded.

B. Summary

In the body of this review gene isolation experiments are grouped according to method of isolation rather than the isolated gene. Table 11 attempts to summarize these studies by gene or gene product. It is intended as a very brief summary of experiments reviewed herein. It gives the main method of preparation, the gene, the organism, the strandedness of the product obtained, approximate degree of purity and the reference. The column on structure indicates whether single or double stranded DNA or mRNA was purified. Messenger RNA is included as we strongly feel that given pure mRNA any gene can be isolated. Purity is arbitrarily designated as high or low. This refers to *our* estimate of the purity of a given gene preparation. The purity says nothing about the yield. In many highly pure gene preparations the yield may be very poor.

Acknowledgements

We are greatly indebted to our colleagues who provided unpublished data. We thank P. GEMSKI for his critical review of parts of the manuscript. Special thanks to ELLEN DOCTOR and to members of the Division of Medicine at Walter Reed Army Institute of Research for typing much of this manuscript. We are grateful to M. BANNING who helped in many aspects of assembling the manuscript.

Responsibility for limiting the scope of this review, and for any omissions and all errors remains solely with the authors.

This review was supported in part by a grant from the National Institutes of Health (GM 19351-01) to one of us (MJF).

References

1. AGARWAL, K. L., BUCHI, H., CARUTHERS, M. H., GUPTA, N., KHORANA, H. G., KLEPPE, K., KUMAR, A., OHTSUKA, E., RAJBHANDARY, U. L., VAN DE SANDE, J. H., SGARAMELLA, V., WEBER, H., YAMADA, T.: Total synthesis of the gene for an alanine transfer ribonucleic acid from yeast. Nature (Lond.) **227**, 27 (1970).
2. ALTMAN, S.: Isolation of tyrosine tRNA precursor molecules. Nature (Lond.) New Biol. **229**, 19 (1971).
3. ALTMAN, S., SMITH, J. D.: Tyrosine tRNA precursor molecule polynucleotide sequence. Nature (Lond.) New Biol. **233**, 35 (1971).

4. Arditti, R. R., Eron, L., Zubay, G., Tocchini-Valentini, G., Connaway, S., Beckwith, J.: *In vitro* transcription of the *lac* operon genes. Cold Spr. Harb. Symp. quant. Biol. **35**, 437 (1970).
4. Ashe, H., Seaman, E., Vunakis, H. V., Levine, L.: Characterization of a deoxyribonuclease of *Mustelus canis* liver. Biochim. biophys. Acta (Amst.) **99**, 298 (1965).
6. Attardi, G., Amaldi, F.: Structure and synthesis of ribosomal RNA. Ann. Rev. Biochem. **39**, 183 (1970).
7. Baltimore, D.: RNA-dependent DNA polymerase in virions of RNA tumor viruses. Nature (Lond.) **226**, 1209 (1970).
8. Baltimore, D., Smoler, D.: Primer requirement and template specificity of the RNA tumor virus DNA polymerase. Proc. nat. Acad. Sci. (Wash.) **68**, 1507 (1971).
9. Baron, L. S., Falkow, S.: Genetic transfer of episomes from *Salmonella typhosa* to *Vibrio cholerae*. Rec. Gen. Soc. Amer. **30**, 59 (1971).
10. Baron, L., Gemski, P., Johnson, E. M., Wohlhieter, J. A.: Intergeneric bacterial matings. Bact. Rev. **32**, 362 (1968).
11. Bautz, E. K. F., Hall, B. D.: The isolation of T4-specific RNA on a DNA-cellulose column. Proc. nat. Acad. Sci. (Wash.) **48**, 400 (1962).
12. Bautz, E. K. F., Reilly, E.: Gene specific messenger RNA: Isolation by the deletion method. Science **151**, 328 (1966).
13. Beckwith, J. R., Signer, E. R.: Transcription of the *lac* region of *Escherichia coli*. I. Inversion of the *lac* operon and transduction of *lac* by ⌀80. J. molec. Biol. **19**, 254 (1966).
14. Beckwith, J. R., Signer, E. R., Epstein, W.: Transcription of the *lac* region of *E. coli*. Cold Spr. Harb. Symp. quant. Biol. **31**, 393 (1966).
15. Bernardi, G.: Chromatography of native deoxyribonucleic acid on calcium phosphate. Biochem. Res. Commun. **6**, 54 (1961).
16. Bernhardt, D., Darnell, J. E.: tRNA synthesis in HeLa cells: a precursor to tRNA and the effects of methionine starvation on tRNA synthesis. J. molec. Biol. **42**, 43 (1969).
17. Birnbaum, L. S., Kaplan, S.: Localization of a portion of the ribosomal RNA genes in *E. coli*. Proc. nat. Acad. Sci. (Wash.) **68**, 925 (1971).
18. Birnstiel, M. L.: Some experiments relating to the homogeneity and arrangement of the ribosomal RNA genes of *Xenopus laevis*. In: Cell differentiation (Reuck, A. V. S., Knight, J., Eds.). Ciba Symposium, p. 178. Little, Brown and Co., 1967.
19. Birnstiel, M., Speirs, J., Purdom, I., Jones, K., Loening, U. E.: Properties and composition of the isolated ribosomal RNA satellite of *Xenopus laevis*. Nature (Lond.) **219**, 454 (1968).
20. Birnstiel, M. L., Wallace, H., Sirin, J. L., Fishberg, M.: Localization of the ribosomal DNA complements in the nuclear organizer region of *Xenopus laevis*. Nat. Cancer Inst. Monogr. **23**, 431 (1966).
21. Bollum, F. J.: DNA polymerizing enzymes from calf thymus glands. In: Methods in enzymology, XII B, 591 (Grossman, L., Moldave, K., Eds.). New York: Academic Press 1968.
22. Bolton, E. T., Britten, R. J., Cowie, D. B., Roberts, R. B., Szafranski, P., Waring, M. J.: Biophysics. Carnegie Inst. Wash. Yearbook **64**, 317 (1965).
23. Borun, T. W., Scharff, M. D., Robbins, E.: Rapidly labeled, polyribosome-associated RNA having the properties of histone messenger. Proc. nat. Acad. Sci. (Wash.) **58**, 1977 (1967).
24. Brenner, D. J., Falkow, S.: Molecular relationships among members of the *Enterobacteriaceae*. Advanc. Genet. **16**, 81 (1971).
25. Brenner, D. J., Fanning, G. R., Johnson, K. E., Citarella, R. V., Falkow, S.: Polynucleotide sequence relationships among members of *Enterobacteriaceae*. J. Bact. **98**, 637 (1969).
26. Brenner, D. J., Fournier, M. J., Doctor, B. P.: Isolation and partial characterization of the tRNA cistrons from *E. coli*. Nature (Lond.) **227**, 448 (1970).
27. Brenner, D. J., Fournier, M. J., Doctor, B. P.: Unpublished data (1972).
28. Britten, R. J.: Repeated DNA and transcription. In: Problems in biology: RNA in development, p. 187 (Hanly, E. W., Ed.). Utah: Univ. Press 1969.

29. Britten, R. J., Davidson, E. H.: Gene regulation for higher cells: A theory. Science **165**, 349 (1969).
30. Britten, R. J., Kohne, D. E.: Nucleotide sequence repetition in DNA. Carnegie Inst. Wash. Yearbook **65**, 78 (1966).
31. Britten, R. J., Kohne, D. E.: Repeated sequences in DNA. Science **161**, 529 (1968).
32. Brown, D. D., Gurdon, J. B.: Absence of ribosomal RNA synthesis in the anucleolate mutant of *Xenopus laevis*. Proc. nat. Acad. Sci. (Wash.) **51**, 139 (1968).
33. Brown, D. D., Reeder, R. H., Roeder, R. G., Suzuki, Y., Wensink, P. C.: Isolation, characterization and control of genes in development. Carnegie Inst. of Wash. Yearbook **68**, 565 (1969).
34. Brown, D. D., Weber, C. S.: Gene linkage by RNA:DNA hybridization. 1. Unique DNA sequences homologous to 4s RNA, 5s RNA and ribosomal RNA. J. molec. Biol. **34**, 661 (1968).
35. Brown, D. D., Wensink, P. C., Jordan, E.: Purification and some characteristics of 5s DNA from *Xenopus laevis*. J. molec. Biol. **51**, 361 (1970).
36. Bun, H., Lingrel, J. B.: Poly A sequences at the 3′ terminus of rabbit globin mRNAs. Nature (Lond.) New Biol. **233**, 41 (1971).
37. Burdon, R. H.: RNA maturation in animal cells. In: Progr. Nucleic Acid. Res. Mol. Biol. **11**, 33 (1971).
38. Burdon, R. H., Clason, A. E.: Intracellular location and molecular characteristics of tumor cell tRNA precursors. J. molec. Biol. **39**, 113 (1969).
39. Chamberlin, M., Berg, P.: Mechanism of RNA polymerase action: Formation of DNA:RNA hybrids with single stranded templates. J. molec. Biol. **8**, 297 (1964).
40. Chantrenne, H., Burny, A., Marbaix, G.: The search for the mRNA of hemoglobin. Progr. Nucleic Acid. Res. Mol. Biol. **7**, 173 (1967).
41. Chattopadhyay, S. K., Kohne, D. E., Dutta, S. K.: Isolation and characterization of ribosomal RNA cistrons from *Neurospora crassa*. Unpublished data, 1972.
42. Chipchase, M. I. H., Birnstiel, M. L.: On the nature of nuclear RNA. Proc. nat. Acad. Sci. (Wash.) **50**, 1101 (1963).
43. Colli, W., Oishi, M.: Purification and analysis of rRNA genes. Fed. Proc. **28**, 531 (1969).
44. Colli, W., Oishi, M.: Ribosomal RNA genes in bacteria: Evidence for the nature of the physical linkage between 16s and 23s RNA genes in *Bacillus subtilis*. Proc. nat. Acad. Sci. (Wash.) **64**, 642 (1969).
45. Colli, W., Oishi, M.: A procedure for gene purification: The purification of ribosomal RNA genes of *Bacillus subtilis* as DNA:RNA hybrids. J. molec. Biol. **51**, 657 (1970).
46. Commerford, S. L.: Iodination of nucleic acids *in vitro*. Biochemistry **19**, 1993 (1971).
47. Coudray, Y., Quetier, F., Guille, E.: New compilation of satellite DNA's. Biochim. biophys. Acta (Amst.) **217**, 259 (1970).
48. Cutler, R. G., Evans, J. E.: Isolation of selected segments from the genome of Hfr *Escherichia coli*. J. molec. Biol. **26**, 81 (1967).
49. Cutler, R. G., Evans, J. E.: Relative transcription activity of different segments of the genome throughout the cell division cycle of *E. coli:* The mapping of ribosomal and transfer RNA and the determination of the direction of replication. J. molec. Biol. **26**, 91 (1967).
50. Daniel, U., Beckmann, J. S., Sarid, S., Grimberg, J. I., Herzberg, M., Littauer, U. Z.: Purification and *in vitro* transcription of a tRNA gene. Proc. nat. Acad. Sci. (Wash.) **68**, 2268 (1971).
51. Daniel, U., Sarid, S., Beckmann, J., Littauer, U. Z.: *In vitro* transcription of tRNA genes. Proc. nat. Acad. Sci. (Wash.) **66**, 1260 (1970).
52. Darnell, J. E., Wall, R., Tushinski, R. J.: An adenylic acid-rich sequence in mRNA of HeLa cells and its possible relationship to reiterated sites in DNA. Proc. nat. Acad. Sci. (Wash.) **68**, 1321 (1971).
53. Davison, P. F.: Isopycnic centrifugation for the isolation of DNA strands coding for ribosomal RNA. Science **152**, 509 (1966).
54. Dawid, I. B., Brown, D. D., Reeder, R. H.: Composition and structure of chromosomal and amplified ribosomal DNA's of *Xenopus laevis*. J. molec. Biol. **51**, 341 (1970).

55. De Kloet, S. R.: The formation of RNA in yeast: Hybridization of high molecular weight RNA species to yeast DNA. Arch. Biochem. biophys. **136**, 402 (1970).
56. De Witt, W., Helinski, D. R.: Characterization of colicinogenic factor E_1 from a non-induced and a mitomycin C-induced *Proteus* strain. J. molec. Biol. **13**, 692 (1965).
57. Doctor, B. P.: Unpublished results (1972).
58. Doctor, B. P., Banning, M. E., Brenner, D. J., Fanning, G. R., Fournier, M. J., Handley, P. S., Miller, W. L., Sodd, M. A., Steigerwalt, A. S.: Isolation, purification and characterization of 5s and tRNA cistrons from *E. coli*. Symp. on Cellular Growth and Differentiation, Atomic Energy Commission, Trombay, India (1972) (in press).
59. Doctor, B. P., Brenner, D. J.: Isolation of 5s RNA cistrons and tRNA cistrons from *E. coli*. Biochem. biophys. Res. Commun. **46**, 449 (1972).
60. Edmonds, M., Vaughan, M. H., Nakazato, H.: Polyadenylic acid sequences in the heterogeneous nuclear RNA and rapidly labeled polyribosomal RNA of HeLa cells: Possible evidence for a precursor relationship. Proc. nat. Acad. Sci. (Wash.) **68**, 1336 (1971).
61. Evans, M. J., Lingrel, J. B.: Hemoglobin messenger ribonucleic acid. Synthesis of 9s and ribosomal RNA during erythroid cell development. Biochemistry **8**, 3000 (1969).
62. Falkow, S., Marmur, J., Carey, W. F., Spilman, W. M., Baron, L. S.: Episomic transfer between *Salmonella typhosa* and *Serratia marcescens*. Genetics **46**, 703 (1961).
63. Falkow, S., Wohlhieter, J. A., Citarella, R. V., Baron, L. S.: Transfer of episomic material to *Proteus*. J. Bact. **87**, 209 (1964).
64. Flamm, W. G., McCallum, M., Walker, M. B.: The isolation of complementary strands from a mouse DNA fraction. Proc. nat. Acad. Sci. (Wash.) **57**, 1729 (1967).
65. Fournier, M. J., Miller, W. L., Doctor, B. P.: Unpublished results (1972).
66. Gaskill, P., Kabat, D.: Unexpectedly large size of globin messenger RNA. Proc. nat. Acad. Sci. (Wash.) **68**, 72 (1971).
67. Gillespie, D., Spiegelman, S.: A quantitative assay for DNA/RNA hybrids with DNA immobilized on a membrane. J. molec. Biol. **12**, 829 (1965).
68. Goodman, H. M., Abelson, J., Landy, A., Brenner, S., Smith, J. D.: Amber suppression: A nucleotide change in the anticodon of a tyrosine tRNA. Nature (Lond.) **217**, 1019 (1968).
69. Gottesman, S., Beckwith, J. R.: Directed transcription of the arabinose operon: A technique for the isolation of specialized transducing bacteriophages for any *E. coli* gene. J. molec. Biol. **44**, 117 (1969).
70. Guha, A., Szybalski, W.: Fractionation of the complementary strands of coliphage T4 DNA based on the asymmetric distribution of the poly U and poly U, G binding sites. Virology **34**, 608 (1968).
71. Hall, R. H.: The modified nucleotides in nucleic acids. New York: Columbia Univ. Press 1971.
72. Haseltine, W. A.: *In vitro* transcription of *Escherichia coli* ribosomal RNA genes. Nature (Lond.) **235**, 329 (1972).
73. Helinski, D. R., Clewell, D. B.: Circular DNA. Ann. Rev. Biochem. **40**, 899 (1971).
74. Hradecna, Z., Szybalski, W.: Fractionation of the complementary strands of coliphage λ DNA based on the asymmetric distribution of the poly I, G binding sites. Virology **32**, 633 (1967).
75. Iida, Y., Kameyama, T.: Regulation of the lactose operon *in vitro*. I. Transcription of the lactose operon. Molec. gen. Genet. **106**, 296 (1970).
76. Ikeda, H.: *In vitro* synthesis of $tRNA^{tyr}$ precursors and their conversion to 4s RNA. Nature (Lond.) New Biol. **234**, 198 (1971).
77. Jacob, F., Adelberg, E. A.: Transfert de caractère génétiques pas incorporation au facteur sexual d'*E. coli*. Compt. Rend. **249**, 189 (1959).
78. Jayaraman, R., Goldberg, E. B.: A genetic assay for mRNA's of phage T4. Proc. nat. Acad. Sci. (Wash.) **64**, 198 (1969).
79. Kacian, D. L., Spiegelman, S., Bank, A., Terada, M., Metafora, S., Dow, L., Marks, P. A.: *In vitro* synthesis of DNA components of human genes for globins. Nature (Lond.) New Biol. **235**, 167 (1972).

80. KASAMUTSU, H., ROWND, R.: Replication of R-factors in *Proteus mirabilis:* Replication under relaxed control. J. molec. Biol. **51**, 473 (1970).
81. KATES, J.: Transcription of the vaccinia virus genome and occurrence of polyriboadenylic acid sequences in messenger RNA. Cold Spr. Harb. Symp. quant. Biol. **35**, 743 (1970).
82. KLAMERTH, O.: Separation of high molecular weight deoxyribonucleic acid and ribonucleic acid. Nature (Lond.) **208**, 1318 (1965).
83. KOHNE, D. E.: Isolation and characterization of bacterial ribosomal RNA cistrons. Biophys. J. **8**, 1104 (1968).
84. KOHNE, D. E.: Evolution of higher organism DNA. Quart. Rev. Biophys. **33**, 327 (1970).
85. KOHNE, D. E.: Personal communication (1972).
86. KOMANO, T., SINSHEIMER, R. L.: Preparation and purification of ∅X-RF component. Biochim. biophys. Acta (Amst.) **155**, 295 (1968).
87. KUECHLER, E., RICH, A.: Two rapidly labeled RNA species in the polysomes of antibody-producing lymphoid tissue. Proc. nat. Acad. Sci. (Wash.) **63**, 520 (1969).
88. LAIRD, C. D., MCCARTHY, B. J.: Magnitude of interspecific nucleotide sequence variability in *Drosophila*. Genetics **60**, 303 (1968).
89. LEE, S. Y., MENDECKI, J., BRAWERMAN, G.: A polynucleotide segment rich in adenylic acid in the rapidly labeled polyribosomal RNA component of mouse sarcoma 180 ascites cells. Proc. nat. Acad. Sci. (Wash.) **68**, 1331 (1971).
90. LENG, M., FELSENFELD, G.: The preferential interactions of polylysine and polyarginine with specific base sequences in DNA. Proc. nat. Acad. Sci. (Wash.) **56**, 1325 (1966).
91. LIM, L., CANELLAKIS, E. S.: Adenine-rich polymer associated with rabbit reticulocyte messenger RNA. Nature (Lond.) **227**, 710 (1970).
92. LINN, S., LEHMAN, J. R.: An endonuclease from *Neurospora crassa* specific for polynucleotides lacking an ordered structure. I. Purification and properties of the enzyme. J. biol. Chem. **240**, 1287 (1965).
93. LINN, S., LEHMAN, J. R.: An endonuclease from *Neurospora crassa* specific for polynucleotides lacking an ordered structure. II. Studies of enzyme specificity. J. biol. Chem. **240**, 1294 (1965).
94. LOW, B.: Formation of merodiploids in matings with a class of Rec^- recipient strains of *E. coli* K-12. Proc. nat. Acad. Sci. (Wash.) **60**, 160 (1968).
95. LUSBY, E. W., DE KLOET, S. R.: A simple procedure for the partial purification of yeast DNA homologous to ribosomal RNA. Biochim. biophys. Acta (Amst.) **209**, 263 (1970).
96. MARGULIES, L., REMEZA, V., RUDNER, R.: Asymmetric template function of microbial deoxyribonucleic acids: Transcription of ribosomal and soluble ribonucleic acids. J. Bact. **103**, 560 (1970).
97. MARKS, A., KEYHANI, E., NAONO, S., GROS, F., SMITH, J. D.: Isolation of a tyrosine-tRNA-tsDNA* hybrid. *tsDNA:DNA sequence complementary to tyrosine tRNA. FEBS Letters **13**, 110 (1971).
98. MARKS, A., SPENCER, J. H.: Isolation and purification of *E. coli* DNA cistron-tRNA hybrids. Fed. Proc. **28**, 531 (1969).
99. MARKS, A., SPENCER, J. H.: Isolation of *Escherichia coli* transfer RNA gene hybrids. J. molec. Biol. **51**, 115 (1970).
100. MARMUR, J., DOTY, P.: Determination of the base composition of deoxyribonucleic acid from its thermal denaturation temperature. J. molec. Biol. **5**, 109 (1962).
101. MARTIN, D., JACOB, F.: Transfert de l'épisome sexuel d'*Escherichia coli* à *Pasteurella pestis*. Compt. Rend. **254**, 3589 (1962).
102. MATSUSHIRO, A.: Specialized transduction of tryptophan markers in *Escherichia coli* K12 by bacteriophage ∅80. Virology **19**, 475 (1963).
103. MAZAITIS, A. J., BAUTZ, E. K. F.: Partial isolation of an rIIb segment of T4 DNA by hybridization with homologous RNA. Proc. nat. Acad. Sci. (Wash.) **57**, 1633 (1967).
104. MCFARLAND, E. S., FRASER, M. J.: Mammalian DNA-sRNA hybrids. Biochem. biophys. Res. Commun. **15**, 351 (1964).
105. MIDGLEY, J. K.: The nucleotide base composition of ribonucleic acid from several microbial species. Biochim. biophys. Acta (Amst.) **61**, 513 (1962).

106. MILLER, R. C., Jr., BESMER, P., KHORANA, H. G., FIANDT, M., SZYBALSKI, W.: Studies on polynucleotides. XCVII. Opposing orientations and location of the su^{+}_{III} gene in the transducing coli phage ∅80 psu^{+}_{III} and ∅80 d su^{+}_{III} su^{-}_{III}. J. molec. Biol. **56**, 363 (1971).
107. MILLER, W. L., DOCTOR, B. P.: Unpublished observations (1972).
108. MIURA, K.: The nucleotide composition of ribonucleic acids of soluble and particle fractions in several species of bacteria. Biochim. biophys. Acta (Amst.) **55**, 62 (1962).
109. MORELL, P., SMITH, I., DUBNAU, D., MARMUR, J.: Isolation and characterization of low molecular weight ribonucleic acid species from *B. subtilis*. Biochemistry **6**, 258 (1967).
110. MORSE, M. L., LEDERBERG, E. M., LEDERBERG, J.: Transduction in *Escherichia coli* K-12. Genetics **41**, 142 (1956).
111. MUDD, J. A., SUMMERS, D. F.: Polysomal ribonucleic acid of vesicular stomatitis virus infected HeLa cells. Virology **42**, 958 (1970).
112. NEMER, M., LINDSAY, D. T.: Evidence that the S-polysomes of early sea urchin embryos may be responsible for the synthesis of chromosomal histones. Biochem. biophys. Res. Commun. **35**, 156 (1969).
113. NORMORE, W. M., BROWN, J. R.: Guanine plus cytosine (G + C) composition of bacteria. In: Handbook of Biochemistry, 2nd Ed., p. H 24 (SOBER, H. A., Ed.). Cleveland: Chemical Rubber Co. 1971.
114. NOVICK, R. P.: Extrachromosomal inheritance in bacteria. Bact. Rev. **33**, 210 (1969).
115. NYGAARD, A. P., HALL, B. D.: A method for the detection of RNA-DNA complexes. Biochem. biophys. Res. Commun. **12**, 98 (1963).
116. PACE, B., PACE, N. R.: Gene dosage for 5s ribosomal ribonucleic acid in *Escherichia coli* and *Bacillus megaterium*. J. Bact. **105**, 142 (1971).
117. PATTERSON, J. B., STAFFORD, D. W.: Sea urchin satellite deoxyribonucleic acid. Its large scale isolation and hybridization with homologous ribosomal ribonucleic acid. Biochemistry. **9**, 1278 (1970).
118. PHILIPSON, L., WALL, R., GLICKMAN, G., DARNELL, J. E.: Addition of polyadenylate sequences to virus-specific RNA during adenovirus replication. Proc. nat. Acad. Sci. (Wash.) **68**, 2806 (1971).
119. PRESS, R., GLANSDORFF, N., MINER, P., DEVRIES, J., KADNER, R., MAAS, W. K.: Isolation of transducing particles of ∅80 bacteriophage that carry different regions of the *Escherichia coli* genome. Proc. nat. Acad. Sci. (Wash.) **68**, 795 (1971).
120. QUAGLIAROTTI, G., RITOSSA, F. M.: On the arrangement of genes for 28s and 18s ribosomal RNA's in *Drosophila melanogaster*. J. molec. Biol. **36**, 57 (1968).
121. RABIN, E. Z., MUSTARD, M., FRASER, M. J.: Specific inhibition by ATP and other properties of an endonuclease of *Neurospora crassa*. Canad. J. Biochem. **46**, 1285 (1968).
122. RADLOFF, R., BAUER, W., VINOGRAD, J.: A dye-buoyant-density method for the detection and isolation of closed circular duplex DNA: The closed circular DNA in HeLa cells. Proc. nat. Acad. Sci. (Wash.) **57**, 1514 (1967).
123. RETEL, J., PLANTA, R. J.: The investigation of the ribosomal RNA sites in yeast DNA by the hybridization technique. Biochim. biophys. Acta (Amst.) **169**, 416 (1968).
124. RITOSSA, F. M., SPIEGELMAN, S.: Localization of DNA complementary to ribosomal RNA in the nucleolus organizer region of *Drosophila melanogaster*. Proc. nat. Acad. Sci. (Wash.) **53**, 737 (1965).
125. ROSS, J., AVIV, H., SCOLNICK, E., LEDER, P.: *In vitro* synthesis of DNA complementary to purified rabbit globin mRNA. Proc. nat. Acad. Sci. (Wash.) **69**, 264 (1972).
126. RUDNER, R., KARKAS, J. P., CHARGAFF, E.: Separation of *B. subtilis* DNA into complementary strands. I. Biological properties. Proc. nat. Acad. Sci. (Wash.) **60**, 630 (1968).
127. RUSSELL, R. L., ABELSON, J. N., LANDY, A., GEFTER, M. L., BRENNER, S., SMITH, J. D.: Duplicate genes for tyrosine transfer RNA in *Escherichia coli*. J. molec. Biol. **47**, 1 (1970).
128. RYAN, J. L., MOROWITZ, H. J.: Partial purification of native rRNA and tRNA cistrons from *Mycoplasma* sp. (Kid). Proc. nat. Acad. Sci. (Wash.) **63**, 1282 (1969).
129. SATO, K., NISHIMUNE, Y., SATO, M., NUMICH, R., MATSUSHIRO, A., ZNOKUCHI, H. OZEKI, H.: Suppressor-sensitive mutants of coliphage ∅80. Virology **34**, 637 (1968).
130. SCHWEIZER, E., MACKECHNIE, C., HALVORSON, H. O.: The redundancy of ribosomal and transfer RNA genes in *Saccharomyces cerevisiae*. J. molec. Biol. **40**, 261 (1969).

131. SGARAMELLA, V., SPADARI, S., FALASCHI, A.: Isolation of the hybrid between ribosomal RNA and DNA of *Bacillus subtilis*. Cold Spr. Harb. Symp. quant. Biol. **33**, 839 (1968).
132. SHAPIRO, H.: Distribution of purines and pyrimidines in deoxyribonucleic acids. In: Handbook of Biochemistry, 2nd Ed., p. H 96 (SOBER, H. A., Ed.). Cleveland: Chemical Rubber Co. 1971.
133. SHAPIRO, J., MACHATTIE, L., ERON, L., IHLER, G., IPPEN, K., BECKWITH, J.: Isolation of pure *lac* operon DNA. Nature (Lond.) **224**, 768 (1969).
134. SKINNER, D. M.: Satellite DNA's in the crabs *Gecarcinus lateralis* and *Cancer pagurus*. Proc. nat. Acad. Sci. (Wash.) **58**, 103 (1967).
135. SKINNER, D. M., KERR, M. S.: Characterization of mitochondrial and nuclear satellite deoxyribonucleic acids of five species of Crustacea. Biochemistry **10**, 1864 (1971).
136. SPIEGELMAN, S., BURNY, A., DAS, M. R., KEYDAR, J., SCHLOM, J., TRAVNICEK, M.. WATSON, K.: Characterization of the products of the RNA-directed DNA polymerases in oncogenic RNA viruses. Nature (Lond.) **227**, 563 (1970).
137. SPIEGELMAN, S., YANKOFSKY, S. A.: The relation of ribosomal RNA to the genome. In: Evolving genes and proteins (BRYSON, V., VOGEL, H. J., Eds.). New York: Acad. Press 1965.
138. STAFFORD, D. W., GUILD, W. R.: Satellite DNA from the sea urchin sperm. Exp. Cell Res. **55**, 347 (1969).
139. SWARTZ, M. N., TRAUTNER, T. A., KORNBERG, A.: Enzymatic synthesis of deoxyribonucleic acid. J. biol. Chem. **237**, 1961 (1962).
140. SZYBALSKI, W.: Use of cesium sulfate for equilibrium density gradient centrifugation. Meth. Enzymol. **12 B**, 330 (1968).
141. TAKAHASHI, H.: An attempt to concentrate ribosomal RNA cistrons in *Bacillus subtilis* by millipore filtration. Biochim. biophys. Acta (Amst.) **190**, 214 (1969).
142. TAYLOR, A. L., TROTTER, C. P.: Revised linkage map of *Escherichia coli*. Bact. Rev. **31**, 353 (1967).
143. TEMIN, H. M., MIZUTANI, S.: RNA-dependent DNA polymerase in virions of Rous sarcoma virus. Nature (Lond.) **226**, 1211 (1970).
144. UDVARDY, A., VENETIANER, P.: Isolation of the ribosomal RNA genes of *Salmonella typhimurium*. Europ. J. Biochem. **20**, 513 (1971).
145. VARMUS, H. E., PERLMAN, R. L., PASTAN, I.: Regulation of *lac* messenger ribonucleic acid synthesis by cyclic adenosine 3,'5'monophosphate and glucose. J. biol. Chem. **245**, 2259 (1970).
146. VERMA, I. M., TEMPLE, G. F., FAN, H., BALTIMORE, D.: *In vitro* synthesis of DNA complementary to rabbit reticulocyte 10s RNA. Nature (Lond.) New Biol. **235**, 163 (1972).
147. VOGT, V.: Breaks in DNA stimulate transcription by core RNA polymerase. Nature (Lond.) **223**, 854 (1969).
148. WALKER, P. M. B., MCLAREN, A.: Fractionation of mouse deoxyribonucleic acid on hydroxyapatite. Nature (Lond.) **208**, 1175 (1965).
149. WALLACE, H., BIRNSTIEL, M. L.: Ribosomal cistrons and the nucleolar organizer. Biochim. biophys. Acta (Amst.) **114**, 296 (1966).
150. WARING, M., BRITTEN, R. J.: Nucleotide sequence: A rapidly reassociating fraction of mouse DNA. Science **154**, 391 (1966).
151. WILLIAMSON, A. R., ASKONAS, B. A.: Biosynthesis of immunoglobulins: The separate classes of polyribosomes synthesizing heavy and light chains. J. molec. Biol. **23**, 201 (1967).
152. WOHLHIETER, J. A., FALKOW, S., CITARELLA, R. V.: Purification of episomal DNA with cellulose nitrate membrane filters. Biochim. biophys. Acta (Amst.) **129**, 475 (1966).
153. WOOD, D. D., LUCK, D. J. L.: The hybridization of mitochondrial rRNA. J. molec. Biol. **41**, 211 (1969).
154. ZEHAVI-WILLNER, T., COMB, D. G.: Studies on the relationship between transfer RNA and transfer-like RNA. J. molec. Biol. **16**, 250 (1966).
155. ZUBAY, G., CHEONG, L., GEFTER, M.: DNA-directed cell-free synthesis of biologically active transfer RNA: su^{+}_{III} tyrosyl-tRNA. Proc. nat. Acad. Sci. (Wash.) **68**, 2195 (1971).

Mechanism of Protein Synthesis and Use of Inhibitors in the Study of Protein Synthesis

AKIRA KAJI

I. Introduction

The biosynthesis of proteins appears to be one of the most complicated synthetic processes catalyzed by biological systems. It is unique in that as many as three to four energy-rich bonds are consumed in the synthesis of a relatively low-energy chemical bond, the peptide linkage. At least nine non-ribosomal soluble protein factors participate in an orderly sequential series of polymerization reactions of aminoacyl tRNA catalyzed by a most complicated biological synthetic machinery — the ribosomes. Protein synthesis can be divided into four major steps, namely, aminoacylation of tRNA, initiation of the polypeptide chain, polypeptide chain elongation, and termination of chain elongation including release of completed chains from the ribosomes. The first step, aminoacylation of tRNA, will not be discussed in this article.

The complexity of the polymerization process of aminoacyl tRNA makes it a target for the action of numerous antibiotics. In this article, I shall attempt to describe the current progress of the studies on the mechanism of protein biosynthesis in relation to the action of antibiotics. Since extensive reviews on the action of antibiotics (PESTKA, 1971) as well as on the mechanisms of protein synthesis (LUCAS-LENARD and LIPMANN, 1971; KAJI, 1970; LENGYEL and SOLL, 1969) have recently been published, we will omit the details of the past developments; emphasis will be placed on the description of current experiments regarding the mechanism of protein synthesis as well as the mode of action of inhibitors in relation to the biosynthesis of protein. The author would like to apologize to those investigators whose contributions despite their importance, could not be covered in this article because of limited space

II. Initiation of Protein Synthesis

1. Association of Ribosomal Subunits with Messenger RNA and the Action of Aurintricarboxylic Acid

It is well accepted that the 30 S ribosomal subunit (in eukaryotic cells, the 40 S ribosomal subunit) participates in the initiation process (NOMURA et al., 1967). Long before this concept was established, 30 S ribosomal subunits were known to combine with synthetic messenger RNA's such as polyuridylic acid (TAKANAMI and OKAMOTO, 1963). The *in-vitro* studies of the initiation process were carried out on the association of the 30 S ribosomal subunit with an AUG codon or f2RNA (NOMURA and LOWRY, 1967). This step is stimulated by two initiation factors, IF_2 and IF_3 [IWASAKI et al.,

1968; Greenshpan and Revel, 1969; Revel et al., 1968 (2)]. These initiation factors were originally discovered among the proteins bound to ribosomes and were required only for the *in-vitro* protein synthesis directed by naturally occurring mRNA such as f2 or MS2 phage RNA. The nomenclature of these factors differs depending on the laboratory. IF_2 is often called F_2, FIII, or C, and IF_3 is called F_3, FII or B. [Revel et al., 1968 (1); Iwasaki et al., 1968; Dubnoff and Maitra, 1969]. However, it was agreed at a recent meeting of workers in this field that they will be called IF_2 and IF_3. We shall discuss the additional factor, IF_1 (F_1, FI or A), in the following section. Despite the fact that IF_2 and IF_3 have been highly purified, the function of these two factors remains somewhat obscure, especially since IF_3 is not required if a synthetic polynucleotide is used to bind smaller subunits [Revel et al., 1968 (1); Lucas-Lenard and Lipmann, 1967; Revel et al., 1968 (2)]. The notion that IF_3 is involved in the ribosomal attachment to the initiation site of messenger RNA [Revel et al., 1970, 1968 (2)] was supported by the finding that this factor changes after virulent phage infection in such a way that the ribosomes specifically seek the phage mRNA rather than host mRNA (Hsu and Weiss, 1969; Schedl et al., 1970; Dube and Rudland, 1970). Furthermore, the synthetic triplet AUG inhibited the binding of 30 S ribosomal subunits to T4 phage mRNA in the absence of IF_3, but this inhibitory effect became negligible in the presence of IF_3 (Revel et al., 1969). Recent evidence indicates that IF_3 can be fractionated into two or more fractions: the function of one fraction is to translate late T4 mRNA, while the other specifically stimulates endogenous *E. coli* protein synthesis as well as MS2 RNA-dependent protein synthesis Grunberg-Manago et al., 1971; Vermeer et al., 1971). The T4-specific IF_3, although present in uninfected cells, increases about fourfold after infection with T4 (Lee-Huang and Ochoa, 1971). In a separate study, it has been shown that initiation factors from T4-infected cells make the normal ribosome bind to a maturation protein cistron of phage RNA which normally does not bind ribosomes unless the RNA is unfolded (Steitz et al., 1970). The fact that IF_3 is localized on the 30S ribosomal subunit is consistent with the notion that the first step in protein synthesis is the binding of this subunit to the initiation triplet of mRNA (Parenti-Rosina et al., 1969; Miller et al., 1969). The actual mechanism through which the 30 S subunit, IF_2 and IF_3, seek the initiation point is not known, but the secondary structure of messenger RNA appears to play an important role. Thus, when the secondary structure of the phage f_2 RNA was changed by formaldehyde treatment (Lodish and Robertson, 1969; Boedtker, 1967), the efficiency of polypeptide synthesis programmed by this RNA increased appreciably, suggesting that the initiation point was exposed by a structural change in the RNA. The same treatment abolished the polar effect of amber mutations in the coat gene on the translation of the RNA polymerase gene. By analysis of the polypeptide formed with the formaldehyde-treated RNA, it was concluded that there are several nucleotide sequences on f_2 RNA which are similar enough for 30 S subunits to initiate peptide synthesis, but these regions are usually hidden because of the three-dimensional structure of the phage RNA (Lodish and Robertson, 1969). In recent studies with formaldehyde-treated phage RNA, IF_3 has been shown to direct ribosomes more to the coat cistron, indicating that not only the structural features of the RNA but also the primary nucleotide sequence play an important role in directing ribosomes to the correct initiation site (Berissi et al., 1971). In this connection, the active role played by ribosomes may also be important. Thus, the ribosomes from *B. subtilis* appear to read efficiently the maturation protein cistron of MS2 phage RNA, while the polymerase and coat-protein cistrons are not translated. The latter two cistrons are efficiently translated by *E. coli* ribosomes (LODISH, 1970).

Recent studies with mammalian systems indicate that the protein factors M_1, M_{2A}, M_{2B}, and M_3 are required for translation of natural mRNA (PRICHARD et al., 1970; SHAFRITZ et al., 1971), while M_3 may not be necessary for the synthetic polynucleotide. In this respect M_3 is similar to IF_3. It is not certain at the moment whether other factors such as M_1 and M_2 (COHEN, 1969; SHAFRITZ et al., 1970) are involved in the binding of messenger RNA to the 40 S ribosomal subunits in a fashion similar to the role of IF_2.

Aurintricarboxylic acid (ATA) has been shown to inhibit specifically the association of ribosomal subunits with messenger RNA. ATA does not dissociate a preformed complex of mRNA, ribosomes, initiator tRNA and initiation factors (GROLLMAN and STEWART, 1968). In the *in-vitro* system for synthesis of phage-related proteins, addition of ATA prior to the addition of phage RNA caused complete inhibition of peptide synthesis. On the other hand, when ATA was added 10 to 15 min after the onset of protein synthesis, it did not inhibit protein synthesis for several minutes [STEWART et al., 1971 (1)]. The observation that labeled ^{3}H ATA was bound by the 30 S subunit but not by the 50 S subunit suggests that the 30 S ribosomal subunit is the target for this inhibitor. Further proof of this concept was provided by the recent observation that a protein fraction isolated from 40 S ribosomal subunits of ascites tumor cells binds polyuridylic acid (poly U), and this binding is sensitive to ATA (ROBERTS and COLEMAN, 1971). The specific inhibitory effect of ATA on the initiation step of the prokaryotic system was used to study the initiation process catalyzed by wheat embryo ribosomes (MARCUS et al., 1970). At 2.5×10^{-5} M ATA the TMV (tobacco mosaic virus) RNA-dependent amino acid incorporation system was completely inhibited, while chain elongation by endogenous polysomes was only mildly inhibited at 5×10^{-5} M. The effect of ATA was markedly diminished if the ribosomes and TMV RNA were pre-incubated with wheat embryo initiation factors prior to the addition of the inhibitor.

2. Binding of Initiator tRNA to the Complex of Messenger RNA and Ribosomal Subunits

The next step in the initiation process is the binding of initiator tRNA to the 30 S subunits. The initiator tRNA is formylmethionyl $tRNA_f$ ($tRNA_f$ is methionyl tRNA which can be formylated) in prokaryotes and methionyl $tRNA_f$ in eukaryote. (CLARK and MARCKER, 1966; SMITH and MARCKER, 1970; LEIS and KELLER, 1970). Evidence that this complex is indeed the obligatory intermediate in protein synthesis was given by an experiment in which ribosomes were labeled with heavy isotopes. In this experiment 70 S ribosomes labeled with ^{15}N and deuterium were mixed with excess light 50 S subunits, poly AUG (random), fMet-tRNA, and factors necessary for the binding of fMet-tRNA and Val-tRNA. It was found that only hybrid 70 S ribosomes contained fMet-tRNA while heavy 70 S ribosomes had Val-tRNA. Thus, 70 S ribosomes must first dissociate in order to bind fMet-tRNA. The resulting 30 S subunits — fMet-tRNA complex would then bind 50 S ribosomal subunits forming the initiation complex of 70 S ribosomes. On the other hand, for the formation of a regular complex of Val-tRNA, messenger RNA, and 70 S ribosomes, the 70 S ribosomes do not have to dissociate in order to participate in this reaction (Guthrie and Nomura, 1968).

The binding of initiator tRNA in prokaryotes is dependent mostly on IF_2 (Wahba et al., 1969). IF_1 participates, but to a lesser extent [Salas et al., 1967 (1)]. Radioactive IF_1 has been shown to bind to 30 S ribosomal subunits in the presence of mRNA, IF_2, GTP, and fMet-tRNA (Hershey et al., 1969). IF_2 has been separated into two fractions, but the exact role of these fractions remains obscure (Parenti-Rosina et al. 1969). The two activities of IF_2, namely the stimulatory effect on the binding of fMet-$tRNA_f$) and the effect on the binding of mRNA to the 30 S subunits, have so far not been assigned to these two IF_2 fractions (Herzberg et al., 1969). These two activities are quite independent of each other because one can inhibit the activity for the template-dependent binding of fMet-tRNA to ribosomes (Mazumder et al., 1969; Wahba et al., 1969) without inhibiting the activity for binding of mRNA (Groner et al., 1970). Although GTP is required for the binding of fMet-tRNA, hydrolysis of the terminal phosphate bond is not necessary. Thus, 5'-guanylyl methylene diphosphate (GMP-PCP) can substitute for GTP in this process (Anderson et al., 1967; Ohta et al., 1967). Other laboratories, however, report that GTP hydrolysis takes place during the formation of the initiation complex with 30 S ribosomal subunits (Lelong et al., 1970; Chae et al., 1969). In a recent experiment, it was shown that, as an analogy to the complex of aminoacyl tRNA, GTP and elongation factor (EFTu), a complex of fMet-$tRNA_f$, IF_2 and GTP may be formed (Lockwood et al., 1971). This complex transfers both fMet-$tRNA_f$ and GTP to 30 S ribosomal subunits in the presence of AUG and initiation factor IF_1. The 30 S initiation complex formed in this manner reacts with 50 S subunits to form a 70 S initiation complex, indicating that the IF_2-GTP-fMet-$tRNA_f$ complex is a true intermediate in the initiation of protein synthesis. When fMet-tRNA is transferred to the 30 S subunits, GTP is also transferred without being hydrolyzed to GDP [Thach and Thach, 1967 (1)].

Studies on the complex of 30 S ribosomal subunits with various polynucleotides showed that the physiological initiation complex, i.e. fMet-tRNA-30 S ribosomal subunits, is most stable. Thus, the initiation factor IF_3 can break up the complex of aminoacyl tRNA, synthetic polynucleotide and 30 S ribosomal subunits while it keeps the physiological initiation complex intact. In the experiment illustrated in Table 1, release of various aminoacyl tRNA's from the 30 S ribosomal complexes was studied in the presence of IF_3. It is clear from this table that as much as 60% of the bound phe-tRNA was released from the complex of phe-tRNA, 30 S subunits and polyuridylic acid by the action of IF_3. In a similar fashion, lysyl-tRNA could be released from the complex of polyadenylic acid and 30 S subunits. On the other hand, no appreciable release of fMet-tRNA took place from the complex of 30 S ribosomal subunits and poly AUG (random). It should be noted that, under identical conditions, Val-tRNA bound to the same complex could be released to an appreciable extent. The poly AUG used in this experiment was a random polymer and contained both AUG (initiation codon) as well as GUA (codon for valine). The initiation factor can apparently distinguish between N-acetyl-phenylalanyl tRNA and fMet-tRNA, because only the former was readily released. This newly discovered function of IF_3 could be very important for living organisms. If, for some reason, 30 S ribosomal subunits happen to bind to the internal region of a cistron and are stabilized by the binding of aminoacyl tRNA, a very undesirable situation would develop. Since 30 S ribosomal subunits cannot move along the messenger RNA, all other 70 S ribosomal movement would halt at this point. Even if this 30 S subunit associates with the 50 S subunit and forms a 70 S ribosome, the protein made by the movement of this 70 S ribosome would be meaningless because it did not start from the correct initiation point. IF_3 can function to prevent formation of a 30 S aminoacyl tRNA-complex in the middle of a cistron, thus avoiding such an undesirable situation. In a separate experiment with labeled synthetic polynucleotides, IF_3 has been shown to release 30 S ribosomal subunits from polynucleotides or triplets other than AUG.

The mechanism of the binding of initiator tRNA to the 40 S ribosomal subunit in eukaryotes is similar to that prokaryotes. A marked difference exists, however, in the

Table 1. Release of aminoacyl tRNA from the 30 S ribosomal subunits. (GUALERZI et al., 1971)

Bound aminoacyl tRNA[a]	mRNA	Aminoacyl tRNA released (% of total bound)
Phe-tRNA	poly U	60
N-acetyl Phe-tRNA	poly U	53
Val-tRNA	poly AUG	27
fMet-tRNA	poly AUG	3
Lys-tRNA	poly A	55

[a] Similar results were obtained when binding of aminoacyl tRNA to 30 S ribosomal subunits was carried out in the presence of GTP and EFTu.

initiator tRNA. In place of formylmethionyl $tRNA_f$, methionyl $tRNA_i$ binds to the 40 S subunit-initiation codon complex. In the eukaryote system, as in the prokaryote, there are two kinds of tRNA specific for methionine, i.e.: $tRNA_f$ and $tRNA_m$. Methionyl $tRNA_f$ can be formylated by the prokaryotic formylating enzyme but methionyl $tRNA_m$ cannot. Since there is no formylating enzyme in the eukaryotic cytoplasm, methionyl $tRNA_f$ functions in its place (SMITH and MARCKER, 1970; LEIS and KELLER, 1970; BROWN and SMITH, 1970; CULP et al., 1970). Methionyl $tRNA_m$ functions to elongate polypeptide chains. Initiation factors M_1, M_{2A} M_{2B} and M_3 have been isolated but the exact functions of these factors are not yet clear. Although an initial assay for M_1 and M_2 was carried out with the polyphenylalanine synthesis system, they seem to function in the rabbit reticulocyte system for the synthesis of hemoglobin (PRICHARD et al., 1970). One of the difficulties in a comparison between M_1, M_{2AB} M_3 and IF_1, IF_2, IF_3 stems from the fact that most of the studies with $M_{1,2,3}$ were carried out with 80 S ribosomes rather than with subunits. Thus, it has been reported that M_1 and M_2 stimulate the binding of Met-$tRNA_f$ to 80 S ribosomes (SHAFRITZ and ANDERSON, 1970). M_2 has been separated into two components; M_{2A} and M_{2B}. In the presence of the 40 S subunit, M_{2A} hydrolyzes γ^{32}P-GTP. In this regard M_{2A} resembles IF_2, because IF_2 also has GTPase activity (SHAFRITZ et al., 1971). In contrast to the well-known GTPase of chain elongation factors, this GTPase activity is not inhibited by fusidic acid (see Section for translocation). As in bacterial systems, GTP is required for the binding of Met-$tRNA_f$ to ribosomes, but GTP hydrolysis is not necessary since GMP-PCP can substitute for GTP. On the other hand, it is believed that at least one molecule of GTP has to be hydrolyzed to GDP for the initiator tRNA to be placed at the puromycin reactive site. In contrast, with *E. coli* 30 S subunits GTP does not appear to be hydrolyzed [LOCKWOOD et al., 1971; THACH and THACH, 1971 (1)].

3. Inhibitors of the Binding of Initiator tRNA to Subunits

a) Streptomycin (SM)

SM appears to act on the initiation step as well as other steps of protein biosynthesis. SM releases ribosome-bound fMet-tRNA from the site corresponding to the donor site (the site where peptidyl tRNA donates its group to the next aminoacyl tRNA). When the binding of fMet-tRNA was carried out in the presence of GMP·PCP, SM did not release the bound fMet-tRNA (MODOLELL and DAVIS, 1970). This report was confirmed by similar studies on the complex of fMet-tRNA, T4-phage messenger RNA, and 70 S ribosomes (LELONG et al., 1971). Even after fMet-tRNA was released, the T4 mRNA still remained attached to the ribosomes. It was observed that during the first minute after the onset of the binding of fMet-tRNA to the ribosomes (the binding of fMet-tRNA during this period reached only 20% of the saturation level), SM caused no release of the bound fMet-tRNA. As fMet-tRNA binding proceeded, the release of fMet-tRNA by SM became apparent. In view of the recent observation that binding of fMet-tRNA to the donor site may consist of at least two steps [THACH and THACH, 1971 (2)], these results may indicate that the first step is not sensitive to SM, but the fMet-tRNA which is already "accommodated" at the donor site is sensitive to SM's releasing action (see next section for "accommodation process"). The observation that fMet-tRNA bound in the presence of GMP ·PCP is not released by SM is also consistent with this hypothesis, because in the presence of this GTP analogue no "accommodation" step takes place.

Even before the effect of SM on the binding of fMet-tRNA had been discovered, evidence from studies on the action of SM *in vivo* had suggested that it may act at some step in initiation. When SM was added to a culture of growing *E. coli*, 30 S and 50 S subunits decreased while 70 S ribosomes increased (SCHLESSINGER et al., 1969; LUZZATO et al., 1969, 1968). The messenger RNA under these conditions appeared to be still attached to the remaining 70 S ribosomes. From these results, it was suggested that an inactive 70 S initiation complex was formed in the presence of SM. These aberrant initiation complexes were thought to be stabilized by SM against dissociation. This is in contrast to the finding that SM releases ribosomes from the initiation complex in the same way as from polysomes. Despite these discrepancies between workers in this field, one can conclude with reasonable assurance that SM interferes with the formation of the initiation complex by either stabilizing aberrant initiation complexes or facilitating the breakdown of the normal complex. The fact that the ribosomal protein responsible for SM resistance is located on the 30 S ribosomal subunit is consistent with the notion that SM inhibits the initiation process which also specifically involves the 30 S subunit.

b) Pactamycin (PA)

Pactamycin acts on both prokaryotes and eukaryotes (BHUYAN, 1967; COHEN and GOLDBERG, 1967; COLOMBO et al., 1966). As a model system to a true protein initiation complex, N-acetylphenylalanyl tRNA binding to 30 S ribosomal subunits is often studied. Like the binding of fMet-tRNA, the binding of N-acetylphe tRNA is dependent on GTP, initiation factors and polyuridylic acid. In this system, pactamycin released the bound N-acetyl Phe-tRNA from 30 S ribosomal subunits (COHEN et al., 1969). The notion that PA may interfere with the initiation step was further supported from studies with the reticulocyte system (MACDONALD and GOLDBERG, 1970). At low concentration (10^{-6} M) PA inhibited globin synthesis by reticulocyte lysates after a lag of 2 min, suggesting that it allows completion of the nascent chain but stops chain initiation. However, at a higher concentration (10^{-5} M) PA also inhibited peptide chain elongation. When reticulocytes are treated with NaF, ribosomes lacking peptidyl tRNA, aminoacyl tRNA and soluble factors accumulate. Such ribosomes appear to bind PA better, probably because the binding site is common to or overlaps with the binding sites of these components. Approximately one molecule of PA was bound per three 30 S subunits. The notion that a low concentration of PA specifically inhibits polypeptide initiation was further strengthened by a recent report on synthesis of hemoglobin by reticulocytes (STEWART-BLAIR et al., 1971). This conclusion was based on three observations. (1) Pactamycin was most active under conditions where there is active formation of new globin chain and is least effective where the system is primarily elongating pre-existing chains. Thus, in lysates of reticulocytes where initiation takes place, Pactamycin inhibits strongly, while in the reconstituted system of isolated ribosomes and soluble enzymes, where no initiation takes place, no appreciable effect was observed. (2) In the presence of pactamycin, incorporation of ^{14}C-valine into the NH_2 terminal position of globin was severely inhibited while incorporation into internal position was much less sensitive. (3) In the presence of 10^{-6} M pactamycin, chain elongation and release of completed globin were still observed. Thus, nascent chains labeled with ^{14}C-amino acids were still released in the presence of pactamycin while the relative amount of monosomes increased. This was interpreted to indicate that the ribosomes released from mRNA at the end of a cistron would accumulate as monosomes because the initiation steps were blocked by PA.

c) Cycloheximide (CH)

Although CH is generally recognized as an inhibitor of the translocation of tRNA, it appears to inhibit the initiation of polyphenylalanine synthesis by the reticulocyte system This was suggested by two earlier observations, First, CH inhibited the binding of unesterified $tRNA^{phe}$ to the complex of reticulocyte ribosomes and polyuridylic acid [MCKEEHAN and HARDESTY, 1969 (1)]. Secondly, synthesis of polyphenylalanine by the reticulocyte system was strictly dependent on the initial binding of $tRNA^{phe}$ to the ribosomes (CULP et al., 1969). The effect of CH on the initiation step for polyphenylalanine synthesis was examined recently in more detail [OBRIG et al., 1971 (1)]. It was found that the "initiation complex", which consists of polyuridylic acid, tRNA specific for phenylalanine, and reticulocyte ribosomes, could not be formed in the presence of CH. The formation of this complex does not require the chain elongation factors and was more sensitive to CH than the chain elongation steps. CH has not been shown to act on the initiation process of other polypeptide synthesis. It should be noted that the binding of unesterified tRNA studied in the reticulocyte system is to 80 S ribosomes and not to 40 S ribosomal subunits. Thus, this system does not seem to have a bearing on the conventional initiation steps where the initiation tRNA binds to the smaller subunit first. Evidence that CH acts after the association of subunits was obtained by studies with NaF which is known to inhibit association of ribosomal subunits. CH apparently acts after the step sensitive to NaF. This is in contrast to the effect of edeine which is discussed in the next section.

d) Kasugamycin (KM), Edeine and Streptogramin A

In recent studies it was reported that kasugamycin, an aminoglycosidic antibiotic inhibits the formation of the 30 S ribosomal initiation complex (Okuyama et al., 1971). The binding of fMet-tRNA to the complex of phage RNA and *E. coli* 30 S ribosomal subunits is sensitive to these antibiotics. KM inhibits binding of phenylalanyl tRNA to the complex of 70 S ribosomes and polyuridylic acid, but has a greater inhibitory effect on the formation of the 30 S initiation complex. Although these data are consistent with the hypothesis that kasugamycin acts specifically on the initiation step, the possibility remains that the binding of aminoacyl tRNA to the 30 S subunit may be generally more sensitive to kasugamycin than the binding of aminoacyl tRNA to 70 S ribosomes. One should perhaps examine the effect of kasugamycin on the binding of phenylalanyl tRNA to the 30 S subunit in comparison with the effect on the formation of the 30 S subunits initiation complex. Strains of *E. coli* resistant to kasugamycin were recently isolated. In parallel with the mutant strain of *S. aureus* resistant to erythromycin (see Section 6 b) the 16 S RNA of this mutant is less methylated than the wild type 16 S RNA. An RNA methylase activity absent from resistant strains, is able to methylate the methyl-deficient 16 S RNA of the mutant as well as control 16 S RNA (Helser et al., 1972).

Edeine affects a broad spectrum of organisms covering both prokaryotes and eukaryotes. Two molecules of edeine bind to one ribosome and inhibit the initiation step at two separate points [Obrig et al., 1971 (2)]. The first step in initiation, the binding of the initiation codon to the 40 S ribosomal subunit, is apparently sensitive. Thus, binding of ^{3}H AUG triplet to the ribosome in the presence of Met-tRNA$_f$ was sensitive to 1 μM edeine. The second step, the binding of met-tRNA$_f$ to the complex of AUG triplet and ribosomes, can also be inhibited. Evidence that edeine inhibits the initiation step specifically was obtained from the studies on the *in vitro* synthesis of globin. Since the NH_2 terminal amino acid of globin is valine, one can follow chain initiation as well as chain elongation by measuring the relative incorporation of valine into the NH_2 terminal end of globin. Edeine (1 μM) caused almost complete inhibition of NH_2 terminal incorporation of valine whereas very little inhibition was observed on the incorporation of valine into the internal chain of globin. The finding that edeine inhibits the binding of AUG triplet to the 40 S subunit is consistent with a separate observation that edeine acts prior to the step sensitive to NaF. Since NaF caused an accumulation of polysomes with unpaired 40 S ribosomal subunits, NaF was regarded as an inhibitor of the association of ribosomal subunits (Hoerz and McCarty, 1969). It is therefore understandable that the ribosomal complex formed in the presence of NaF is insensitive to edeine.

Streptogramin A and B have been suggested as inhibitors of the initiation process (ENNIS, 1970). When these antibiotics were added to a culture of growing bacteria, polysomes rapidly disappeared with a concomitant increase of 30 S and 50 S subunits. Binding of fMet-tRNA to ribosomes was inhibited by these antibiotics. Since Streptogramin A specifically inhibits the binding of aminoacyl tRNA at the acceptor site (see later section), these observations may indicate that distortion of the acceptor site may influence the donor site, which is presumably the binding site for fMet-tRNA.

4. Association of 50 S Ribosomes with the 30 S Subunit Initiation Complex

Long before the discovery that formylmethionyl tRNA binds to the 30 S ribosomal subunit, it was known that 30 S subunits can bind phenylalanyl tRNA in the presence of polyuridylic acid (MATTHAEI et al., 1964; SUZUKA et al., 1965; KAJI et al., 1966; PESTKA and NIRENBERG, 1966). From the NH_2-terminal analysis of the polyphenylalanine which was synthesized from the 30 S subunit bound phenylalanyl tRNA, it was concluded that it has only one binding site (IGARASHI and KAJI, 1969). The concept that the 50 S subunit can, upon association with the 30 S ribosomal subunit, create the second aminoacyl tRNA binding site was formed on the basis of two separate observations. Firstly, the amount of phenylalanyl tRNA bound on the 30 S subunit is doubled upon association with the 50 S ribosomal subunit (SUZUKA et al., 1966). Similarly, when a complex of the 30 S subunit fMet-$tRNA_f$ and poly AUG was mixed with methionyl $tRNA_m$ a 70 S ribosomal complex containing both met-tRNA's was formed (GHOSH and KHORANA, 1967). Secondly, NH_2-terminal analysis of C^{14}-polyphenylalanine formed from ^{14}C-phenylalanyl tRNA bound to 70 S ribosomes indicated that there are two binding sites on the 70 S ribosome. Since the 50 S subunit by itself cannot bind aminoacyl tRNA, and only one aminoacyl tRNA is bound to the 30 S subunit, the association of 50 S and 30 S subunits must then create the second site (IGARASHI and KAJI, 1967). If the 30 S subunit binds one aminoacyl tRNA in the presence of synthetic polynucleotide, to where does it bind? Does it bind to the site which becomes, upon association with the 50 S subunit, the acceptor site or the donor site of the 70 S ribosome? In order to answer this question, the experiment shown in Table 2 was performed. In this experiment 30 S ribosomal subunits were mixed with polyuridylic acid and ^{14}C phe-tRNA. Because of the excess of 30 S ribosomal subunits, most of the ^{14}C-phe-tRNA in the mixture was bound to the 30 S ribosomal subunits. To this mixture 50 S ribosomal subunits were added and 70 S ribosomes having ^{14}C-phe-tRNA were formed. Puromycin was added to this complex and the amount of puromycin derivative formed from the bound ^{14}C-phe-tRNA was measured. As described later, if aminoacyl tRNA is bound at the donor site (D site, see Section III-1), it will be reactive with puromycin while the acceptor site-bound (A site, see Section III-1) aminoacyl tRNA would not react with puromycin. On the other hand translocase factor G (EFG, see Section III-5, and GTP move peptidyl tRNA from the acceptor site to the donor site, making it available to puromycin. As shown in this table, the major portion of the 30 S subunit-bound ^{14}C-phe-tRNA was at the donor site because it was reactive with puromycin in the absence of EFG and GTP. If ^{12}C-phe-tRNA was given during the association reaction with 50 S subunits, ^{14}C-phenylalanyl-^{12}C-phenylalanyl-tRNA was formed (SUZUKA and KAJI, 1967) and bound to the acceptor site. Thus, the reaction of this bound diphe-tRNA with puromycin was dependent largely on the addition of EFG and GTP.

A similar situation exists with regard to the binding site of fMet-tRNA on the 30 S subunits. Analogous to the complex of phe-tRNA and the 30 S subunits, formylmethionyl-tRNA binds to the 30 S subunit at the site which corresponds to the donor site of the 70 S ribosome [THACH and THACH, 1971 (2)]. Previously, it was thought that formylmethionyl tRNA would bind to the 30 S ribosomal site, which corresponds to the A site. According to this concept, upon association with the 50 S ribosomal subunit, the fMet-tRNA would be translocated from the acceptor site to the donor site by IF_2 with concomitant hydrolysis of GTP (KOLAKOFSKY et al., 1968; review: LENGYEL and SOLL, 1969). This possible translocation was examined by measuring mRNA movement on ribosomes. It is possible to determine the position of mRNA on the ribosomes because ribosomes protect the bound mRNA from attack by RNase. No movement of mRNA was observed in the presence of IF_2 and GTP, while clear-cut evidence for movement of mRNA was observed with the conventional translocation by GTP and EFG. Thus, one can conclude that during the association of the 50 S ribosomal subunit, with the 30 S initiation complex, fMet-tRNA is *not* translocated. Since fMet-tRNA is bound to the D site, without translocation, it must bind to the D site directly. A similar conclusion was obtained from an experiment involving a RNase-treated initiation complex of 70 S ribosomes (KUECHLER, 1971). It has been demonstrated that, after exposure of an initiation complex containing ribosomes, fMet-$tRNA_f$, and R17 RNA, to RNase the ribosome could still translate the protected RNA fragment. The product of this translation process was fMet Ala-Ser-Asn-Phe (the NH_2 terminal pentapeptide of the coat protein). The initiation complex made in the presence of GMPPCP also yielded the same N-terminal pentapeptide. These results show that the distance between the fmet-$tRNA_f$ binding site and the point of entry of the messenger RNA corresponds to four codons, regardless of the presence of GTP. This indicates that fMet-tRNA binds to the D site directly.

Table 2. Evidence that the aminoacyl tRNA binding site of the 30 S subunit corresponds to the donor site of the 70 S ribosome. (IGARASHI et al., 1971)

Experiment	Additions during association with 50 S subunits	Additions during the puromycin reaction	Puromycin derivative formed (% of total bound)
1	none	none	71
1	none	EFG, GTP	70
2	^{12}C-Phe-tRNA	none	18
2	^{12}C-Phe-tRNA	EFG, GTP	67

The complex of 30 S subunits, ^{14}C-Phe-tRNA and polyuridylic acid was made and 50 S subunits were added in the presence (Expt. 2) and absence (Expt. 1) of ^{12}C-phenylalanyl tRNA. The 70 S ribosomal complex thus formed was then incubated with puromycin either in the presence or absence of GTP and EFG.

Evidence has been presented which indicates that binding of fMet-tRNA to the donor site may involve two separate steps. The first is binding *per se*, which can be accomplished by either GMPPCP (β-γ-methylene GTP) or GTP. The second step, called "accommodation", properly situates the fMet-tRNA at the donor site. It appears that the accommodation step requires the GTP, which cannot be substituted by GMPPCP. Although these results suggest that the "accommodation" step may require the terminal phosphate energy of GTP, a recent surprising observation cast some doubt on this conclusion. When γ-^{32}P labeled GTP is used in this process, one can show that it binds to the 30 S subunit initiation complex. Upon association with the 50 S subunit the GTP^{32} was released *without* hydrolysis. The fMet-tRNA must have been "accommodated" during this process because it was reactive with puromycin. One therefore is led to conclude that the terminal energy of GTP is *not* dispensed at all during the formation of the 70 S ribosomal initiation complex (LOCKWOOD and MAITRA, 1972).

The fate of the initiation factors during the association of the 30 S initiation complex with the 50 S ribosomal subunit was studied with the use of labeled initiation factors. When ^{3}H labeled IF_1 was used, the radioactive protein remained bound to the 30 S subunit in the presence of AUG, IF_2, fMet-tRNA and GTP. The bound IF_1 was then released upon association with the 50 S subunit (HERSHEY et al., 1969) Similarly, ^{35}S-labeled IF_3 was shown to bind to 30 S ribosomal subunits in the absence of messenger RNA, fMet-tRNA, IF_1 and IF_2. Upon association with the 50 S subunit, whether brought about by increasing the Mg^{++} concentration or through formation of a 70 S initiation complex, IF_3 was released from ribosomes (LEE-HUANG and OCHOA, 1971).

III. Chain Elongation

1. Binding of Aminoacyl tRNA to the Acceptor Site of Ribosomes

As indicated in the previous section, upon association of the 50 S subunit with the initiation complex of the 30 S ribosomal subunit, the second site for the binding of aminoacyl tRNA is created. The binding of aminoacyl tRNA programmed by the second triplet of the cistron is therefore the next step in the polypeptide synthesis. The earlier work on the binding of aminoacyl tRNA to ribosomes started from the observation that the complex of polyuridylic acid and ribosomes can bind phenylalanyl tRNA (KAJI and KAJI, 1963, 1964; SPYRIDES and LIPMANN, 1964; ARLINGHAUS et al., 1964). Most of the work during this period was centered on the non enzymatic binding of aminoacyl tRNA to the complex of ribosomes and synthetic polynucleotides. One of the important contributions from these studies was the elucidation of the genetic code by this method (NIRENBERG and LEDER, 1964). Earlier, it had been found that certain synthetic polynucleotides would stimulate amino acid incorporation into acid-insoluble material (NIRENBERG and MATHAEI, 1961). From these results a rough idea of the genetic code was obtained, but it was only after the discovery of the binding of aminoacyl tRNA to the complex of synthetic nucleotides and ribosomes that the complete elucidation of the genetic code *in vitro* became possible.

The concept that there are two ribosomal sites for the binding of aminoacyl tRNA emerged from the early studies on the non-enzymatic binding of aminoacyl tRNA. The complex of ^{14}C-phenylalanyl tRNA, polyuridylic acid, and ribosomes was prepared and isolated from the unbound ^{14}C-phenylalanyl tRNA by sucrose density gradient centrifugation. To this complex, an excess amount of ^{12}C-phenylalanyl tRNA, soluble factors, and GTP were added to produce polyphenylalanine. Since the direction of chain elongation is from the NH_2-terminal end to the COOH-terminal end (DINTZIS, 1961), the ^{14}C-phenylalanine would be located mostly at the amino-terminal end and the rest of the polyphenylalanine would then be composed of ^{12}C-phenylalanine under these experimental conditions. If there is only one site for the binding of aminoacyl tRNA on the ribosome, one would expect that *all* the ^{14}C-phenylalanine in the polyphenylalanine would be at the NH_2^2 terminal. On the other hand, if there are two sites for the binding of aminoacyl tRNA, only 50% of ^{14}C-phenylalanine in the polyphenylalanine would be expected at the amino-terminal end. The experimental evidence clearly supported the latter view (IGARASHI and KAJI, 1967). Experiments with P^{32} tRNA have also revealed that there are two sites for the binding of aminoacyl tRNA on mammalian ribosomes (WARNER and RICH, 1964). Earlier, SCHWEET and his associates studied the role of soluble factors in peptide synthesis and suggested a two-site model for the ribosome (HEINTZ et al., 1966) These two sites were called donor (D) and acceptor (A) site (see Section II-4). They are often called peptidyl tRNA site (P site) and aminoacyl tRNA site, respectively. This, however, is a misnomer, since peptidyl tRNA shuttles between these two sites and does not remain on the site called the "peptidyl site". On the other hand, the term "donor site" is more appropriate, because it is the site where the peptidyl group is "donated" to the incoming aminoacyl tRNA located at the "acceptor site". In addition to the evidence cited above, clear-cut data supporting the two-site model were given recently (ROUFA et al., 1970). This experiment took advantage of the known sequence of the initiation regions of f2 phage RNA. The initiation region of the coat-protein cistron has the nucleotide sequence AUG·GCU·UCU·AAC·UUU, which corresponds to the sequence of the NH_2-terminal five amino acids of this protein: fMet, Ala, Ser, Asn, Phe. The RNA polymerase cistron of this RNA has the sequence AUG·UCG·AAG·ACA·ACA·AAG, corresponding to the NH_2-terminal sequence of this protein: fMet, Ser, Lys, Thr, Thr, Lys. The third cistron of this RNA which codes for the maturation protein has the sequence AUG·CGA·GCU·UUU·AGU, corresponding to the sequence fMet, Arg, Ala, Phe, Ser (STEITZ, 1969). From these known sequences, it is noted that two degenerate serine code words, UCU and UCG occur in the third and second positions of the coat and RNA polymerase cistrons respectively. To avoid complications due to possible translocation of tRNA, antibody against purified EFG was added to the system. Under these conditions, only the NH_2-terminal peptides of three proteins, fMet-Ala, fMet-Arg and fMet-Ser, were synthesized. When the binding of aminoacyl tRNA to the complex of f2 RNA and ribosomes was examined, fMet-, Ala-, and Ser-tRNA were bound. In order to decide whether the Ser-tRNA binding is in response to the third codon of the coat cistron (UCU) or the second codon (UCG) of the polymerase cistron, the $tRNA^{ser}$ was fractionated by the countercurrent method. One species of $tRNA^{ser}$ responds only to UCU while the other $tRNA^{ser}$ responds to UCG, UGA and UCA. It was found that, of these two tRNA's, only Ser-tRNA for UCG, UGA and UCA was bound, indicating that this binding is in response to the second codon of the polymerase and not to the third codon of the coat protein. This is in confirmation of the two-site model proposed above. Results of a similar experiment with $Q\beta$ phage RNA are also consistent with the two-site model (SKOGERSON et al., 1971). Various other experiments have also led to this same conclusion (NAKAMOTO, 1967; BRETCHER and MARCKER, 1966; KOLAKOFSKY et al., 1968; GILBERT, 1963; SISLER and MOLDAVE,

1969; Bretcher, 1968; Spirin, 1968). On the other hand, some laboratories have suggested multi-site models (for example, Wettstein and Noll, 1965). The most recent evidence for a multi-site model was obtained from studies with RNA from bacteriophage fr (Swan and Matthaei, 1971). In this experiment, fr-RNA (similar to R17 RNA) was found to stimulate the binding of equimolar amounts of Ala-, Aspn-, fMet-, and Ser-tRNA's. The aminoacyl tRNA's bound corresponded to the first four amino acids of the amino-terminal sequence of the fr coat protein: fMet-Ala-Ser-Aspn-Phe-Glu. An important point of this experiment is that no tripeptide was synthesized, and all the bound ^{14}C-serine was identified as free serine. From these observations, it was concluded that the ribosome has four sites for the binding of aminoacyl tRNA. One striking fact in this experiment is that the binding of Aspn tRNA was strongly inhibited by fusidic acid, (an inhibitor of translocation), despite the absence of translocase in the system. It is possible that the discrepancies between these two schools of thought may be due to the absence or presence of translocation during the binding of aminoacyl tRNA.

Working with a mammalian system, one laboratory is strongly in favor of a three-site model (Culp et al., 1969). Based on the observation that the initiation of polyphenylalanine synthesis at low $MgCl_2$ concentrations in the reticulocyte system is dependent on the addition of unesterified tRNA specific for phenylalanine, a third site called the "entry site" was proposed. The experimental evidence indicates that unesterified tRNA specific for phenylalanine binds to reticulocyte ribosomes in the presence of polyuridylic acid. In order to localize the site of this $tRNA^{phe}$, reticulocyte ribosomes, polyuridylic acid, and unesterified $tRNA^{phe}$ were incubated and the complex of these components was formed. Unlabeled tRNA charged with ^{14}C-phenylalanine was then added, and the mixture was incubated. It was found from this experiment that non-enzymatic and enzymatic binding of ^{14}C-phenylalanyl-tRNA takes place at a site other than the site at which the unesterified $tRNA^{phe}$ is bound. Furthermore, diphenylalanine was formed in the presence of GTP and factor T1 (binding enzyme). These data indicate that a ribosome may bind two molecules of phenylalanyl tRNA. Since the binding site of the unesterified $tRNA^{phe}$ is different from those of phenylalanyl tRNA, one is forced to conclude that there are three sites for the binding of tRNA on the reticulocyte ribosome. A similar experiment with *E. coli* ribosomes led to the conclusion that sites occupied by unesterified tRNA are identical to those sites occupied by aminoacyl tRNA (Ishitsuka et al., 1970). It should be pointed out, however, that these results are obtained in the experimental conditions where 70 S or 80 S ribosomes with no bound tRNA were exposed to tRNA or aminoacyl tRNA. Such a situation would never occur under physiological conditions. During polypeptide synthesis, the donor site would never be empty. It would be occupied by either peptidyl tRNA or unesterified tRNA. Thus, the binding of aminoacyl tRNA would take place only at the acceptor site under normal physiological conditions.

In bacterial cells at least two soluble factors have been assigned to catalyze the binding of aminoacyl tRNA. These factors have been called EFTs (elongation factor Ts) and EFTu (elongation factor Tu), based on their sensitivity to heating at 50 °C, EFTu being unstable (Lucas-Lenard and Lipmann, 1966). In cell extracts, these two factors exist in the form of a complex called EFT [Nishizuka and Lipmann, 1966 (2)]. Although different laboratories called these factors by different names (S_1, S_2,

S_3, Flu, Fl_s, etc.; Skoultchi et al., 1968; Shorey et al., 1969), workers in this field have agreed to call them EFTu, EFTs and EFG (EF standing for elongation factor). A relatively large portion (3%) of the entire soluble protein of growing *E. coli* is EFT (Gordon, 1970) and the relative amount of EFT to ribosomes is constant. Despite the fact that the amount of EFT is regulated by the amount of ribosomes, the genetic locus for EFT is not linked to any one of the known ribosomal protein genes (cited in Review by Lucas-Lenard and Lipmann, 1971). Both of these factors have been purified and crystallized [Miller and Weissbach, 1970; Parmeggiani and Gottschalk, 1969 (2)]. EFTu has a molecular weight of about 40,000 and binds GDP strongly (Kdiss $= 3 \times 10^{-9}$ M) through an SH group of the enzyme (Miller et al., 1971). On the other hand, the binding constant of EFTu to GTP is much lower. EFTs has a high an affinity to EFTu as to GDP (Lucas-Lenard et al., 1969; Miller and Weissbach, 1969), and therefore functions to displace GDP from EFTu. In the presence of GTP, GDP, EFTs and aminoacyl tRNA, EFTu forms a complex with GTP and aminoacyl tRNA (Ravel et al., 1968), but loses its affinity for GDP and Ts (Cooper and Gordon, 1969; Miller and Weissbach, 1969). The specificity for aminoacyl tRNA in this reaction is rather strict in that unesterified tRNA, N-acetyl aminoacyl tRNA, and initiator tRNA (Ono et al., 1968) could not replace regular aminoacyl tRNA. It therefore appears that there is a built-in safeguard mechanism against fMet-tRNA binding during chain elongation. In fact, non-formylated Met-$tRNA_f$ does not react with EFTu. In the ternary of GTP-EFTu-aatRNA, each component exists in an equimolar ratio (1:1:1) but GTP can be substituted by the analogue GMPPCP [Ono et al., 1969 (1); Skoultchi et al., 1970]. The complex, when mixed with a complex of a ribosome and a messenger RNA, transfers its aminoacyl tRNA to the ribosome. During this process a concomitant, stoichiometric loss of the terminal phosphate of GTP takes place [Ono et al., 1969 (2); Gordon, 1969]. In the presence of GMPPCP, in place of GTP, the complex of GMPPCP, aminoacyl tRNA and EFTu remains intact and stays on the ribosome. In the presence of GTP, both EFTu and GTP are quickly released as an EFTu-GDP complex upon binding of aminoacyl tRNA to ribosomes (Skoulchi et al., 1969; Ravel et al., 1969). The resulting EFTu-GDP complex then reacts with EFTs to free the EFTu of GDP (Waterson et al., 1970; Hachmann et al., 1971). The overall process can be summarized in the following equations (Weissbach et al., 1970):

$$\text{EFTu-GDP} + \text{GTP} \xrightarrow{\text{EFTs}} \text{EFTu-GTP} + \text{GDP}$$

$$\text{EFTu-GTP} + \text{aatRNA} \longrightarrow \text{aatRNA-EFTu-GTP}$$

$$\text{Ribosome-mRNA} + \text{aatRNA-EFTu-GTP} \longrightarrow$$

$$\longrightarrow \text{aatRNA-ribosome-mRNA} + \text{EFTu-GDP} + \text{Pi}$$

The fact that EFTs may play a relatively unimportant role was suggested by the observation that EFTs dependency is observed only in the presence of low levels of EFTu [Weissbach, 1971 (2)]. The EFTs dependency obtained under these conditions can be replaced 60 to 80% by phosphoenol pyruvate and pyruvate kinase. This suggests that pyruvate kinase may phosphorylate EFTu-GDP, and therefore EFTu-GTP is made without EFTs (Weissbach et al., 1970). In an attempt to localize the active site of tRNA which interacts with EFTu, $tRNA^{phe}$ was oxidized by periodate and reduced by sodium borohydride. This modified $tRNA^{phe}$ was phenylalanylated but the phenylalanyl tRNA did not react with EFTu (CHEN and OFENGAND, 1970).

The aminoacyl tRNA-EFTu-GTP complex can transfer its aminoacyl tRNA moiety not only to the 70 S ribosome but also to the 30 S subunit. GTP hydrolysis, however, does not take place with the 30 S subunit (BROT et al., 1970). Since the 30 S ribosomal subunit has the tRNA binding site which corresponds to the donor site of the 70 S ribosome, these observations may indicate that EFTu interacts with the donor site despite the fact that its main function is to bring aminoacyl tRNA to the acceptor site of the ribosome. This view was further strengthened by the recent experiment in which the peptidyl transferase (which is presumably at the donor site) was shown to be involved in the interaction of EFTu-GTP-aatRNA with ribosomes (RAVEL et al., 1970). The ribosome-poly U-N-acetyl phenylalanyl tRNA complex prepared in the presence of NH_4^+ and K^+ has active peptidyl transferase. The EFTu-aminoacyl tRNA-GTP complex binds aminoacyl tRNA to this active ribosome, resulting in the formation of a peptide bond and the hydrolysis of approximately one GTP for each aminoacyl tRNA bound. If the complex of ribosome-poly U-N-acetyl phenylalanyl tRNA was isolated in the absence of NH_4^+ and K^+, the peptidyl transferase of this ribosomal complex was inactivated, binding much less aminoacyl tRNA. When the peptidyl transferase activity was restored by incubation with NH_4^+ or K^+, the binding capacity of the ribosome was also restored in a parallel fashion. Similarly, inactivation of ribosomes by sparsomycin (a specific inhibitor of peptidyl transferase, see Section III-4-c) resulted in a parallel inhibition of the aminoacyl tRNA binding capacity. In addition to the EFTu factor, a presumably new protein factor has been reported to stimulate the binding of the tRNA to the 30 S subunits. This factor stimulated polyphenylalanine synthesis by the presence of the 30 S and 50 S subunits (KAN et al., 1970). The role of this factor may be to "accommodate" peptidyl tRNA to the donor site properly in cooperation with EFG factor (see section on translocation III-5).

The enzyme responsible for the binding of aminoacyl tRNA to ribosomes of eukaryotes is called transferase 1 or T1. It is equivalent to the EFT of *E. coli*. This enzyme has been isolated from rabbit reticulocytes and has a molecular weight of 186,000 [MCKEEHAN and HARDESTY, 1969 (2)]. It consists of three identical subunits of 62,000 molecular weight. T1 has also been isolated from rat liver and has a molecular weight between 100,000 and 300,000 (SCHNEIR and MOLDAVE, 1968). The original finding of SCHWEET indicated that T1 together with GTP is involved in the binding of aminoacyl tRNA (ARLINGHAUS et al., 1964). From the early observation that T1 from rat liver was stabilized against heat inactivation by the presence of aminoacyl tRNA and GTP, a complex of T1, aminoacyl tRNA, and GTP was suggested (IBUKI and MOLDAVE, 1968). Further studies indicated that such a complex indeed exists (RAO and MOLDAVE, 1967). In the yeast system (RICHTER, 1970) as well as in wheat embryo (JEREZ et al., 1969), a similar complex has been observed using Sephadex column chromatography. The relationship between T1 and the EFT factor of the *E. coli* system has been strenghtened by the discovery that the EFT factor can substitute for T1 in polypeptide formation as well as in the binding of aminoacyl tRNA (KRISCO et al., 1969). On the other hand, T1 could not substitute for EFT in the *E. coli* system.

T1 from rat liver binds to 40 S ribosomal subunits in the presence of GTP and aminoacyl tRNA (IBUKI and MOLDAVE, 1968; RAO and MOLDAVE, 1969). In parallel with the observation that EFTu factor may interact with the binding site of the 30 S ribosomal subunit (which corresponds to the donor site of the 70 S ribosome), these observations may indicate again that T1 may interact with the donor site of ribosomes. On the other hand, it has been reported that T1 factor and GTP do not significantly increase the amount of phenylalanyl tRNA binding to the 40 S subunit. In the presence of both subunits, T1 stimulated the binding of phenylalanyl tRNA fivefold, indicating that *both* 40 S and 60 S subunits are necessary for the stimulation of aminoacyl tRNA binding (BUSIELLO et al., 1971). By analogy with the 30 S subunit, one can assume that the aminoacyl tRNA binding site of the 40 S subunit corresponds to the donor site of the 80 S ribosomes. The observation that T1 cannot interact with the 40 S subunit implies that T1 factor cannot interact with the donor site of the 80 S ribosome. This is consistent with the observation that in the mammalian system, in contrast to the bacterial system, aminoacyl tRNA cannot bind to the donor site without translocation (SISLER and MOLDAVE, 1969). A recent report indicates that T1 from pig liver can be separated into two complimentary fractions. Whether these two fractions have roles similar to EFTs and EFTu of *E. coli* remains to be elucidated (IWASAKI et al., 1971).

2. Inhibitors of the Binding of Aminoacyl tRNA

a) Tetracycline (TC)

The demonstration that non-enzymatic binding of N-acetylphenylalanyl tRNA to 70 S ribosomes was reduced approximately 50% by TC established that this antibiotic inhibits the binding of aminoacyl tRNA to one of the two ribosomal sites (SUAREZ and NATHANS, 1965). A later finding that the aminoacyl tRNA bound in the presence of tetracycline can react with puromycin led to the concept that tetracycline specifically inhibits at the acceptor site (SARKER and THACH, 1968). Studies with the binding of polylysyl tRNA to ribosomes revealed that this binding is relatively insensitive to tetracycline inhibition, while lysyl tRNA binding to the complex of polyadenylic acid and ribosomes was sensitive to TC (GOTTESMAN, 1967). Assuming that polylysyl tRNA may prefer to bind to the peptidyl site (donor site), these data were taken to support the specific action of TC on the acceptor site. However, there is some evidence that peptidyl tRNA may also bind to the acceptor site (SUAREZ and NATHANS, 1965; TANAKA et al., 1972). Additional evidence that TC acts mostly on the acceptor site was provided by the Mg^{++} concentration optimum studies on the TC effect. It is known that in the absence of EFT, at low (5 mM) Mg^{++}, phe-tRNA binds almost exclusively to the donor site, while at high (13 mM) Mg^{++}, both sites are occupied by phe-tRNA. Thus, TC has very little effect on the binding of phe-tRNA at 5 mM Mg^{++}, while it inhibits the binding at 13 mM Mg^{++} (IGARASHI and KAJI, 1970). The relative insensitivity of the donor site to TC has been utilized in determining the relationship between the ribosomal sites defined by the NH_2 terminal analysis of the polyphenylalanine and the acceptor and donor sites. In this experiment, the complex of ribosomes with ^{14}C-phenylalanyl tRNA was prepared in the presence of TC. Starting from this complex, polyphenylalanine was made with ^{12}C-phenylalanine. The resulting polyphenylalanine had ^{14}C-phenylalanine mostly at the

NH_2-terminal end, indicating that the ^{14}C-phe tRNA which became the NH_2 terminus was at the donor site. The observations discussed above tend to support the relatively specific nature of TC action on the acceptor site. However, there is some evidence that TC's action is not *exclusive* to the acceptor site. First of all, TC inhibits the binding of fMet-tRNA to the complex of AUG and 70 S ribosomes (SARKER and THACH, 1968). It is now clear that fMet-tRNA does not bind to the acceptor site (III-1, II-2). Furthermore, binding of fMet-tRNA to the complex of R17 phage RNA and ribosome in the presence of GTP and initiation factors was stimulated by a low (4×10^{-4} M) concentration of TC but inhibited by higher concentrations of TC (MODOLELL, 1970). Secondly, TC clearly inhibits the binding of aminoacyl tRNA to the 30 S ribosome subunit (SUZUKA et al., 1966). As described in the preceding section, the aminoacyl

Table 3. Tetracycline inhibition of the binding of phenylalanyl tRNA to two ribosomal sites

Expt.	Mg^{++}	EFT	Tetracycline	% inhibition of the binding to	
				donor site	acceptor site
1	6 mM	—	5×10^{-4} M	4	69
2	6 mM	+	5×10^{-4} M	18	97
3	13 mM	—	5×10^{-4} M	31	82

The data were obtained from the amount of diphenylalanyl puromycin and monophenylalanyl puromycin formed from the complex of ribosomes, polyuridylic acid and phenylalanyl tRNA prepared under the condition described in the Table. The principles used in this calculation are (a) phenylalanyl puromycin formed in the absence of translocation of bound tRNA was located at the donor site; (b) diphenylalanyl puromycin was derived from phenylalanyl tRNA bound to two sites, thus the radioactivity of this fraction was divided by two and equally distributed to the acceptor and donor site; (c) phenylalanyl puromycin dependent on translocation was alloted to the acceptor site.

tRNA binding site of the 30 S subunit becomes, upon association with the 50 S subunit, the donor site of the 70 S ribosome (IGARASHI et al., 1971).

In order to obtain more quantitative information on the specificity of TC's action on the two ribosomal sites, a Sephadex G-15 column was devised which separates phenylalanyl puromycin from diphenylalanyl puromycin. This column permitted a quantitative determination of the amount of phenylalanyl tRNA bound to the donor and acceptor site. A complex of ^{14}C-phenylalanyl tRNA, polyuridylic acid and ribosomes was prepared in the presence or absence of TC. The puromycin derivative was formed from these complexes in the presence or absence of GTP and EFG, and was analyzed by the Sephadex G-15 column. As is shown in Table 3, TC's action is specific to the acceptor site in the presence of EFT at low Mg^{++}. This specificity becomes weaker when the binding of Phe-tRNA was carried out at high Mg^{++} in the absence of EFT. This action of TC on the donor site, however, need not be considered under physiological conditions for protein synthesis. During polypeptide synthesis, the donor site is occupied by either peptidyl tRNA or unesterified tRNA, so that only the binding of aminoacyl tRNA to the acceptor site is possible. Only under such an artificial situation where ribosomes with no tRNA and mRNA on them are presented

to mRNA and aminoacyl tRNA, is the binding of aminoacyl tRNA to the donor site possible.

Binding of labeled tetracycline has been shown to take place on 30 S as well as 70 S ribosomes (DAY, 1966; CONNAMACHER and MANDELL, 1968; MAXWELL, 1968). This suggests that the site on the 30 S subunit which would become the acceptor site upon association with the 50 S subunit may be the binding site for TC. Taking advantage of the fact that the fluorescence of TC is enhanced when it binds to ribosomes, it has been possible to show that Mg^{++} plays an important role in this binding (WHITE and CANTOR, 1971).

The effect of TC *in vivo* is more complicated than its effect *in vitro*. TC (2×10^{-4} M) induced breakdown of polysomes [GURGO et al., 1969 (1); CUNDLIFFE, 1968], but association of new messenger RNA with ribosomes was not inhibited under similar conditions. TC (2×10^{-4} M) may not inhibit the binding of aminoacyl tRNA *in vivo* because the ribosomes isolated in this experiment carried aminoacyl tRNA. On the other hand, TC (8×10^{-4} M) reduced the amount of bound aminoacyl tRNA, and under these conditions polysomes were preserved. It appears that low concentrations of TC permit the run-off of ribosomes by allowing movement of ribosomes along the mRNA but do not permit reinitiation of protein synthesis. This would result in a decrease of polysomes. At high concentrations, TC freezes the ribosomes on the mRNA by inhibiting the binding of aminoacyl tRNA to ribosomes.

b) Streptomycin (SM)

The observation that SM influences the binding of aminoacyl tRNA to 70 S ribosomes preceded the discovery of its action on the chain initiation process. From genetic (GORINI and KATAJA, 1964) as well as *in-vitro* studies (DAVIES et al., 1964; VAN KNIPPENBERG et al., 1965) it became clear that SM influences protein synthesis by causing misreading of the genetic message, so that miscoded amino acids are incorporated into protein (OLD and GORINI, 1965; BISSEL, 1965; BODLEY and DAVIE 1966). The fact that SM causes misreading at the step of binding of aminoacyl tRNA to 70 S ribosomes was discovered independently by two laboratories (KAJI and KAJI, 1965; PESTKA et al., 1965). In these studies synthetic polynucleotides were used and non-enzymatic binding of aminoacyl tRNA to the complex of ribosome and RNA was examined. It was found, for example, that isoleucyl tRNA binding to the complex of ribosomes and polyuridylic acid was stimulated by SM, while the binding of phenylalanyl tRNA to the same complex was markedly inhibited. With the use of synthetic polynucleotides of known sequences, SM was found to stimulate the misreading of the 5' terminal and internal nucleotide of the triplet code word. Thus, SM caused (1) the misreading of only two pyrimidine bases, (2) the misreading of pyrimidine bases in the 5' terminal position of a triplet codon as other pyrimidines, and (3) the misreading of internal pyrimidines as both pyrimidines and purines (DAVIES et al., 1966). Earlier studies showed that ribosomes, in particular, the 30 S subunits, are the site of action of streptomycin; this conclusion was supported by the direct demonstration that SM inhibited the codon-specific binding of aminoacyl tRNA to the 30 S subunit (KAJI et al., 1966; PESTKA, 1967). It should be pointed out, however, that the aminoacyl tRNA binding site of the 30 S ribosomal subunit corresponds to the donor site of 70 S ribosomes. This site, therefore, is not the decoding site for the

70 S ribosome. The decoding site for the 70 S ribosome is the acceptor site which can be created only through the combination of 30 S and 50 S ribosomal subunits. It is therefore understandable that, although one can demonstrate the inhibitory effect of SM on the binding of aminoacyl tRNA to the 30 S ribosomal subunit, one cannot demonstrate the *miscoding* effect of SM *unless* the 50 S ribosomal subunit is present (KAJI, 1967).

Approximately one molecule of dihydro-SM is bound to one 30 S subunit at the concentration of dihydro-SM which influences the binding of aminoacyl tRNA (KAJI and TANAKA, 1968). The 30 S subunit which had been reversibly inactivated did not bind dihydro-SM (VOGEL et al., 1970). This binding of dihydro-SM did not take place with 30 S subunits from a SM-resistant strain. The demonstration that SM resistance is caused by the change in a 30 S subunit protein came from an experiment involving the binding of dihydro-SM (TANAKA and KAJI, 1968; OZAKI et al., 1969). It has been shown that upon sedimentation through a CsCl solution, the 30 S subunit yields a core particle and split protein (HOSOKAWA et al., 1966). The split protein and core particle can be reconstituted into the 30 S ribosomal subunit. When the binding of dihydro-SM to a reconstituted particle was studied, a maximum binding of labeled dihydro-SM was observed when the core particle from the SM-sensitive strain of *E. coli* was used. This indicated that SM resistance resides in the core particle. One peculiar finding in this experiment was that, despite the fact thet the SM-sensitive protein resides in the core particle of the 30 S ribosomal subunit, the binding of dihydro-SM did *not* take place with the core particle alone. It was dependent on the presence of split protein, suggesting that the structural integrity of the entire 30 S subunit is responsible for the binding of dihydro-SM. A more direct demonstration that proteins of the core particle were responsible for the binding of dihydro-SM was made possible by the total reconstitution of 30 S particles from proteins and 16 S ribosomal RNA (TRAUB and NOMURA, 1968; NOMURA et al., 1969). Thus, when 30 S ribosomal subunits were reconstituted from proteins of sensitive 30 S subunits and 16 S RNA of resistant ribosomal subunits, the reconstituted particle bound ^{3}H-dihydro-SM while the particles composed of 16 S RNA of sensitive ribosomes and proteins of resistant ribosomes did not. Other biological activities of the reconstituted 30 S ribosomal subunits and the effect of SM on them were examined. All data from these experiments were consistent with the conclusion that a protein of the 30 S ribosomal subunit is responsible for the sensitivity or resistance of this subunit to SM. After fractionation of 30 S ribosomal proteins into 21 pure fractions, it was shown that a specific protein named P10 (or S12) (for this nomenclature see WITTMANN et al., 1971) was responsible. In confirmation of the earlier results with CsCl core particles, the protein S12 by itself could not bind SM, whereas the completed 30 S subunits could.

The fact that S12 does not play a main role in chain elongation was demonstrated by the observation that S12-deficient particles could carry out polyphenylalanine synthesis to as much as 50 to 80% of the control. On the other hand, the S12-deficient particles were only 20% active, compared to the control 30 S subunits, in MS2 RNA-dependent protein synthesis, suggesting that S12 is indispensible for physiological translation of natural messenger RNA. Since the S12-deficient particle has partially lost the capacity to bind fMet-tRNA, S12 may be closely related to the chain initiation event. This fact is consistent with the finding that SM acts at the step of initiation of protein synthesis. Ribosomes (70 S) which are made up of S12-deficient 30 S particles exhibit a marked decrease of SM-stimulated misreading. Furthermore, they are resistant to other agents which induce miscoding. Thus, one can regard S12 as a protein which is responsible for the infidelity of ribosomal translation. S12 is also responsible for SM dependency which is characteristic of a mutant whose growth is dependent on SM (Birge and Kurland, 1969; Apirion et al., 1969).

For the reconstitution of 30 S ribosomal subunits from purified 30 S ribosomal proteins and 16 S RNA, proteins have to react with 16 S RNA in an orderly fashion. Based on this phenomenon an "assembly map" of reconstitution of 30 S ribosomal subunits has been constructed (Mizushima and Nomura, 1970). Using sensitivity of ribosomal proteins to trypsin, similar topography of ribosomal proteins was obtained (Chang and Flaks, 1970). Although the "assembly" maps indicate no definite position of S12, the trypsin topography method suggests that S12 is situated well inside of the subunit. However, one has to be cautious in deciding protein topography of ribosomes based on the sensitivity of each ribosomal protein to trypsin. The slight denaturation of a protein can greatly alter its sensitivity to trypsin (Kaji, 1965).

While S12 acts as an "infidelity protein", a 30 S subunit protein "ram" (ribosomal ambiguity) (see Gorini, 1970, for review), may be regarded as a "fidelity protein". Discovery of the ram mutant originated from a search for a ribosomal mutation which causes suppression in the absence of streptomycin. This is a mutant in which all three types of nonsense codons were suppressed. Furthermore, the ribosomes isolated from the ram mutant showed extensive misreading of synthetic mRNA's and this altered property was shown to reside in the 30 S sububit. It therefore appears that a 30 S ribosomal component controlled by the ram gene must play an important role in maintaining translational fidelity (Rosset and Gorini, 1969). The ambiguity caused by the ram mutation was antagonized by a mutation of S12. The protein altered by the ram mutation was identified as the protein S4 of the 30 S ribosomal subunit [Zimmerman et al., 1971 (1)].

The concept that SM inhibits protein synthesis at the stage of the binding of aminoacyl tRNA to ribosomes has been temporarily forgotten because of the notion that it may inhibit initiation of polypeptide chain synthesis. However, recent experiments "re-discovered" SM inhibition at the step of the binding of aminoacyl tRNA to ribosomes. Thus, studies on the SM effect on the f2 RNA-directed polypeptide synthesis revealed that the addition of SM (3.4×10^{-5} M) rapidly inhibits chain extension (Modolell and Davis, 1968). Similar results were obtained with endogenous messenger RNA. Under the experimental conditions used, just before the addition of SM more than 50% of the ribosomes were in the form of polysomes (mainly dimers and trimers). Since puromycin rapidly and completely converted the polysome to 70 S ribosomes, these polysomes appeared to be active. Yet, no appreciable polysome run-off was seen upon addition of streptomycin. From these observations, it was concluded that SM rapidly freezes the ribosomes and their attached polypeptides in polysomes. This would be expected if the binding of aminoacyl tRNA to the acceptor site was primarily inhibited by SM. Although SM stabilized polysomes as indicated above, slow breakdown of "stabilized" polysome was observed in the presence of SM (Modolell and Davis, 1969). When polysomes with labeled nascent polypeptides were used in these experiments, one could observe that the SM-induced breakdown was accompanied by the release of the nascent peptidyl tRNA. Prior to the release by SM, the peptidyl tRNA was bound to the donor site. For the release to take place, aminoacyl tRNA and subsequent peptidyl transfer were essential. From these observations it was concluded that SM distorts the binding of aminoacyl tRNA to the acceptor site. The rationale behind this conclusion was that the aminoacyl tRNA which is unstably bound to the acceptor site would, after peptide bond formation, produce peptidyl tRNA unstably bound at the acceptor site. This unstable peptidyl tRNA would be gradually released.

The observation that SM "freezes" ribosomes on mRNA and causes immediate cessation of polypeptide chain elongation explains the dominance of SM sensitivity over SM resistance (SPARLING et al., 1968). If the sensitive ribosome is stopped, normal flow of resistant ribosomes on the messenger RNA must also stop. It has been found that the SM-sensitive/SM-resistant heteroploids can produce colonies after brief treatment with SM. This is a bacteriostatic effect of SM in contrast to the well-known bacteriocidal effect. During a short exposure to SM, protein synthesis by the heteroploids stops. Thus, SM sensitivity is dominant over SM resistance with respect to its bacteriostatic effect also (BRECKENRIDGE and GORINI, 1969). The bacteriostatic effect of SM could be observed not only in heteroploids of resistant and sensitive strain but also in a strain which carries the ram mutation (ROSSET and GORINI, 1969). This effect was seen at a streptomycin concentration lower than that required for killing. In separate studies, it was suggested that SM's bacteriocidal action is due to its induction of synthesis of a certain class of RNA (STERN and COHEN, 1964). These observations suggest that the bacteriocidal and bacteriostatic effects of SM may be completely separate. The miscoding effect of SM is perhaps related to the bacteriostatic effect and not to the bacteriocidal effect.

The notion that the miscoding effect of SM is separate from the bacteriocidal effect was further supported by isolation of a mutant which exhibits a high degree of misreading *in vitro* and yet is resistant to killings by SM (BROWNSTEIN and LEWANDOWSKI, 1967; LEWANDOWSKI and BROWNSTEIN, 1966). Analysis of this mutant revealed that a 30 S subunit protein, which is not S12, was changed (KREIDER and BROWNSTEIN, 1971). The affinity of the 30 S subunit of this mutant for dihydro-SM was lower than that of normal subunits. Thus, not only S12 but also other proteins of the 30 S subunit contribute to the affinity for SM. Additional evidence separating the bacteriocidal effect of SM from the miscoding effect is the fact that a derivative of streptomycin, deoxystreptamine, has *in vitro* miscoding effect but has no bacteriocidal effect (TANAKA et al., 1967). It is known that the SM concentration at which misreading occurs is less than the concentration which causes stoppage of chain elongation or lethality. From these observations it has been suggested that SM may interact with the ribosome in two steps. In the first and reversible step, SM causes misreading and in the second step the reversible binding changes into irreversible binding by the process of peptide bond formation. It is this second step which may eventually lead to the killing of the cell (DAVIS, 1968).

On the other hand, some parallelism between the killing and the miscoding effects of SM has been observed. Streptomycin-dependent mutants require any one of several factors for growth; i.e., streptomycin, paromomycin or ethanol. Streptomycin and paromomycin in combination killed these mutants. The amount of misreading by ribosomes extracted from this strain is small when induced by either SM or paromomycin, but is quite large in the presence of both. These observations are at least consistent with the notion that killing by SM and paromomycin in *in vivo* is related to high levels of misreading *in vitro* (ZIMMERMAN et al., 1972).

c) Thiopeptin and Related Compounds

A sulfur-containing antibiotic, thiopeptin, has been reported to inhibit the binding of aminoacyl tRNA to ribosomes during the elongation steps. Thus, 5×10^{-7} M thiopeptin inhibited about 45% the binding of ^{14}C-phenylalanyl-tRNA in the presence of EFT. This inhibitory effect can be compared with that of 5×10^{-5} M tetracycline, which produced 50% inhibition of the binding of Phe-tRNA. In contrast to tetracycline, however, thiopeptin did not have any effect on non-enzymatic binding of aminoacyl tRNA (compare with siomycin effect below). Another noteworthy difference between tetracycline and thiopeptin is their effect on GTP hydrolysis during the binding of aminoacyl tRNA. While tetracycline had no effect, thiopeptin inhibited GTP hydrolysis strongly. Neither of these antibiotics had an effect on the EFG-dependent GTPase (KINOSHITA et al., 1971).

Thiostrepton, which is a close chemical derivative of thiopeptin (ANDERSON et al., 1970), has recently been shown to inhibit protein synthesis *in vivo* at the step of binding of aminoacyl tRNA to ribosomes (CUNDLIFFE, 1971). In this experiment, a relatively high concentration (300 μg/ml) of thiostrepton was given to protoplasts of *B. megaterium* which had been labeled with ^{32}P for a few generations and pulse-labeled with ^{3}H leucine for 30 sec. A rapid stoppage of leucine incorporation into proteins was observed, but no degradation of polysomes took place. Subsequent addition of puromycin was followed by rapid and quantitative release of nascent peptides from ribosomes and extensive breakdown of polyribosomes. These observations indicate that thiostrepton stops the movement of ribosomes on mRNA, and keeps most, if not all, of the bound nascent polypeptidyl tRNA at the donor site which is sensitive to the action of puromycin. In a separate experiment it was found that thiostrepton does not inhibit peptide bond formation or the translocation step *in vivo*. Thus, the puromycin reaction of the bound peptidyl tRNA is insensitive to thiostrepton regardless of the presence of EFG and GTP. This conclusion may appear contradictory to the *in vitro* inhibitory effect of thiostrepton on translocation as discussed in III-6 C. In view of the fact that thiostrepton affects translocation of tRNA by inhibiting the binding of EFG to ribosomes, this effect on translocation may not be so important *in vivo* where EFG is abundant. For comparison an identical experiment was performed with tetracycline — a known inhibitor of the binding of aminoacyl tRNA. The polysomes, frozen by the addition of tetracycline, had exactly the same characteristics as those produced by thiostrepton.

Siomycin A (SA) is an antibiotic closely related to thiostrepton and has an inhibitory action on bacterial protein synthesis but not on mammalian protein synthesis [TANAKA et al., 1970 (2)]. SA acts on ribosomes rather than on the soluble components of protein synthesis. Therefore, the inhibitory effect of SA was much more influenced by the amount of ribosomes in the *in vitro* system than other components such as tRNA, messenger RNA, and soluble proteins. When ribosomes were treated with an excess of SA, the treated ribosomes showed almost no activity in synthesizing polyphenylalanine [TANAKA et al., 1970 (1)]. The evidence that SA inhibits the binding of aminoacyl tRNA to the acceptor site of the ribosome came from studies with ribosomes containing tRNA at the donor site. In this experiment, ribosomes, polyuridylic acid and tRNA (not aminoacylated) were incubated so that the ribosomal sites for tRNA binding were occupied with unesterified tRNA. ^{14}C-phenylalanyl tRNA binding to such a ribosomal complex was inhibited by SA as well as by tetracycline. It is known that under these conditions ^{14}C-phe-tRNA does not bind to the D site, while binding to the A site takes place through exchange of pre-bound tRNA with ^{14}C-phe-tRNA (Ishitsuka and Kaji, 1972). In addition, the aminoacyl tRNA bound under these conditions was not reactive with puromycin (Watanabe and Tanaka, 1971). One can therefore conclude that SA inhibits the binding of Phe-tRNA at the A site. Similar results were obtained with the complex of natural messenger RNA and ribosomes [Modolell et al., 1971 (1,2)]. In the protein-synthesizing system directed by f2-phage RNA, most of the ribosomes initiate at the coat-protein cistron (Lodish, 1968). Since the first two triplets of the coat cistron of f2 phage correspond to fMet-Ala, binding of alanyl tRNA would then be at the A site. This alanyl tRNA binding was dependent on EFT factor, GTP, fMet-tRNA and R17RNA, suggesting that the binding site of alanyl tRNA becomes available only after the initiator tRNA binds to the ribosomes. Under the experimental conditions used, at least 75% of the bound alanine became formylmethionylalanine, indicating that the binding of alanyl tRNA in this experiment represents mostly binding at the A site. Siomycin A strongly inhibited this alanyl tRNA binding. An important point in this experiment is that SA did not inhibit the binding of fMet tRNA. Similar results were obtained with TC suggesting that the mode of action of SA is similar to that of TC. The fact that SA does not inhibit the binding of fMet-tRNA suggests that SA is specific for the A site, since fMet-tRNA, as discussed before, binds to the donor site rather than to the acceptor site. Additional evidence that SA is a specific inhibitor to the A site was obtained from studies with the binding of phenylalanyl tRNA to the complex of polyuridylic acid and 70 S ribosomes at various Mg^{++} concentrations. It is known that, in the absence of EFT, phenylalanyl tRNA binds exclusively to the D site at 5 mM Mg^{++} while both sites are occupied with phe tRNA at 13 mM Mg^{++}. When SA was added in this system, the inhibitory effect was observed only at 13 mM Mg^{++}, and not at 5 mM Mg^{++}. Thus, the SA effect is observable only under conditions where the binding of Phe-tRNA to the A site takes place.

Siomycin A, like thiopeptin, inhibited GTP hydrolysis coupled with the binding of aminoacyl tRNA (Modolell et al., 1971). Inhibition was observed at a concentration of SA which corresponds to less than two molecules per ribosome. SA acts on 50 S subunits but not on 30 S subunits [Tanaka et al., 1970 (1)]. Since the A site of the 70 S ribosome is created by the association of the 30 S subunit with the 50 S subunit, the binding site of SA may be the site on the 50 S subunit which becomes, upon association with the 30 S subunit, the acceptor site of the 70 S ribosome. This is in contrast to the tetracycline binding site, which is probably the site on the 30 S subunit which becomes, upon association with the 50 S subunit, the acceptor site of the 70 S ribosome.

d) Aurintricarboxylic Acid (ATA)

In addition to its action on the binding of messenger RNA to the 30 S subunit (Grollman and Stewart, 1968), ATA inhibits the binding of aminoacyl tRNA to ribosome, but a higher concentration of ATA is necessary for this effect (Marcus et al., 1970). The locus of ATA's action on the chain-elongation step appears to be at the reaction involving EFTs. As described earlier, EFTs catalyzes the exchange between GDP and GTP on EFTu. It can also catalyze the exchange between GDP with ^{3}H labeled GDP on EFTu. Since EFTu-GDP (^{3}H-labeled) can be retained on Millipore filter, the exchange of GDP (^{3}H) with non-labeled GDP on EFTu can be measured by this method. EFTs stimulated this exchange at 0° as well as 37 °C, and ATA almost completely abolished this stimulatory effect at 0 °C (Weissbach and Brot, 1970). There was exchange of GDP which was not dependent on Ts at 37 °C, but this exchange was not inhibited by ATA. It should be pointed out, however, that the inhibitory effect of ATA on chain elongation cannot be explained by this observation alone because attempts to reverse the ATA effect by the addition of excess amounts of EFTs have failed. It appears that the inhibition by ATA of chain elongation may be due to the sum of various effects, and not to inhibition of a specific step. Polyphenylalanine synthesis was inhibited maximally at 300 μM ATA. Under these conditions binding of polyuridylic acid to 30 S ribosomal subunits, aminoacylation of tRNA and the GTPase activity of EFG were all inhibited (Siegelman and Apiron, 1971). It should be emphasized that at a lower concentration of ATA the polypeptide chain initiation steps are specifically inhibited.

e) Edeine

The discovery that edeine (Hettinger and Craig, 1970) inhibits protein synthesis at the step of aminoacyl tRNA binding to ribosomes (Hierowski and Kurylo-Borowska, 1965) preceded the studies of its action on the initiation of protein synthesis (see II-3 d). Edeine does not inhibit the puromycin reaction of the ribosome-bound fMet-tRNA, indicating that peptide bond formation is insensitive to edeine (Monro and Vazquez, 1967). In addition to the effect on the binding of aminoacyl tRNA to ribosomes, edeine inhibits the dissociation of ribosomes. When ribosomes are exposed to edeine, they lose their normal characteristic of dissociating into subunits at 10^{-4} M Mg^{++} (Kurylo-Borowska and Hierowski, 1965). The 30 S and 50 S subunit reassociated in the presence of edeine even at low (10^{-4} M) Mg^{++}. Since labeled ^{14}C-edeine binds to 30 S and 50 S subunits, it may act as a connecting link between these two subunits (Kurylo-Borowska and Hierowski, 1965).

Several lines of evidence are available which point to the specific action of edeine on the donor site of the 70 S ribosome. The inhibitory effect of edeine on the non-enzymatic binding of phenylalanyl tRNA decreases as the Mg^{++} concentration increases (Szer and Kurylo-Borowska, 1970). At low concentrations of Mg^{++}, Phe-tRNA binds almost exclusively to the donor site, and edeine inhibits this binding completely. At 18 to 20 mM Mg^{++}, both acceptor and donor sites bind Phe-tRNA, and edeine inhibits this binding by only 40%. Secondly, the inhibitory effect of edeine is additive to that of tetracycline. Since tetracycline is relatively specific to the acceptor site, edeine must act on the donor site. Further support for this conclusion came from the observation that aminoacyl tRNA bound in the presence of edeine did not react with puromycin. This indicated that these aminoacyl tRNA's were not at the donor site. The third piece of evidence is that edeine inhibits the binding of fMet-tRNA to the complex of ribosomes and poly AUG or MS2 RNA. As described earlier, fMet-tRNA binds to the donor site without going through the acceptor site. The final piece of evidence for the specific nature of edeine's action on the donor site is the fact that it inhibits the binding of aminoacyl tRNA to the 30 S subunit [WEISSBACH et al., 1971 (1)]. As described earlier, the aminoacyl tRNA binding site on the 30 S subunit corresponds to the donor site of the 70 S ribosome. Since the binding of aminoacyl tRNA to the donor site does not take place during polypeptide synthesis, edeine's specific action on this process merely suggests that this antibiotic acts on the donor site, but it does not indicate that edeine inhibits protein synthesis by inhibiting the binding of aminoacyl tRNA at the donor site. It would be of interest to examine the effect of edeine on the translocation step, which may be inhibited by edeine's specific action on the donor site. In contrast to this overwhelming piece of evidence pointing to the specific action of edeine on the D site, edeine has recently been reported to inhibit the binding of aminoacyl tRNA in the presence of EFTu. Since EFTu presumably places aminoacyl tRNA on the acceptor site [WEISSBACH et al., 1971 (3)], edeine may also influence the A site of ribosomes [WEISSBACH et al., 1971 (1)]. The effect of edeine on the initiation of globin synthesis has been discussed earlier [OBRIG et al., 1971 (2), see II-3 d].

f) Virginiamycin Group

Antibiotics belonging to this group consist of two components. For example, Virginiamycin (or Osterogrycin), an antibiotic which is produced by a mutant of *Streptomyces virginiae* (VANDERHAEGHE et al., 1957), is composed of two components, Virginiamycin M (VM) and Virginiamycin S (VS). The common feature of this group of antibiotics (which includes mikamycin, PA 114 factor, pristinamycin or pyostacin, streptogramin and vernamycin) is that they consist of components which, though exerting different degrees of inhibition on various microorganisms, posses a synergistic action in a given bacterium [VAZQUEZ, 1966 (1)]. The combination of VM and VS results in bacteriocidal effects while each alone is a bacteriostatic agent (COCITO, 1969). These antibiotics do not influence aminoacylation of tRNA or binding of messenger RNA to ribosomes (ENNIS, 1970; YAMAGUCHI and TANAKA, 1967), but they inhibit di- and tripeptide formation as well as the binding of aminoacyl tRNA to ribosomes [PESTKA, 1970 (2)]. Polyphenylalanine formation programmed by polyuridylic acid is sensitive to streptogramin A, especially when the antibiotic is added prior to the onset of the reaction.

Direct evidence that Vernamycin A (VA) acts on the 50 S ribosomal subunit was provided by an experiment in which ^{3}H-vernamycin A was shown to bind to the 50 S subunit and the 70 S ribosome (ENNIS, 1971). This binding requires K^+ or NH_4^+, and Mg^{++}. Optimum binding was observed at 100 to 200 mM K^+ or NH_4^+ and 1.5 to 10 mM Mg^{++}. Binding of VA is temperature-dependent. No binding was observed at 0 °C while it was complete within 5 min at 37 °C. However, if the ribosomes were preincubated at 37 °C in the presence of K^+, they could bind VA even at 0 °C. This is reminiscent of the binding of dihydro-SM which takes place with "activated ribosomes". In a similar fashion, affinity of ribosomes for erythromycin was influenced by K^+ which presumably influences the ribosomal configuration (TERAOKA, 1970; see Section III-6 b). These observations suggest that for maximum antibiotic binding, ribosomes have to be in a proper configuration. The observation that the removal of K^+ released the bound VA is consistent with the concept. The binding site of VA is probably common to or overlaps with that of antibiotics which act on the 50 S subunit (Erythromycin, synergistin A and spiramycin).

At least one possible mechanism for the synergistic action of the two components of Vernamycin was offered by the stimulatory effect of Vernamycin Ba (VB) on the binding of VA. In the presence of VB, the binding of VA was stimulated and became irreversible even at a low K^+ concentration. In fact, the K^+ requirement for the binding of VA was partially eliminated in the presence of VB. It was proposed that VB changes (or activates) the ribosomal configuration in a manner similar to K^+ ions so that VA binds irreversibly (Ennis, 1971).

In a recent experiment it was found that Virginiamycin M (VM) strongly inhibited polyphenylalalanine synthesis as well as phage protein synthesis *in vitro* (Cocito and Kaji, 1971). When Virginiamycin M was tested on the non-enzymatic binding of

Table 4. Puromycin reaction of phenylalanyl tRNA bound in the presence of virginiamycin

Ribosome complex	Puromycin derivative of phenylalanine cpm/100µg of ribosomes			
	with EFG		without EFG	
	− VM	+ VM	− VM	+ VM
Ribosome, poly U, ^{14}C-Phe-tRNA	399	315	224	193
Ribosome, poly U, ^{14}C-Phe-tRNA and Virginiamycin M	0	0	0	0

The reaction mixture for the formation of puromycin derivative of bound phenylalanine contained, per ml, 382 µg of ribosomal complex with 6.5×10^3 cpm of bound ^{14}C-Phe-tRNA. Where indicated, 13.4 µg of EFG and/or 1 µg virginiamycin M were added to 1 ml of the reaction mixture. The mixture was incubated for 15 min at 30 °C and the radioactive puromycin derivatives in a 50 µl aliquot were measured.

Phe-tRNA, approximately 50% inhibition was observed at 15 mM Mg^{++}. At this concentration of Mg^{++} both the acceptor and the donor sites are occupied by aminoacyl tRNA. In the presence of 5 mM Mg^{++} which permits the binding of aminoacyl tRNA only to the donor site, very little inhibition by Virginiamycin M was observed. A striking finding was, however, that the complex prepared in the presence of Virginiamycin M at 15 mM Mg^{++} was completely inactive for the puromycin reaction (Table 4). It is noted in this table that VM only slightly inhibited the translocation dependent on the EFG factor or the puromycin reaction if VM was added after the complex was made. From these observations it was concluded that VM specifically inhibits the binding of aminoacyl tRNA to the acceptor site and the same time makes the donor site-bound peptidyl tRNA unavailable to puromycin. Perhaps VM binds to the acceptor site and exerts this dual effect. The possibility remains, however, that VM may act directly on the donor site but cannot bind to ribosomes at lower Mg^{++} concentration (6 mM).

3. Peptide Bond Formation

When both acceptor site and donor site are filled with aminoacyl tRNA and peptidyl tRNA, respectively, the next reaction to take place is the formation of the peptide bond. This step is catalyzed by an enzyme which apparently resides in the 50 S (60 S in eukaryotes) ribosomal subunit (MADEN et al., 1968; MONRO, 1967). The enzyme which catalyzes this reaction is called peptidyl transferase and does not require any other soluble factors. This reaction has been traditionally studied with the use of puromycin. The reaction of puromycin with the ribosome-bound aminoacyl tRNA or peptidyl tRNA is regarded as an analogue of natural peptide bond formation.

The enzyme, peptidyl transferase, resides in the 50 S ribosomal subunit and therefore fMet-puromycin can be formed by the 50 S subunit alone. Although this system is quite artificial in that it requires 33% alcohol, the fact remains that the puromycin reaction can be catalyzed solely by the 50 S ribosomal subunit (MONRO and MARCKER, 1967). In a recent study, however, the concept that peptide bond formation is entirely the function of 50 S subunits has been challenged. Thus, in the absence of messenger RNA and polynucleotides, the 30 S subunits were shown to have a stimulatory effect on the formation of N-acetyl aminoacyl puromycin from several N-acetyl aminoacyl tRNA's. Although the addition of 30 S subunits to the reaction mixture for the puromycin reaction stimulated peptide bond formation as much as fivefold, the stimulatory effect could partially be due to the stimulation of non-specific binding of aminoacyl tRNA. The stimulation of peptidyl transferase activity by 30 S subunits is most likely due to the configurational change of 50 S subunits upon association with 30 S subunits rather than direct participation of 30 S ribosomal subunits in the peptidyl transferase reaction (BERMAN and MONIER, 1971).

During the studies of effects of monovalent cations such as NH_4^+, Rb^+, K^+ and Cs^+ on the activities of ribosomes, it became clear that the peptidyl transferase reaction was absolutely dependent on the continued presence of one of these ions (MISKIN et al., 1970), their effectiveness being in the order given above. The activity of the peptidyl transferase was lost if the cation was removed, and the addition of the cation brought back the activity. This activation process was greatly facilitated by incubation at relatively high temperatures (39 to 40 °C). There was no appreciable reactivation below 15 °C. Although this requirement suggested that a ribosomal conformational change may take place during the activation, no such change was detected, at least by the sedimentation analysis of the inactive and activated ribosomes. The reactivation of peptidyl transferase appeared to be first-order with respect to the appearance of active ribosomes. The activation energy calculated from the kinetics of the reactivation varied from about 40 kcal/mole at 14 to 30 °C to about 20 kcal/mole at 39 to 45 °C. It should be pointed out that in these studies the peptidyl transferase activity was assayed for its ability to catalyze the formation of an ester bond between the carboxyl group of fMet-tRNA and alcohol. These data have been confirmed with the use of the conventional assay for peptidyl transferase, namely, fMet-puromycin formation. In order to examine the effect of bound fMet-tRNA on this activation of ribosomes, fMet-tRNA was bound in the presence of initiation factors to the complex of inactive ribosomes and AUG at low temperature. Controls were those ribosomes without fMet-tRNA. It was found that under marginal conditions of reactivation of ribosomes, only the ones carrying fMet-tRNA were activated, suggesting that bound tRNA stimulated the activation process.

To identify the protein or proteins which are responsible for peptide bond formation, 50 S ribosomal proteins were partially dissociated from the particle, and the peptidyl transferase activity of the core particle was tested. This attempt has met with only partial success. Neither core nor dissociated proteins alone are active, but combining together restored the peptidyl transferase activity (Staehelin et al., 1969). It was possible, however, to eliminate participation of certain ribosomal proteins in the peptidyl transferase reaction because particles which lacked all of the acidic proteins were able to catalyze peptide bond formation.

In the peptidyl transferase reaction, the molecule which donates is peptidyl, aminoacyl, or N-blocked aminoacyl tRNA. To establish which part of the tRNA molecule is required for the peptidyl transferase reaction, a fragment of formylmethionyl tRNA was tested for its activity in the puromycin reaction. It was found that the fragment having the composition of CAACCA-f methionine can donate the f methionine group to puromycin in the presence of methanol. Upon further degradation it was found that CCA-f methionine still reacts with puromycin while CA-f methionine or A-f methionine did not. From these observations it appears that a minimum requirement for the peptidyl transferase is CCA at the 3' end of tRNA (Monro et al., 1968).

In a similar attempt to dissect the peptidyl transferase reaction, binding of oligonucleotidyl phenylalanine to 70 S ribosomes was studied [Pestka, 1969 (2)]. The rationale for regarding the binding of oligonucleotidyl phenylalanine as a part of the transferase reaction is as follows: The peptidyl tRNA on the donor site and aminoacyl tRNA (the acceptor) must come into contact with each other on the surface of the peptidyl transferase. The binding of acceptor tRNA to the peptidyl transferase can be regarded as one of the important steps of this reaction. The binding of oligonucleotidyl amino acid presumably represents this process independent of the interaction of the other portion of tRNA (such as the anticodon loop) with ribosomes. After this step, the ternary complex, i.e. the complex of the peptidyl tRNA, aminoacyl tRNA, and the peptidyl transferase, will become the complex of tRNA and peptidyl tRNA (one amino acid longer). These studies are based on two assumptions. The first is that the binding of oligonucleotidyl amino acid (oligonucleotidyl phenylalanine was used mostly) to the ribosome takes place at the peptidyl transferase site exclusively. The second is that the peptidyl transferase reaction goes through two steps; the binding of the acceptor aminoacyl tRNA to the enzyme, followed by the transfer of the peptidyl group to the acceptor molecule. Perhaps one can imagine that the ribosome-bound aminoacyl tRNA is bound to the ribosome at two spots, one being on the 30 S subunit, where the binding is helped by the presence of mRNA, and another being on the 50 S ribosomal subunit where the peptidyl transferase is located. The binding to the latter spot probably involves the CCA end of the tRNA and the aminoacyl moiety. As supporting evidence for these assumptions the following observations are given: (1) There is an absolute requirement for K^+ or NH_4^+. It should be recalled that peptidyl transferase is dependent on one of these ions. (2) Deacylated ^{32}P oligonucleotide binds to ribosomes much less than the aminoacyl oligonucleotide, indicating that a bound amino acid is required. (3) The amino group of the oligonucleotidyl amino acid has to be free. N-acetyl aminoacyl tRNA does not bind. This is what one expects of the peptidyl transferase site for the acceptor aminoacyl tRNA. (4) Many antibiotics which inhibit peptide bond formation inhibit this reaction.

In separate studies, the efficiencies of various donor peptidyl tRNA's were examined in relation to the length and kind of peptidyl group (Panet et al., 1970). The rates of the puromycin reaction with aminoacyl phenylalanyl tRNA and diaminoacyl phenylalanyl tRNA were compared. Surprisingly, the dipeptidyl tRNA reacted slower than tripeptidyl tRNA at 4 °C. On the other hand, acetyl phenylalanyl tRNA reacted faster than dipeptidyl tRNA but slower than the tripeptidyl tRNA. Because the free NH_2 group of dipeptidyl tRNA is closer to the ester linkage than that of triphenylalanyl tRNA, it was suggested that a free NH_2 group may have an inhibitory action on the peptidyl transferase. The bond energy of the ester linkage between peptide and tRNA would become less as the chain length of the peptide lengthens. The fact that the tripeptidyl tRNA reacted faster than dipeptidyl tRNA indicates that factors other than bond energy play an important role in the rate of peptide bond formation.

The peptidyl transferase requires two substrates, i.e. peptidyl tRNA (donor) and the aminoacyl tRNA (the acceptor). The structural requirement for the acceptor is even simpler than that for the donor (CERNA et al., 1970; WALLER et al., 1966; NATHANS and NEIDLE, 1963). For example, 2'3'-O-aminoacyl adenosine can be a substrate. However, the acceptor activity of these compounds was dependent on the nature of the side chain of the amino acid residue bound to adenosine. Aminoacyl adenosines of aliphatic amino acids such as methionyl and phenylalanyl adenosine had a high acceptor capacity, while that of polar amino acids such as seryl, sulfuryl methionyl adenosine had poor activity. This suggests that a hydrophobic interaction plays a role in the peptidyl transferase reaction. In addition to the hydrophobic forces, electrostatic interactions may be important, as evidenced by the high acceptor capacity of lysyl adenosine and benzoyl histidyl adenosine. On the other hand, even the NH_2 group of the acceptor substance can be changed to an OH group without hampering the reaction. When a modified puromycin which has the OH group instead of the NH_2 group was given to the complex of ribosomes and aminoacyl tRNA, an ester bond was formed between the carboxyl group of the amino acid and the OH group (FAHNSTOCK et al., 1970). This is consistent with the fact that ethanol can be an acceptor in the peptidyl transferase reaction. The acceptor activity was also affected by the nature of the donor tRNA derivative. For example, 2'(3')-O-L-3-(1-benzyl-4-imidazolyl)-alanyl adenosine was a weak acceptor of lysine peptides from polylysyl tRNA while it was a good acceptor for the pentanucleotide fragment of acetyl phenylalanyl tRNA (RYCHLIK et al., 1970). The transfer reaction catalyzed by peptidyl transferase has a stereospecificity with respect to the amino acid of the acceptor substrate. Thus, phenylalanyl adenosine is a good acceptor while D-phenylalanyl adenosine did not accept the aminoacyl residue from any of the donors, such as polylysyl tRNA, acetyl phenylalanyl tRNA, and acetyl leucyl pentanucleotides.

4. Inhibitors of the Peptidyl Transferase

a) Chloramphenicol (CM)

If an antibiotic inhibits the puromycin reaction of donor site-bound aminoacyl or NH_2-blocked aminoacyl tRNA, this antibiotic is regarded as an inhibitor of peptidyl transferase. Chloramphenicol is a representative of this category. It is a relatively simple, well-known bacteriostatic agent and only one of its four stereoisomers is active (MAXWELL and NICKEL, 1954). One of the early suggestions that CM may inhibits peptide bond formation came from an experiment in which CM inhibited the formation of peptidyl puromycin *in vitro* (NATHANS et al., 1962) as well as *in vivo*

(Nathans, 1964). Numerous experiments, mostly involving the puromycin reaction in the absence of translocation, established that CM is an inhibitor of the peptidyl transferase. This was further supported by the fact that CM inhibits peptide formation between formyl methionyl oligonucleotide (hexanucleotide) and puromycin in the presence of alcohol (Monro and Marcker, 1967).

One peculiar finding which has not been explained is that the effect of CM on polypeptide synthesis programmed by synthetic polynucleotide varies depending on the kind of polynucleotide. Thus, incorporation of leucine and phenylalanine with poly UA and poly U respectively is more resistant to CM than incorporation of lysine with poly A. Polypeptide synthesis directed by MS2-phage RNA or *E. coli* mRNA is much more sensitive to chloramphenicol than most of the amino acid incorporation systems programmed by synthetic polynucleotides [Kucan and Lipmann, 1964; Vazquez, 1966 (2)]. This was thought to be due to CM's action of inhibiting the binding of certain messenger RNA's to ribosomes. It was further suggested that polynucleotides high in uridylate have less secondary structure and bind better to ribosomes, making them insensitive to CM. Since it has been shown that CM inhibits peptide bond formation, this view is no longer tenable. The differential sensitivity to CM may be related to the kind of amino acid which is being polymerized rather than the polynucleotide. It is possible that the sensitivity to chloramphenicol may depend on the amino acid which is participating in peptide bond formation. Those amino acids which have a stronger affinity for the peptide synthetase may be more resistant to the action of chlroamphenicol.

When ribosomes with nascent polypeptides were incubated in the presence of puromycin, about 50% of the nascent chain were released. Surprisingly, only 50% of this release was sensitive to CM (Weber and Demoss, 1969). From this experiment it was proposed that CM inhibited only those peptidyl tRNA's which are not properly situated at the donor site. Another possible interpretation is that those peptidyl tRNA's which have a stronger affinity to the peptidyl transferase are relatively insensitive, while those that have a weaker affinity are sensitive to CM. The observation that CM inhibited all sizes of proline peptides programmed by poly C, while inhibiting only longer polylysine peptides programmed by poly A (Julian, 1965; Irvin and Julian, 1970) can be interpreted in a similar way on the assumption that the affinity of polylysyl tRNA to the peptidyl transferase varies depending on the size of polylysine while that of polyprolyl-tRNA does not. In a similar fashion, the rate of oligophenylalanine synthesis was inhibited by CM, while the extent of formation of diphenylalanine was either unchanged or stimulated [Pestka, 1970 (1)]. This can be understood by assuming that phenylalanyl tRNA has a stronger affinity to the peptidyl transferase than the diphenylalanyl tRNA.

The effect of chloramphenicol was studied with a system in which binding of aminoacyl oligonucleotide to ribosomes was examined [Pestka, 1969 (1); Pestka et al., 1970). As discussed earlier, this was assumed to represent a partial reaction of the peptidyl transferase. The binding of aminoacyl oligonucleotide to peptidyl transferase takes place probably at the site of the acceptor molecule. Together with other inhibitors on the peptidyl transferase, CM has a strong inhibitory action on this system. One may wonder, then, why CM does not inhibit the binding of aminoacyl tRNA to 70 S ribosomes. This is perhaps due to the fact that, with 70 S ribosomes, a major contribution to the binding force of aminoacyl tRNA to the peptidyl transferase comes from the 30 S subunit. In fact, it is so strong that aminoacyl tRNA binds to the 30 S subunit alone at the site corresponding to the donor site. The inhibitory effect of the CM on the binding of phenylalanyl oligonucleotide appears to have some correlation with its *in vivo* activity, because only the bacteriostatically active isomer significantly inhibited this binding.

Studies with labeled CM showed that it binds to ribosomes (VAZQUEZ, 1963; WOLFE and HAHN, 1965). Of the two subunits, CM binds to the 50 S subunit [VAZQUEZ, 1966 (3)] and the bound CM can be washed off the ribosomes (HURWITZ and BRAUN, 1967). From the observation that streptogramin A, lincomycin, and macrolide antibiotics inhibit the binding of CM, it was suggested that the binding sites of these antibiotics may overlap.

However, one should be cautious about the interpretation of these data. It should be emphasized that ribosomes can change configuration when they bind antibiotics (SHERMAN and SIMPSON, 1969) and such configurational changes may cause the inhibition of the binding of other antibiotics.

Turning now to the effect of CM *in vivo*, the most noteworthy effect, besides stopping amino acid incorporation into protein, is its stabilization of polysomes (DAS et al., 1966; WEBER and DEMOSS, 1966; DRESDEN and HOAGLAND, 1967; FLESSEL, 1968). It was first suggested that CM causes this effect by stopping chain elongation, thereby freezing the ribosomes on the mRNA. Under these conditions run-off of ribosomes would also be prevented and one would expect the stabilization of the polysome. This concept that CM preserves the polysome by stopping the movement of ribosomes was challenged by the suggestion that continued binding of new ribosomes to messenger RNA takes place in the presence of 3×10^{-4} M CM for as long as 30 min at almost a normal rate, while protein synthesis is inhibited almost completely [GURGO et al., 1969 (2)]. For comparison, another inhibitor of polypeptide chain elongation, fusidic acid, was added to the growing bacteria under identical conditions. In contrast to CM, fusidic acid (an inhibitor of translocation) inhibited the entry of ribosomes into polysomes, indicating that movement of ribosomes was indeed stopped under the influence of fusidic acid. Although the data are consistent with the notion that ribosomes can move along mRNA, run off, and bind to mRNA in the absence of peptide bond formation, the actual mechanism of this "uncoupled" translocation is hard to visualize. In the presence of CM, most of the ribosomes should contain peptidyl tRNA on the donor site and aminoacyl tRNA at the acceptor site. Translocation of such a complex *in vitro* has not been possible because of the inability of EFG factor (translocase) to release aminoacyl tRNA from the ribosome (ISHITSUKA et al., 1970). One possible interpretation is that the continued binding of mRNA to ribosomes in the presence of CM does not represent normal initiation, movement, and run off of ribosomes from messenger RNA. It has been reported that CM even increased the amount of polysomes through the formation of *non*-functional polysomes brought about by physical association of ribosomes with messenger RNA (CAMERON and JULIAN, 1968; WEBER and DEMOSS, 1966).

The idea that CM preserves polysomes (LEVINTHAL et al., 1963; MANGIAROTTI and SCHLESSINGER, 1966) by "freezing" ribosomal movement due to its inhibitory effect on peptide bond formation was supported by an experiment with tryptophan-operon messenger RNA. It was found in this experiment that the addition of CM after depression immediately prevented the accumulation of messenger RNA of more distal genes of the tryptophan operon (MORSE, 1971). In a normal situation, a cluster of ribosomes would closely follow the RNA polymerase so that nascent messenger RNA will be immediately translated. The closely packed cluster of ribosomes is then followed by 5' exonucleolytic degradation of the messenger RNA. When the progress of ribosomes is interrupted by a nonsense codon, the ribosomes are rapidly discharged from the messenger RNA (WEBSTER and ZINDER, 1969). The nascent mRNA is then exposed, unprotected by ribosomes, to endonucleolytic attack and rapid degradation of mRNA ensues. This degradation of mRNA distal to the nonsense codon was regarded as one of the reasons why the cistrons distal to the nonsense mutation are not efficiently expressed (the polar effect). If endonucleolytic attack is initiated when the mRNA is prematurely exposed by discharge of ribosomes at the nonsense codon, one should be able to get the same effect by artifically stopping the movement of ribosomes on mRNA. Conversely, if some agent can simulate the effect of the nonsense codon (at the mRNA level), then one can say that the agent must have stopped ribosomal movement on mRNA. This is exactly what was observed when CM was added to this system. Addition of CM to *E. coli* while ribosomes are translating the early portion of nascent tryptophan mRNA immediately prevented accumulation of mRNA for the more distal genes of the tryp operon. This is a situation almost identical to the case where a nonsense codon prevents accumulation of mRNA for the portion more distal to itself. The polar effect caused by a nonsense codon is relieved by a mutation called suA (SCAIFE and BECKWITH, 1966). This mutation was shown to permit detection of the untranslatable mRNA immediately distal to a nonsense codon in the tryptophan operon. It was therefore proposed that suA may make a non-functional endonuclease. In a similar fashion, the suA mutation prevented the degradation of mRNA exposed by the addition of CM.

One of the well-known *in-vivo* effects of CM is its ability to counteract the bacteriocidal effect of SM (ANAND and DAVIS, 1960; PLOTZ and DAVIS, 1962) as well as the bacteriostatic effect of low concentrations of SM (KIRSCHMANN and DAVIS, 1969). These effects are not due to the competition for the binding sites of CM and SM, because they bind to different subunits. Although the actual mechanism of CM's antagonism to SM action is not known, it may be related to the slow breakdown of polysomes by SM, which requires peptide bond formation.

b) Puromycin

Puromycin (PM), though not a direct inhibitor of peptidyl-transferase, inhibits normal peptide chain elongation by participating in the peptidyl transferase reaction. It is for this reason, that this antibiotic is discussed in the section on peptidyl transferase. It is well known that puromycin releases the nascent peptide chain from ribosomes by virtue of its capacity to be a substrate (acceptor) of peptidyl transferase (ALLEN and ZAMECNIK, 1962), one puromycin molecule being incorporated for each chain released. Despite the fact that the puromycin reaction with the ribosome-bound peptidyl tRNA or aminoacyl tRNA has been widely used as a model for peptide bond formation, the evidence that this compound attacks the donor-site-bound aminoacyl tRNA from the direction of the A site became available only recently. In order to examine this point, an experiment described in Table 5 was performed. In this experiment a complex of ribosomes with ^{14}C-phenylalanyl tRNA only on the donor site was prepared. This complex was then incubated with and without N-acetyl ^{12}C-phenylalanyl tRNA or ^{12}C-phenylalanyl tRNA at 13 mM Mg^{++} which permitted the binding of these aminoacyl tRNA's to the acceptor site. Puromycin reactions using these complexes were then studied. It can be seen from this table that the amount of puromycin derivative of ^{14}C-phenylalanine formed in the absence of the translocase (EFG) was reduced when ^{12}C-phenylalanyl tRNA or N-acetyl ^{12}C-phenylalanyl tRNA was at the acceptor site. Since the ^{14}C-phenylalanyl tRNA was located almost exclusively at the donor site, the puromycin reaction of this bound aminoacyl tRNA was not influenced by the presence of EFG, provided the acceptor site remained empty. However, if the acceptor site was occupied with ^{12}C-phenylalanyl tRNA, ^{14}C-phenylalanyl-^{12}C-phenylalanyl tRNA was formed and was located at the acceptor

Table 5. Effect of acceptor site-bound aminoacyl tRNA on the reaction of donor site-bound ^{14}C-phenylalanyl tRNA with puromycin (TANAKA et al., 1972)

Aminoacyl tRNA for additional binding at 13 mM Mg^{++} (second step)	Bound ^{14}C phenylalanyl tRNA (after the second step) CPM	^{14}C-phenylalanyl puromycin derivatives formed (third step)	
		without EFG	with EFG
No aminoacyl tRNA	736	483	541
(^{12}C)-phenylalanyl tRNA	644	107	468
N-acetyl-^{12}C-phenylalanyl tRNA	634	111	136

The experiment was performed in three steps. In the first step, the binding of ^{14}C-phenylalanyl tRNA to the donor site of the complex of ribosomes and polyuridylic acid was carried out at 5 mM Mg^{++}. The second step is the additional binding of ^{12}C-phenylalanyl tRNA and N-acetyl-^{12}C-phenylalanyl tRNA at 13 mM Mg^{++} to the acceptor site of the complex prepared in the first step. The third step is the puromycin reaction of the ribosome bound ^{14}C-phenylalanine in the presence or absence of EFG (translocase).

site. Thus, the bound ^{14}C-phenylalanine is unavailable for the puromycin reaction unless EFG was added to the reaction mixture. When N-acetyl ^{12}C phenylalanyl tRNA was bound at the acceptor site, no diphenylalanyl tRNA could be formed, and EFG had no appreciable effect on the puromycin reaction of the bound ^{14}C-phenylalanyl tRNA. An important point in this experiment is the fact that N-acetyl-^{12}C-phenylalanyl tRNA at the acceptor site reduced the reactivity of the donor-site-bound ^{14}C-phenylalanyl tRNA. This observation strongly suggests that puromycin attacks the donor site from the direction of the acceptor site.

In a series of studies with oligonucleotidyl phenylalanine, puromycin was shown to inhibit the binding of the aminoacyl end of aminoacyl tRNA to the ribosome (PESTKA, 1969). Since puromycin attacks the donor-site-bound peptidyl tRNA from the direction of the acceptor site, this is consistent with the notion that the binding of the oligonucleotidyl amino acid to the 50 S subunit represents the binding to the peptidyl transferase from the direction of the acceptor site. This binding is supposed to represent the binding of acceptor tRNA but not peptidyl tRNA (donor tRNA). One would therefore expect that the binding of the oligonucleotidyl amino acid would take place at or near the A site. The inhibition by puromycin supports this notion.

The structural requirements for the puromycin-like action have been investigated using analogues of puromycin (Symons et al., 1969). It has been found that the amino-acyl analogues of puromycin which have puromycin-like activity were those in which the p-methoxyphenylalanine was substituted by either α-phenylalanine or tyrosine (Nathans and Neidle, 1963), a single benzene ring in the amino acid side chain being necessary for the activity. Thus, substitution with S-benzyl-L-cysteine or benzyl-L-histidine gave only moderate activity. Substitution at the 5' hydroxyl group of the nucleotide portion of puromycin resulted in a heightened activity of the analogue which otherwise was inactive. Thus, it was found that the cytidine 3' phosphoryl group substitution of 3'-N-glyceryl aminonucleoside resulted in a compound almost as active as puromycin itself. The effect is specific to cytidine, or guanosine substitution, because phosphoryl, adenosine-3' phosphoryl, or uridine 3'-phosphoryl substitution did not give active compounds (Rychlik et al., 1967).

c) Sparsomycin (SPM)

Sparsomycin (SPM) (Wiley and MacKellar, 1970) is regarded as one of the most specific inhibitors of peptide bond formation. After the action of this antibiotic was localized to the protein synthetic machinery (Slechta, 1965), studies with in vitro *E. coli* systems established that this compound specifically inhibits peptide bond formation [Goldberg and Mitsugi, 1967 (1, 2)]. This antibiotic influences both bacterial and mammalian protein synthesis (Colombo et al., 1966). The sensitivity of polypeptide synthesis to sparsomycin varies depending on the base composition of the mRNA used in the *in-vitro* protein-synthesizing system. Thus, polyproline formation with poly C is most sensitive to sparsomycin, followed by polylysine formation with poly A and polyphenylalanine formation with poly U. Furthermore, polypeptide synthesis programmed by polynucleotide copolymers containing uridine plus other nucleotides was more sensitive than polyphenylalanine synthesis with poly U. This is reminiscent of CM's action on the polypeptide synthesis. The molecular basis for the messenger RNA specificity of SPM action remains obscure. Although polyphenylalanine synthesis is relatively insensitive to SPM, it was with this system that SPM was established as an inhibitor of the peptidyl transferase. When ribosomes charged with ^{14}C-polyphenylalanyl tRNA were incubated with puromycin, ^{14}C-polyphenylalanine was released from the ribosomes. The release can be detected by sucrose density gradient centrifugation of the reaction mixture. SPM inhibited this release, as well as the release of polylysine from the complex of polylysyl tRNA and ribosomes.

The notion that SPM acts at the step of peptidyl transferase was further supported by its inhibitory effect on the "fragment" reaction (Monro and Vazquez, 1967). In the presence of alcohol, 50 S subunits from *E. coli* catalyzed the puromycin reaction of various N-acyl aminoacyl oligonucleotide fragments from tRNA to give the corresponding N'-acyl aminoacyl puromycin (Monro, 1967; Monro et al., 1968). This fragment reaction should not be confused with the binding of aminoacyl oligonucleotidyl amino acid, which presumably represents the binding of the *acceptor aminoacyl* tRNA to the peptidyl transferase. The reaction of N'-acyl aminoacyl oligonucleotide with puromycin in alcohol presumably proceeds through the binding to the peptidyl transferase at the site for the *donor peptidyl* tRNA. SPM's inhibitory effect on the puromycin reaction of peptidyl tRNA diminishes as the concentration of puromycin increases, indicating that these two antibiotics would interact with the peptidyl transferase at the same point — probably the site for acceptor molecules [PESTKA, 1970 (1); GOLDBERG and MITSUGI, 1967 (2)].

SPM appears to induce an inert complex of peptidyl donor (presumably at the donor site) and the 50 S ribosomal subunit. In fact, in the presence of alcohol, SPM stimulates the binding of N-blocked aminoacyl nucleotides to 70 S ribosomes as well as to 50 S subunits (MONRO et al., 1969). This binding induced by SPM does not involve a covalent linkage, since this complex disintegrates by dissolving the complex in a buffer without alcohol. Blocking of the amino group of aminoacyl tRNA was essential for the stimulation to occur. This is in sharp contrast to the binding of the aminoacyl fragment tRNA where the blocking is inhibitory. The SPM-induced ribosomal complex of amino-blocked fragment tRNA does not react with puromycin. One possible explanation for this observation is that the ribosomal conformation is changed by SPM in such a manner that it is no longer available for puromycin. Another possibility is that SPM binds tightly to the site of peptidyl transferase for the acceptor molecule so that puromycin is unable to approach the donor molecule. This possibility is consistent with two separate observations. The first observation is that SPM inhibited the binding of acceptor aminoacyl tRNA to the peptidyl transferase [PESTKA, 1969 (2)]. Secondly, despite an earlier report to the contrary (VAZQUEZ and MONRO, 1967), SPM appears to inhibit the binding of ^{14}C-chloramphenicol (in review by WEISBLUM and DAVIES, 1968). Since CM inhibits the binding of the acceptor tRNA to the peptidyl transferase, these results indicate that SPM may bind to the peptidyl transferase at the site for the acceptor tRNA. Furthermore, the SPM-induced binding of N-blocked aminoacyl oligonucleotide is inhibited by several antibiotics which inhibit peptide bond formation (such as CM, spiramycin IV, carbomycin and lincomycin, MONRO et al., 1969).

d) Lincomycin (LM)

LM (HOEKSEMA et al., 1964) inhibits polypeptide synthesis *in vitro* (TERAOKA et al., 1969) and appears to bind to the 50 S subunit (CHANG and WEISBLUM, 1967). The dissociation constant of the ^{14}C-LM-ribosome complex was 340 μM (FERNANDEZ-MUNOZ et al., 1971). Three pieces of evidence suggest that LM binds to the peptidyl transferase at or near the site for the acceptor tRNA. First of all the 50 S subunit which contains peptidyl transferase is responsible for resistance or sensitivity to LM (APIRION, 1967; KREMBEL and APIRION, 1968). The second piece of evidence is that the binding site of LM overlaps with that of CM [VAZQUEZ, 1966 (2); VAZQUEZ and MONRO, 1967]. CM binding to ribosomes is inhibited by LM, and, conversely, N-methyl ^{14}C-LM binding to ribosomes is inhibited by CM (FERNANDEZ-MUNOZ et al., 1971). As discussed in the section on CM, this antibiotic perhaps binds to the peptidyl-transferase at or near the A site. The third piece of evidence, which is perhaps strongest, is that the LM binding to ribosomes is inhibited by puromycin which, by definition, must associate with peptidyl transferase at the site for the acceptor molecule. Although these observations suggest that the binding site of LM may be closely related to the site for the acceptor, the following two observations suggest that the binding site of LM is *not* identical to the site for the acceptor tRNA. First, in contrast to the binding of CM, ethanol faciliates LM binding to ribosomes, indicating that the LM and CM binding sites which are presumably at the site for the acceptor molecule, are not identical. Secondly, erythromycin (EM) inhibits the binding of LM. Since EM does not inhibit the binding of the aminoacyl fragment tRNA, this may be another piece of evidence that LM's binding site is not identical with the site for the acceptor molecule.

Although it was first suggested that LM may inhibit the binding of aminoacyl tRNA to ribosomes, this concept appears to be incorrect in view of the evidence indicating that LM is a specific inhibitor of peptidyl transferase. Thus, LM can inhibit the puromycin reaction of the 70 S ribosome-bound formylmethionyl hexanucleotide, causing about 50% inhibition at 1.8 to 3.7 μM LM (Monro and Vazquez, 1967). Inhibitory effects of LM on the peptidyl transferase were confirmed in a separate system involving the puromycin reaction of the donor-site-bound phenylalanyl tRNA. Polyphenylalanine formation was also sensitive to LM, but the sensitivity was much less than that of the puromycin reaction. The lack of inhibition of the translocation step by LM was demonstrated by its inability to stop the EFG dependent release of tRNA from the ribosome (Igarashi et al., 1969).

In spite of the clear-cut evidence obtained from *in-vitro* studies, the *in-vivo* action of LM cannot be completely explained on the basis of its peptidyl transferase inhibition. When protoplasts were incubated in the presence of 5×10^{-4} M LM, polyribosomes were degraded. Increasing the drug concentration beyond this value did not protect polyribosomes, while a complete inhibition of protein synthesis was observed. It is therefore clear that breakdown of polysomes under these conditions was not due to incomplete inhibition of protein synthesis which would have allowed run-off of ribosomes. In the presence of LM, polysomes were converted to subunits (50 S and 30 S), and not to 70 S ribosomes (Cundliffe, 1969). It should be recalled that chloramphenicol preserved polyribosomes by virtue of its action on the peptidyl transferase. Why, then, doesn't LM act similarly? The answer to this question is not known. A recent report which indicates that LM may not interfere with peptidyl transferase *in vivo* but instead inhibits the initiation process may have bearing on this question (Pestka, 1971).

In eukaryotic cells, both mitochondrial and chloroplast protein syntheses are inhibited by LM, while cytoplasmic protein synthesis is insensitive. This characteristic of LM has been used in studying protein synthesis in these organelles (Ellis and Hartley, 1971). In this experiment pea apices were illuminated in the presence of LM and the increase of various enzyme activities was followed. Many enzymes, except ribulose diphosphate carboxylase, showed an increase of specific activity even though chlorophyll synthesis was reduced by 75% by LM. From these observations it was concluded that chloroplast ribosomal activity is necessary for the synthesis of ribulose diphosphate carboxylase.

e) *Gougerotin*

Gougerotin is an aminohexose pyrimidine nucleoside belonging to a group of antibiotics which includes amicetin and blasticidin. These antibiotics, together with sparsomycin, are effective inhibitors of both eukaryotic and prokaryotic systems (Vazquez et al., 1969; Clark and Chang, 1965). Gorgerotin is regarded as a peptidyl transferase inhibitor on the basis of the regular puromycin assay. Thus, in the absence of translocation, 10^{-3} M gougerotin caused 70% inhibition of the formation of N-acetyl phenylalanyl puromycin [PESTKA, 1970 (1)]. Likewise, puromycin-dependent polylysine release from polylysyl tRNA was blocked by this antibiotic [GOLDBERG and MITSUGI, 1967 (2)]. The fragment reaction with fMet-oligonucleotide (MONRO and VAZQUEZ, 1967) and acetyl leucyl oligonucleotide (NETH et al., 1970) is also sensitive to 10^{-4} M gougerotin. Like SPM and CM, gougerotin's inhibitory effect on the puromycin reaction decreases as the concentration of puromycin increases (CASJENS and MORRIS, 1965). The inhibitory effect of gougerotin on the *in vitro* polypeptide synthesis is dependent on the kind of synthetic polynucleotide used. For example, polyproline formation programmed by poly C is more sensitive than poly U-dependent polyphenylalanine synthesis. As described earlier, such differential sensitivity could be attributed not only to the difference in the polynucleotide, but also to the difference in the peptidyl group being synthesized.

Several analogues of gougerotin have recently been synthesized and their effect on the peptidyl transferase tested [CERNA et al., 1971 (1)]. Replacement of the -$CONH_2$ group of gougerotin by a CH_2OH group decreased the inhibitory activity by 90%. Peptidyl transferase was inhibited only by compounds which had a sarcosyl D-seryl amide residue on the glucose moiety. At a concentration which inhibits the peptidyl transferase, gougerotin stimulated the binding of the donor substrate N-acetyl leucyl CACCA in the presence of alcohol. On the other hand, gougerotin inhibited the binding of the acceptor substrate phenylalanyl CACCA to the ribosomes. From these observations it has been suggested that gougerotin and its derivative inhibit peptidyl transferase by competing with the acceptor substrate at the ribosomal acceptor site [CERNA et al., 1971 (1)]. It should be pointed out that the binding of aminoacyl tRNA to the acceptor site of the ribosome is not inhibited by this antibiotic. If the acceptor site of the peptidyl transferase is identical to the acceptor site of the ribosomes, these results suggest that the interaction between codon and anti-codon on the ribosomes is strong enough to keep the aminoacyl tRNA at the acceptor site of the 70 S ribosome, even if gougerotin inhibits the interaction of the CCA end of the tRNA with the 50 S subunit at the peptidyl transferase site. Alternatively, these results may indicate that the site of the peptidyl transferase for the acceptor molecule is different from the acceptor site of the ribosome.

5. Translocation

After the peptide bond is formed between the donor-site-bound peptidyl tRNA and the acceptor-site-bound aminoacyl tRNA, the newly elongated peptidyl tRNA is bound at the acceptor site and the unesterified tRNA is bound to the donor site. The next step is to translocate the newly elongated peptidyl tRNA from the acceptor site to the donor site and to release unesterified tRNA from the donor site. This step is called translocation. The translocation step has been shown to be dependent on GTP and EFG in bacteria or T2 in the mammalian system. In contrast to the relationship between T1 (of mammalian system) and EFT, T2 and EFG are not interchangeable. However, eukaryotic mitochondrial translocase (mitochondrial T2) is interchangeable with EFG in a polypeptide-synthesizing system by E. coli ribosomes (KRISCO et al., 1969; KUNTZEL, 1969). The EFG from *E. coli* has been crystallized (PARMEGGIANI,

1968; Kaziro et al., 1969), has a molecular weight of 72,000 and is heat-stable at 50 °C for a least 10 min. T2 also has been purified and has a molecular weight of approximately 65,000, which is similar to that of EFG (Galasinski and Moldave, 1969). In growing *E. coli*, EFG can be as much as several percent of the total protein. In growing HeLa cells, there are approximately 10^7 molecules of T2, or 3 to 4 molecules per ribosome [Gill et al., 1969 (2)].

There are five methods of measuring the translocation of tRNA. The first and most commonly used method is to examine the availability of bound aminoacyl tRNA to the puromycin reaction. Thus, the aminoacyl tRNA bound at the acceptor site is not available for the puromycin reaction unless it is translocated to the donor site by the action of translocase. For example, at high (13 mM) Mg^{++}, both the acceptor and the donor sites are occupied, and peptide bond formation between these two bound aminoacyl tRNA's ensues. The dipeptidyl tRNA thus formed is bound to the acceptor site and unesterified tRNA is bound to the donor site. In the presence of EFG, GTP and puromycin, diphenylalanyl puromycin is formed.

The second method is to examine tripeptide synthesis dependent on EFG and GTP. In the presence of a synthetic polymer such as AUG UUU UUU it has been shown that EFG is required for the synthesis of fMet-Phe-Phe, while fMet-Phe can be synthesized without this factor. This indicates that for the translation of the third codon, movement of mRNA and bound tRNA is required (Erbe et al., 1969). With RNA from phage, reading of the first three codons depends on the presence of EFG, while the first dipeptide can be synthesized without EFG.

The third method of measuring translocation is based on the fact that simultaneous release of unesterified tRNA takes place from the donor site of the ribosomes. For demonstration of the release of tRNA from ribosomes, a complex of tRNA specific for phenylalanine, polyuridylic acid and ribosomes was prepared under the condition where the acceptor and the donor site are occupied with unesterified tRNA. When such a complex was exposed to EFG and GTP, release of tRNA took place (Ishitsuka et al., 1970). The release of tRNA is probably not the primary action of EFG, but the consequence of translocation of tRNA from the acceptor site to the donor site. Thus, if one prepares the complex of ribosomes which contain tRNA mostly on the donor site, EFG and GTP do not release this tRNA. If the ribosomal complex with tRNA is prepared in the presence of an excess of ribosomes, most of the complex would contain only one tRNA. EFG does not release tRNA from such a complex either. These observations strongly support the notion that release of tRNA takes place because the donor site tRNA is *pushed out* by the translocating tRNA. It should be recalled that the natural substrate for EFG is the ribosomal complex having tRNA and peptidyl tRNA at the donor and the acceptor sites, respectively. The experiments cited above show that EFG not only acts on the natural substrate (complex), but also on an artificial complex having unesterified tRNA on both sites. Similar experiments with labeled tRNA further indicated that tRNA release takes place simultaneously with translocation (Lucas-Lenard and Haenni, 1969; Lucas-Lenard et al., 1969). In this experiment phe-^{3}H-tRNA was prepared from a uracil-requiring mutant of *E. coli*. The complex of ribosomes having N-acetyl-^{14}C-phe-^{3}H tRNA presumably at the donor site was prepared, and ^{12}C-phenylalanyl tRNA was added to this complex in the presence of EFT and GTP. Under these conditions, ^{12}C-phenylalanyl tRNA was bound to the acceptor site and formed N-acetyl ^{14}C-phenylalanyl ^{12}C-phenylalanyl tRNA. The unesterified ^{3}H tRNA was bound to the donor site. Incubation of this complex with GTP and EFG resulted in release of ^{3}H tRNA. Simultaneous with this release, the N-acetyl ^{14}C-phenylalanyl ^{12}C-phenylalanyl tRNA became available for the puromycin reaction, indicating that translocation of the peptidyl tRNA took place.

One important point about the EFG-dependent release of tRNA is that aminoacyl tRNA or peptidyl tRNA cannot be released from the donor site. This is an important feature of the translocase, since release of the peptidyl tRNA from the donor site by the translocase would end chain elongation. Because EFG cannot release

Table 6a. Inability of ribosome-bound tRNA to accept amino acid. (Ishitsuka and Kaji, 1972)

Substrate	Phenylalanyl tRNA formed (% of total)
A) Ribosome with tRNA on A and D sites	5.1
B) Ribosome with tRNA only on D site	2.8
C) Unbound tRNA	97.3

Ribosomal complexes and free tRNA were separately incubated with the regular mixture for aminoacylation for 2 min at 37 °C. Total amounts of tRNA on the ribosomes were separately measured by assaying the acceptor capacity of the tRNA's after they were released from the ribosomes.

Table 6b. Selective exchange of A-site-bound tRNA^{Phe} with free N-acetyl ^{14}C phenylalanyl tRNA. (Ishitsuka and Kaji, 1972)

Incubation time	Distribution of exchanged aminoacyl tRNA (% of total exchanged) acceptor site	donor site
30′	99.63	0.37
60′	92.44	7.56

The ribosomal complex of polyuridylic acid having tRNA^{Phe} on both the A and D sites was incubated with N-acetyl ^{14}C-phenylalanyl tRNA for 30 and 60 min at 37 °C. The complex of N-acetyl ^{14}C-phenylalanyl tRNA with ribosomes thus formed was subjected to the puromycin reaction to decide which of the two sites was occupied by N-acetyl ^{14}C-phenylalanyl RNA.

aminoacyl tRNA, occupation of the donor site by aminoacyl tRNA would immediately stop chain elongation. During polypeptide synthesis such a situation can occur in three ways. The first is aminoacylation of the donor site tRNA before it is released by the translocation of peptidyl tRNA from the acceptor site. The second way is to exchange the donor-site tRNA with aminoacyl tRNA. A third conceivable way is a binding of aminoacyl tRNA to the donor site. However, this can happen only when the empty ribosome is exposed to aminoacyl tRNA which does not occur *in vivo* because the donor site is always occupied by either tRNA or peptidyl tRNA during polypeptide synthesis. The experiments indicated in Table 6 show that there is a built-in mechanism in the ribosome to prevent occupation of the donor site during polypeptide synthesis. As shown in Table 6 a, aminoacylation of ribosome-bound tRNA is much slower than that of unbound tRNA. In experiments A and B, the ribosomal complexes with $tRNA^{phe}$ on both sites, and those with $tRNA^{phe}$ only on the donor site were used. It is clear from these experiments that tRNA on either the D or A site is prevented from aminoacylation. Table 6 b shows that exchange of tRNA in the donor site with free aminoacyl tRNA does not occur. In this experiment, ribosomes each with the two molecules of bound tRNA were exposed to N-acetyl ^{14}C-phenylalanyl tRNA. The only way for the aminoacyl tRNA to bind to the ribosome is through exchange with the bound tRNA. The labeled N-acetyl-Phe-tRNA thus bound was tested for its reactivity with puromycin. As can be seen from Table 6 b less than 1% of the aminoacyl tRNA's exchanged during a 30-min incubation period was reactive with puromycin, indicating that very little, if any, exchange takes place at the donor site. Even after a one-hour exposure of the complex to N-acetyl Phe-tRNA, the major portion (90%) of the exchange between bound unesterified tRNA and free N-acetyl ^{14}C-phe tRNA was at the acceptor site.

The fourth assay method of translocation is unusual in that it measures the movement of mRNA rather than tRNA. This method depends on the fact that the portion of mRNA bound to the ribosome is protected from digestion by RNase. For this assay, ribosomes, initiation factors, EFTu, and a mixture of tRNA containing fMet-tRNA, valyl tRNA, GTP and synthetic polynucleotide AUG $GC(U)_{30}$ were mixed. Analysis of the complex formed in this mixture indicated that fMet-Val-tRNA is bound to the acceptor site. Addition of EFG to this complex translocated the bound fMet-Val-tRNA to the puromycin reactive site. The complex was then treated with RNase and analyzed for the chain length of the ribosome-bound messenger fragments. It was found that ApUGG $(U)_{15}$ was bound to the ribosome and was RNase-resistant. On the other hand, in the absence of added EFG, the RNase-resistant, ribosome-bound RNA was $AUGG(U)_{12}$. These results suggest that the addition of EFG has caused the movement (relative to the ribosome) of RNA about three nucleotides in the 5' direction [THACH and THACH, 1971 (2)]. In addition to these four assays mentioned above, one can measure the translocation of acceptor tRNA by measuring the availability of the acceptor site for further binding of aminoacyl tRNA. This assay depends on the evacuation of the acceptor site by translocation. One measures EFG-dependent binding of aminoacyl tRNA to the acceptor site.

Although the mode of action of EFG and the mechanism through which the chemical energy of GTP is converted to kinetic energy are not known, sequential events on the uncoupled hydrolysis of GTP are understood. EFG, together with ribosomes hydrolyzes GTP in the absence of apparent translocation of the tRNA or messenger RNA. This is the reason why the translocase was originally named "G" factor. The hydrolysis of GTP by EFG proceeds as follows [BODLEY et al., 1970 (2)]:

$$\text{Ribosome} + \text{EFG} + \text{GTP} \longrightarrow \text{Ribosome} - \text{EFG} - \text{GTP} \quad (1)$$

$$\text{Ribosome} - \text{EFG} - \text{GTP} \longrightarrow \text{Ribosome} + \text{EFG} - \text{GDP} + \text{Pi} \quad (2)$$

$$\text{Ribosome} - \text{EFG} - \text{GDP} \longrightarrow \text{Ribosome} + \text{EFG} + \text{GDP} \quad (3)$$

As indicated in (1), EFG makes a complex with ribosomes in the presence of GTP [KAZIRO et al., 1969; BROT et al., 1969; PARMEGGIANI and GOTTSCHALK, 1969 (2); BODLEY and LIN, 1970]. GTP can be replaced with GDP or GMPPCP [PARMEGGIANI and GOTTSCHALK, 1969 (2); BROT et al, 1969]. All the components of reaction (1) are necessary for the complex formation because no stable binding has been observed between nucleotide and either ribosomes or EFG alone. In the mammalian system, most of T2 is bound to the ribosomes, but the presence of GTP or GDP in this complex has not been examined [GILL et al., 1969 (2)]. The ribosome-bound T2 is released only after the ribosomes are dissociated from mRNA at the termination step. Although the original report indicated that both ribosomal subunits are required for the binding of EFG to ribosomes, the 50 S subunit alone appears to be sufficient for this binding [BODLEY and LIN, 1970; MODOLELL et al., 1971 (1)]. Reaction (2) is irreversible and therefore the exchange reaction between Pi-GTP or GDP-GTP does not take place [NISHIZUKA and LIPMANN, 1966 (2)]. The product of this reaction, i.e. the complex of GDP, EFG and ribosome, is stable and can be isolated by either sucrose density gradient centrifugation or Millipore filtration [BODLEY et al., 1970 (1); BODLEY et al., 1969; BROT et al., 1969]. Despite these fine studies the most crucial question as to how these reactions are related to actual translocation remains to be answered.

Recent studies with hydrogen-exchange techniques indicate that configurational changes of ribosomal structure take place during translocation (CHUANG and SIMPSON, 1971). In this experiment, ribosomes carrying acetyldiphenylalanyl tRNA in the acceptor site were incubated with EFG and GTP. These ribosomes had been equilibrated with ^{3}H-H_2O. Although ^{3}H radioactivity exchanged out upon incubation with non-labeled buffer solution, the EFG and GTP addition caused an immediate increase in the exchange rate, inducing the exchange-out of about 350 more ^{3}H atoms per ribosome per unit period, or about 6.6% of the ^{3}H atoms present at the time of EFG and GTP addition. GMPPCP produced only one third of the increase. The configurational change of ribosomes was demonstrated by the change of sedimentation value also. The pre- and post-translational complex sedimented with the speed of 72 S and 69 S, respectively. The sedimentation behavior of the post-translational complex increased when it was converted back to the pre-translocational state by addition of EFT and aminoacyl tRNA. Conformational change of ribosomes was also suggested earlier by a separate but similar study with synthetic polynucleotide (SCHREIER and NOLL, 1971). Under appropriate conditions, formation of an "initiation complex" with synthetic polynucleotide has been demonstrated and the following sequence of events suggested:

$$\text{poly U} + 30\text{ S} \rightleftarrows \text{poly U} - 30\text{ S} \quad (1)$$

$$(\text{poly U} - 30\text{ S}) + \text{tRNA}^{\text{phe}} \rightleftarrows 30\text{ S} - \text{poly U} - \text{tRNA}^{\text{phe}} \quad (2)$$

$$(30\text{ S} - \text{poly U} - \text{tRNA}^{\text{phe}}) + 50\text{ S} \rightarrow 60\text{ S} - \text{poly U} - \text{tRNA}^{\text{phe}} \quad (3)$$

The initiation complex (60 S) will then change its configuration to form 70 S upon association with aminoacyl tRNA. This form is called pre-translocational form:

$$60\text{ S} - \text{poly U} - \text{tRNA}^{\text{phe}} + \text{Phe-tRNA EFTu} - \text{GTP} \longrightarrow$$
$$\longrightarrow \text{poly U} - \text{Phe-tRNA } 70\text{ S} - \text{tRNA}^{\text{phe}} + \text{EFTu} + \text{GDP} + \text{Pi} \quad (4)$$

The 70 S pre-translocational form would then undergo configurational change by EFG and GTP back into the 60 S form.

$$\text{poly U} - \text{Phe-tRNA } 70\text{ S} - \text{tRNA}^{\text{phe}} + \text{EFG} + \text{GTP} \longrightarrow$$
$$\longrightarrow \text{tRNA}^{\text{phe}} + (\text{poly U} - \text{Phe-tRNA} - 60\text{ S}) + \text{EFG} + \text{GDP} \quad (5)$$

The 60 S in (5) would be called "post-translocational form". The 70 S complex in (4) corresponds to the pre-translocational ribosome which was designated as 72 S ribosome in the preceding experiment. The 60 S ribosome in this experiment corresponds to the 68 S ribosome designated as post-translocational ribosome. The apparent difference in the sedimentation values was due to the different experimental conditions during the sedimentation analysis of various ribosomal complexes.

In spite of these clear-cut observations on the change of sedimentation behavior of ribosomes during the translocation, the interpretation of the slight change in S-value has become somewhat difficult because of a recent report concerning the pressure exerted on ribosomes during sedimentation analysis. Thus, sedimentation behavior of 75 S ribosomes from unfertilized sea urchin eggs varied with rotor speed and the duration of centrifugation (Infante and Baierlein, 1971). As the ribosomes sedimented and were exposed to increasing pressure, equilibrium between ribosomes and subunits was shifted toward dissociation, giving an absorbance peak sedimenting more slowly than 75 S particles. Considering these findings, it appears possible that the 60 S "post-translational form" behaves as 60 S because only one $tRNA^{phe}$ is holding the subunits together, while the "pre-translocational" form is 70 S because two tRNA's hold both subunits together, making the associated form more resistant to pressure.

6. Inhibitors of Translocation

a) Fusidic Acid (FA)

FA inhibits the GTPase catalyzed by EFG (or T2) and ribosomes and has been accepted as one of the typical inhibitors of the translocation step. The fact that FA acts on EFG is established by the isolation of an FA-resistant mutant of *E. coli* which has an altered EFG (the GTPase activity of which is insensitive to FA) (Tocchini-Valentini et al., 1969; Kinoshita et al., 1968). The genetic locus for fusidic acid resistance is located near the SM resistance locus [Tanaka et al., 1971 (1)]. Of the three steps of the GTPase reaction, FA inhibits the last step, namely, the dissociation of the complex of ribosomes, EFG and GDP (Brot et al., 1971). In the presence of FA, this complex becomes stable and can be isolated. FA had no effect on the second reaction (the formation of this complex) and thus a single round of translocation may not be sensitive to FA [Bodley et al., 1970 (2)]. In accordance with this view, FA's effect on translocation appeared to decrease with increasing amounts of EFG. However, from the fact that FA inhibits the GTPase reaction *after* the hydrolysis of GTP, another possibility arises. That is, translocation may occur *after* the splitting of GTP. It is possible, although unlikely, that actual translocation of tRNA may occur at the time GDP and EFG dissociate from the ribosome, and it is this step that is inhibited by FA. If this hypothesis is correct, only one GTP would be necessary for each peptide bond formation, because the GDP created by EFTu during the binding of aminoacyl tRNA may be used for translocation. This point will be discussed further in the section dealing with thiostrepton.

FA binds to EFG at a 1:1 molar ratio, the association constant being 1.2×10^5 M^{-1}, and this binding is stimulated by GTP or GDP. This association does not take place with EFG from FA-resistant *E. coli*. A similar binding of FA to mammalian translocase (T2) has been found (Malkin and Lipmann, 1969). In the presence of ribosomes a stable complex of GDP, ribosome, FA and EFG is formed with a molar ratio of 1:1:1:1 (OKURA et al., 1970). The association constant of FA to the complex of GDP, ribosome and EFG is 2×10^6 M^{-1}, indicating that this complex is more stable than the EFG-FA complex. A similar complex is formed with eukaryotic translocase (RICHTER et al., 1971).

The inhibitory effect of FA on translocation can be demonstrated with the various methods of assay for translocation as described in III-5. Thus, with a ribosomal complex having tRNA and aminoacyl tRNA on the D and A sites, respectively, FA inhibits the release of D-site tRNA during translocation (see the section on erythromycin).

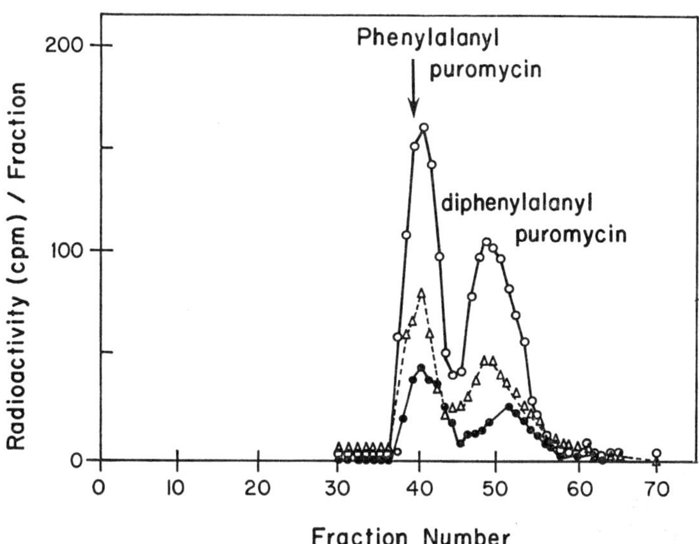

Fig. 1. Stimulation of phenylalanyl puromycin formation by EFG and its inhibition by fusidic acid (TANAKA and KAJI, 1972). The complex of ^{14}C-Phe-tRNA, poly U and the ribosome was made with EFT, and this complex was incubated with puromycin, with and without EFG or fusidic acid. ○—○, 2,100 cpm of the puromycin derivatives formed with EFG were placed on the Sephadex column; △··△, 1,140 cpm of those formed with EFG and 5×10^{-4} M fusidic acid were analyzed; ●—●, 850 cpm of those formed in the absence of EFG were placed on the Sephadex column

FA can therefore inhibit translocation in two conceivable ways. The first possibility is that FA "fixes" unesterified tRNA on the donor site and thus prevents the movement of aminoacyl or peptidyl tRNA from the A site to the D site. The second possible mechanism is that FA stops the movement of aminoacyl or peptidyl-tRNA from the A to the D site, even if the D site is empty. To distinguish these two possibilities, translocation of aminoacyl-tRNA from the A site to the empty D site was studied in the presence and absence of FA. In the experiment illustrated in Fig. 1, the complex of Phe-tRNA, polyuridylic acid and ribosomes was prepared in the presence of EFT. Among the complexes formed, some carried Phe-tRNA on both sites, some only on the D site, and some only on the A site. The puromycin reaction was performed with these complexes in the presence and absence of EFG. The puromycin derivative thus formed was analyzed by a Sephadex column which separates diphenylalanyl puromycin from phenylalanyl puromycin. It is noted in this figure that EFG markedly stimulated *mono* as well as diphe-puromycin. The stimulation of *mono* phe-puromycin formation indicates that EFG can translocate Phe-tRNA in the absence of tRNA on the D site. A separate experiment established that the D sites of these complexes were indeed empty. As shown in this figure, FA inhibited the EFG-stimulated phe-puromycin formation, indicating that the presence of tRNA at the donor site is *not* necessary for the inhibitory action of FA on translocation. We therefore eliminated the first possibility and concluded that FA stops movement of aminoacyl tRNA *per se*.

b) Erythromycin (EM)

As a representative of macrolides, EM shall be discussed in this section. EM is active against gram-positive bacteria, but higher concentrations of EM also inhibit the growth of *E. coli*. Although EM has been shown to inhibit the translocation step, other antibiotics belonging to the macrolides inhibit other steps of protein synthesis. The *in-vitro* inhibitory effect of EM can be observed with the phage MS2 RNA-directed protein-synthesizing system. On the other hand, the effect of EM on polyphenylalanine formation by the *E. coli* system depends on the Mg^{++} concentration (Tanaka and Kaji, unpublished observation). At a higher Mg^{++} (13 mM) concentration EM even stimulated polyphenylalanine synthesis, while at a lower (5 mM) Mg^{++} concentration it inhibited the synthesis (Wolfe and Hahn, 1964). Similarly, with *B. subtilis*, EM's inhibitory effect on polyphenylalanine synthesis is observable at 12 mM Mg^{++} (Wilhelm and Corcoran, 1967). Although polylysine formation is sensitive to EM, di- and trilysine formation is insensitive (Teraoka et al., 1969).

There are two schools of thought concerning the mode of action of EM, one claiming that EM inhibits peptide bond synthesis and the other believing that EM acts on the translocation step. We shall review the evidence for these two opposing views and propose a hypothesis which accommodates most of the available data. There are several pieces of evidence which suggest that EM acts on the translocation step, but not on peptide bond synthesis. First of all, a typical peptidyl transferase reaction, N-acetyl phenylalanyl puromycin or phenylalanyl puromycin formation, was insensitive to EM, while release of tRNA during translocation was sensitive (Igarashi et al., 1969). The puromycin reaction of fMet-tRNA was not inhibited by EM (Mao and Robishaw, 1971), and occasionally as much as a twofold stimulation was observed (Vogel et al., 1971). The second piece of evidence comes from the *in-vivo* studies of EM. In protoplasts of *B. megaterium*, EM inhibited the puromycin release of nascent peptides, but did not inhibit the release if tetracycline was also present. This is not due to a competition of these two antibiotics for binding to ribosomes, because the binding sites are clearly different. In the presence of tetracycline, with its specific inhibitory effect on the binding of aminoacyl tRNA, the ribosomes carry peptidyl tRNA on the donor site but have an empty acceptor site. If EM inhibited peptide bond formation, the puromycin reaction with the peptidyl tRNA on these ribosomes would have been inhibited (Cundliffe and McQuillen, 1967). The lack of inhibition by EM in this case was taken as evidence that EM inhibits translocation. Thirdly, the binding of an acceptor molecule to the peptidyl transferase was not inhibited by EM. Thus, 10^{-3} M EM (the concentration which is sufficient to inhibit translocation) did not inhibit the binding of CACCA-phenylalanine to the 50 S subunit [PESTKA, 1969 (2)]. The fourth piece of evidence is that the puromycin reaction of acetyl leucyl pentanucleotide was stimulated by 1 μM EM, indicating that EM does not inhibit the peptidyl transferase reaction [CERNA et al., 1971 (2)].

The large body of evidence suggesting that EM does not act on the peptidyl transferase is contrasted by some data which suggest that EM may inhibit peptide bond formation. Firstly, 10^{-6} M EM inhibited the reaction of polylysyl tRNA with puromycin in the absence of translocation and prevented the addition of a single lysine unit to polylysyl tRNA (CERNA et al., 1969; JAYARAMAN and GOLDBERG, 1968). Similarly, diphenylalanyl puromycin formation was sensitive to EM even in the absence of EFG and GTP [CERNA et al., 1969; TANAKA et al., 1971 (3); and TERAOKA et al., 1970]. Since phenylalanyl and N-acetyl phenylalanyl puromycin formation was not sensitive to EM, the inhibitory action of EM on peptide bond formation appears to depend on the nature of the donor group. The second piece of evidence is the correlation of peptidyl transferase activity with the EM resistant mutation. The peptidyl transferase activity of EM-resistant ribosomes is less than that of sensitive ribosomes. This turned out to be a reversible change in the 50 S subunit [CERNA et al., 1971 (2)]. Thus, when the 70 S ribosomes were reconstituted from the subunits, the peptidyl transferase activity of reconstituted ribosomes of the sensitive strain was the same as that of the resistant strain. Despite these facts, one can at least conclude that resistance to EM influences the peptidyl transferase. The third piece of evidence is circumstantial. Antibiotics belonging to the macrolide group include, in addition to EM, niddamycin, carbomycin, spiramycin and tylosin. All of these antibiotics inhibit peptide bond formation (MAO and ROBISHAW, 1971). It was argued that EM cannot be a translocation inhibitor because it permits oligolysine formation. From these considerations, it was concluded that EM is probably a peptidyl transferase inhibitor.

These seemingly contradictory reports may be reconciled by a hypothesis which postulates that EM inhibits the translocation step by inhibiting proper accommodation of the donor molecule (peptidyl tRNA) to the ribosomal D site. This hypothesis was derived from experiments in which the translocation step was measured by four different procedures as discussed in III-5. In the experiment shown in Table 7a the EFG-dependent puromycin reaction of ribosome-bound aminoacyl tRNA was studied. It is clear from this table that this reaction was sensitive to fusidic acid regardless of the ribosomal complex used, while EM inhibited only the puromycin reaction of diphenylalanyl tRNA (Complex B). Table 7 b shows translocation measured by the release of tRNA and the effect of EM on this process. Even with complex A, which was not susceptible to EM by the puromycin assay, EM has a clear-cut effect on translocation measured by this method. The hypothesis which is consistent with these and other observations is illustrated in Fig. 2. By binding to or near the peptidyl transferase, EM inhibits translocation at the step of proper positioning of peptidyl tRNA to the D site. This step may be analogous to the "accommodation" step of the initiation of protein synthesis (see II-2). Thus, translocation in the presence of EM would result in peptidyl tRNA located near, but not on the donor site. Since the peptidyl tRNA is not quite on the donor site, no release of D-site tRNA takes place. It is postulated that the position of peptidyl tRNA translocated in the presence of EM is so close to the D site that the puromycin reaction can take place with those peptidyl tRNA's having a higher affinity for the peptidyl transferase, but not with those having a lower affinity. N-acetyl Phe-tRNA, Phe-tRNA and fMet-tRNA all have a high affinity to the transferase and can react with puromycin even at a position slightly away from the D site. On the other hand, diPhe-tRNA and polylysyl-tRNA have a lower affinity for the peptidyl transferase and can react with puromycin only when they are properly positioned on the D site, and EM inhibits this positioning. The

Table 7a. Effect of erythromycin A on the EFG-dependent reaction of bound ^{14}C aminoacyl- and ^{14}C-peptidyl-tRNA with puromycin. (TANAKA and KAJI, unpublished observations)

Complex	EFG-dependent puromycin reaction (cpm)			% inhibition by	
	control	fusidic acid	erythromycin	fusidic acid	erythromycin
A	1399	408	1404	71	—1
B	1166	0	292	100	75
C	1755	166	1620	90	8

Complexes A, B and C have N-acetyl ^{14}C-phenylalanyl tRNA, ^{14}C-diphenylalanyl tRNA and N-acetyl ^{14}C-diphenylalanyl tRNA on the acceptor site, respectively. The donor sites of all complexes are occupied with tRNA specific for phenylalanine. The values are expressed as the total amount of puromycin derivatives formed in the reaction mixture (0.4 ml). Where indicated, 5×10^{-4} M fusidic acid and 10^{-5} M erythromycin were added.

Table 7b. Effect of erythromycin A on the EFG-dependent release of $tRNA^{Phe}$ from ribosomes. (TANAKA and KAJI, unpublished)

Complex	Erythromycin	EFG dependent release of $tRNA^{Phe}$ (cpm)	% inhibition
A	none	1,100	59
	10^{-5} M	452	
B	none	2,168	53
	10^{-5} M	1,028	

The complexes A and B are as in Table 7a. The values represent the acceptor activity of $tRNA^{Phe}$ released, expressed as its acceptor capacity for ^{14}C-phenylalanine.

variable results with the puromycin reaction may then be due to the variable affinity of different peptidyl tRNA's for the peptidyl transferase.

The hypothesis proposed above is supported by two additional pieces of evidence. Since translocation is accompanied by *release* of tRNA from the donor site, this can be followed by the *decrease* of the ribosome-bound tRNA. This EFG-dependent decrease of bound tRNA was sensitive to EM. The second piece of evidence is that EM did *not* inhibit the EFG-dependent binding of aminoacyl tRNA. When a ribosomal complex with tRNA on the D site and aminoacyl tRNA on the A site was exposed to ^{14}C-aminoacyl tRNA, very little binding of ^{14}C-aminoacyl tRNA took place because the binding sites were all occupied. Upon translocation, however, the A site became empty and ^{14}C-aminoacyl tRNA bound to this evacuated A site.

According to the hypothesis, EM allows the movement of aminoacyl or peptidyl tRNA to a site near donor site, thus making the acceptor site empty. Therefore, no inhibition of EFG-dependent binding of aminoacyl tRNA by EM was expected.

Attempts to get information on the ribosomal binding site of EM by studying the effect of various antibiotics on binding of EM to ribosomes gave somewhat confusing results. The fact that EM inhibits the binding of CM [Wolfe and Hahn, 1965; Vazquez, 1966 (2)], and that LM inhibits the binding of EM (Fernandez-Munoz

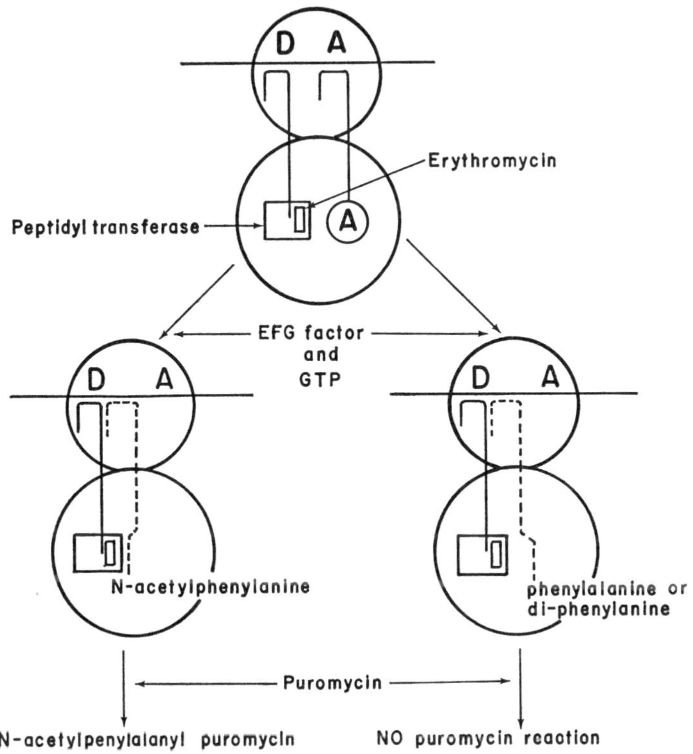

Fig. 2. Possible mechanism of EM action. (Tanaka and Kaji, unpublished)

et al., 1971) is consistent with the hypothesis that EM binds to the peptidyl transferase near or at the site for the acceptor molecule. Yet, with *B. subtilis* ribosomes it has been reported that CM did not inhibit the binding of EM (Oleinick et al., 1968). The reverse effect, that of EM inhibition on CM binding could be observed only when a high concentration (10^{-3} M) of CM was used. It was concluded from these results that the binding sites for EM and CM are clearly different; it is only at a high concentration that CM may weakly bind to the EM site. Recent reports on the binding of EM to *E. coli* ribosomes suggest that CM has only a partial effect on EM binding. Furthermore, puromycin, which presumably interacts at the site for the acceptor molecule did not appreciably inhibit the binding of EM (Fernandez-Munoz et al.,

1971). The binding of EM to the ribosome is dependent on NH_4^+ or K^+ and each ribosome binds one molecule of EM at 4×10^{-7} M EM (OLEINICK and CORCORAN, 1969). Of the two subunits, the action of EM has been localized to the 50 S subunit (WILHELM and CORCORAN, 1967).

A mutant strain of bacteria resistant to EM has been isolated and the protein responsible for this mutation was identified as one of the 50 S subunit proteins (OTAKA et al., 1970). A different type of resistant mutant has recently been isolated which had altered 23 S ribosomal RNA (LAI and WEISBLUM, 1971). In certain strains of *S. aureus*, EM induces resistance to certain inhibitors which act on the 50 S ribosomal subunits. The 50 S ribosomal subunits obtained from the induced and constitutively resistant cells have a decreased capacity to bind EM and lincomycin. For the induction of resistance by EM, synthesis of protein and RNA, but not of DNA was required. The 23 S ribosomal RNA from the EM-resistant cells contained N^6-dimethyladenine, which did not exist in the control 23 S RNA. Methylated adenine was also found in the 16 S ribosomal RNA from *S. aureus* 30 S ribosomes, but this was not affected by EM. These observations represent the first case of drug resistance caused by modification of RNA.

c) Thiostrepton and Related Compounds

In addition to their inhibitory action on the binding of aminoacyl tRNA to the acceptor site, this group of antibiotics also inhibits the translocation step. Polypeptide formation and EFG-dependent GTPase are inhibited by thiostrepton or siomycin. This is due to the inhibitory action of these antibiotics on the binding of EFG to ribosomes [BODLEY et al., 1970 (3)]. Thus, among the sequential events related to the GTPase action of EFG, thiostrepton inhibited formation of the GTP-ribosome-EFG-complex. Therefore, thiostrepton and fusidic acid appear to act in opposite ways. Fusidic acid stabilizes the complex GDP-ribosome-EFG, whereas thiostrepton (10^{-6} M) prevents the formation of this or the GTP-ribosome-EFG complex [WEISBLUM and DEMOHN, 1970; BODLEY et al., 1970 (3)]. As indicated in the previous section, siomycin and thiostrepton inhibit the binding of aminoacyl tRNA to the acceptor site of ribosomes as well as translocation. This dual action suggests that the binding site of the complex EFTu-GTP-aminoacyl tRNA may overlap with the ribosomal binding site for EFG. In an attempt to characterize these binding sites, GTP hydrolysis by EFG was carried out with ribosomal subunits instead of 70 S ribosomes. The results suggest that the primary interaction in GTP hydrolysis takes place between EFG and the 50 S subunit. The 50 S subunit-mediated GTPase was inhibited by siomycin (0.5 μM) (MODOLELL et al., 1971). These results show that the 50 S subunit contains the siomycin as well as the EFG binding site. Are they identical to the peptidyl transferase center (which is also on the 50 S subunit)? The answer to this question was given by the finding that siomycin does *not* inhibit the fragment reaction with puromycin. Additional evidence that the peptidyl transferase center is *not* involved in the binding of siomycin to ribosomes is the observation that the ribosomal binding of CM, LM and EM is not inhibited by siomycin. As described before, CM, LM and EM presumably bind to the peptidyl transferase at mutually exclusive sites or in allosterically linked positions.

If the binding site of the EFG and the site for the complex EFTu-GTP-aminoacyl tRNA are identical, this could be the site where GTP hydrolysis for the aminoacyl tRNA binding may be coupled to the translocation step. Careful analysis of the amount of GTP consumed in the overall polypeptide synthesis indicates that only one GTP is used per one polypeptide synthesis [NISHIZUKA and LIPMANN, 1966 (1); cited in LUCAS-LENARD and LIPMANN, 1971). On the other hand, studies on translocation and binding of aminoacyl tRNA clearly indicate that at least two GTP molecules are necessary per one peptide bond. One would therefore have to assume that GTP hydrolysis of these two steps must be coupled somehow. Although experimental data indicate that there is no such coupling of GTP hydrolysis [ONO et al., 1969 (2)], the proximity or identity of the EFG site and aminoacyl tRNA site of the ribosome strongly suggests that such coupling may indeed take place during polypeptide synthesis.

d) Bottromycin and Streptomycin

Bottromycin A_2 has been shown to act on the 50 S ribosomal subunit and specifically inhibit the step of translocation (LIN et al., 1968; KINOSHITA and TANAKA, 1970). Bottromycin A_2 does not inhibit the binding of lysyl tRNA nor does it influence the puromycin reaction of ribosome bound fMet-tRNA, while the EFG-dependent puromycin reaction of polylysyl tRNA was inhibited strongly by this antibiotic. However, simultaneous release of tRNA from ribosomes was not sensitive to bottromycin [TANAKA et al., 1971 (2)]. From these observations it was suggested that EFG's main function is to release the bound tRNA. According to this view, translocation of peptidyl tRNA and mRNA to the vacated donor site is catalyzed by the 50 S ribosomal subunit and not by EFG. It postulates further that bottromycin inhibits the translocation catalyzed by the sububit but not the tRNA released, which is catalyzed by EFG. Recent finding, however, indicates that this finding is in error and Bottromycin is actually an inhibitor to the peptide bond formation (OTAKA and KAJI, unpublished).

In addition to its action on the initiation step and the binding of aminoacyl RNA to the acceptor site, *streptomycin* also inhibits translocation. In the presence of 0.5 mM SM, about 90% of the EFG dependent puromycin reaction of diphenylalanyl tRNA was inhibited. The EFG-dependent release of tRNA from the complex of tRNA, ribosomes and polyuridylic acid was also sensitive to 0.5 mM streptomycin (IGARASHI et al., 1969). In addition, indirect, but suggestive evidence has been presented for SM's action at the translocation step. Treatment of sensitive *E. coli* strains with SM causes the accumulation of up to 50% of the ribosomes as monosomes. These streptomycin monosomes are attached to relatively long mRNA molecules but are inactive in protein synthesis. If these ribosomes are blocked *at the initiation step* after the binding of initiator tRNA, they should contain mostly fMet-tRNA, while other tRNA's would be bound if they are blocked near the initiation site. In order to distinguish between these two possibilities, the relative amount of fMet-tRNA on the monosome was determined and compared with the amount of ribosome-bound Ala- and Leu-tRNA (alanine is often found at the NH_2 terminal of *E. coli* proteins). Although SM-induced monosomes contained twice as much fMet-tRNA when compared with regular monosomes, one would expect at least a six-fold increase if SM had blocked at the step immediately after the binding of fMet-tRNA. From these considerations it was suggested that SM allows the ribosomes to move only a few codons on the mRNA after the initiation step, indicating that translocation after initiation is specifically inhibited (Lennette and Apirion, 1970). Other interpretations for this observation are possible. As described before, an attempt has been made to explain the multiple action of SM on various steps of protein synthesis. In short, SM distorts the acceptor site of ribosomes (Modolell and Davis, 1969). This distortion impairs effective binding of aminoacyl tRNA to the acceptor (decoding) site, resulting in misreading. It makes binding of peptidyl tRNA at the acceptor site unstable, resulting in the slow release of peptidyl tRNA, leading in turn to a slow breakdown of polysomes. The distortion at the A site may make it difficult for translocation to occur. This unitary hypothesis is also consistent with the conclusion that SM changes ribosomal configuration. When ^3H-equilibrated ribosomes were placed in a buffer in the presence of SM, the ^3H exchange-out rate was increased compared to the control without SM. This effect was observed at a concentration of SM which is enough to inhibit polypeptide synthesis (Sherman and Simpson, 1969).

e) Showdomycin and Cycloheximide

Showdomycin (Nishimura et al., 1964) inhibits protein synthesis in both gram-positive and gram-negative bacteria and in eukaryotic cells (Matsua et al., 1964). When the translocase (T2) from a mammalian system was preincubated with showdomycin, the enzyme was completely inactivated, whereas preincubation of ribosomes with this antibiotic resulted in less inhibition. It was concluded that showdomycin is a specific inhibitor of translocation.

Cycloheximide is one of the antibiotics which act exclusively on eukaryotic systems. A resistant mutant of yeast has been isolated whose resistance was due to changes in the 60 S ribosomal subunits. The heterozygous diploid gave a mixture of sensitive and resistant ribosomes, but the sensitive phenotype was dominant (Cooper et al., 1967; Rao and Grollman, 1967).

Both *in-vivo* and *in-vitro* studies support the notion that CH inhibits translocation. First of all, cycloheximide prevents GTP-dependent breakdown of polysomes (Wettstein et al., 1964). Similarly, CH inhibited protein synthesis in reticulocytes and caused a slight increase in the polyribosome content of the cells (Godchaux et al., 1967). The conversion of reticulocyte polysomes into monosomes in the presence of NaF was inhibited by 4×10^{-6} M CH. NaF was added in this experiment to prevent reinitiation of polypeptide chain synthesis by the run-off ribosomes. These experiments show that CH prevents movement of ribosomes (chain elongation) but do not prove that CH is a translocation inhibitor. They are, however, consistent with this notion which was later proven by *in-vitro* experiments (Baliga et al., 1970). Freshly harvested rat-liver polyribosomes were mixed with ^3H puromycin and the incorporation of ^3H radioactivity into polypeptide was measured. These polysomes had bound peptidyl tRNA, 80% of which was located at the acceptor site. Thus, CH inhibited approximately 80% of the puromycin reaction in the presence of T2 and GTP. If the polysomes were preincubated with GTP and T2 before the addition of CH, no inhibitory effect was observed. Similar results were obtained with fusidic acid, indicating that CH acts identically to a typical translocation inhibitor. Furthermore, the inhibitory efiect of CH can be overcome by the addition of excess T2 and GTP. The second piece of evidence is the demonstration that CH inhibits release of tRNA which accompanies translocation [Obrig et al., 1971 (1)]. The dual action of CH on translocation and the initiation step can be due to CH's inhibitory effct on the movement of tRNA on ribosomes in general.

f) Pederin and Diphtheria Toxin

Pederin is a poisonous substance produced by the insect *Paederus fuscipes* (PAVAN and BO, 1953; CARDANI et al., 1965). It inhibits the growth of cultured mammalian cells at concentrations of 1.5 nanogram/ml. In cell-free amino acid polymerizing system from human cell culture, 0.01 μg of peridin inhibited the incorporation of amino acid into polypeptide (PERANI et al., 1968). The release of nascent chains from polysomes by puromycin was inhibited by about 50% and this was taken as suggestive evidence for inhibition of the translocation step (JACOBS-LORENA et al., 1971). The rationale behind this is that the puromycin reaction of those peptidyl tRNA's on the donor site (representing 50% of total) is insensitive but the puromycin reaction of those on the donor site would be sensitive to prederin. More experiments would be necessary before we can definitely assign this compound as a translocase inhibitor. Pederin inhibits cytoplasmic protein synthesis very effectively, while mitochondrial protein synthesis is much less sensitive.

Diphtheria toxin is another inhibitor of translocation specific for eukaryotic cells. It inhibits protein synthesis by inactivating T2 (translocase). This inactivation is dependent on NAD. The following reaction is catalyzed by the toxin [HONJO et al., 1968; GILL et al., 1969 (1)].

$$\underset{\text{(active)}}{NAD^{+} + T2} \longrightarrow \underset{\text{(inactive)}}{ADPR - T2}$$

(ADPR: adenosine diphosphate ribose)

T2 is inactivated by toxin and NAD with respect to its GTPase activity as well as its translocase activity (RAEBURN et al., 1968). It has been proposed that a ternary intermediate composed of toxin, NAD, and T2 is formed. In this complex, toxin is bound to the adenosyl moiety of NAD, as well as to T2, which, in turn, is bound to the NMN moiety of NAD (GOOR and MAXWELL, 1970). The specificity of this reaction to T2 has been used to measure quantitatively the amount of T2 in tissue extracts. Radioactive NAD bound to T2 in the presence of toxin can be measured in tissue extracts without purification of T2. However, this method gives the amount of T2 in tissue not including that bound to ribosomes. When T2 is bound to ribosomes, toxin cannot inactivate T2 (RAEBURN et al., 1968). This is regarded as the reason why inactivation of T2 *in vivo* is prevented by cycloheximide. By virtue of its action to stop translocation, cycloheximide "freezes" polysomes, and most of the T2 stays bound to these polysomes. It would be of interest to see if T2 inactivation is prevented if polysomes are frozen at different steps such as peptide bond formation or binding of aminoacyl tRNA.

The process of inactivation described in the above equation takes place *in vivo* as well as *in vitro*. Thus, a high-speed supernatant fluid prepared from homogenates of HeLa cells previously exposed to toxin *in vivo* can support relatively little amino acid incorporation [GILL et al., 1969 (2)]. The activity of this supernatant fluid can be restored to the level of the control if the supernatant is first incubated with both nicotinamide and toxin to convert ADPR T2 back to T2. The reaction is thus reversible, toxin catalyzing the back reaction also. Since the toxin acts catalytically, only a few molecules of it are required to kill a single HeLa cell. It has been suggested that toxin adsorbs to a limited number of receptor sites at or near the surface of the sensitive cell and catalyzes the ADP ribosylation of T2 near the surface.

IV. Chain Termination

1. Release of Completed Chains and Inhibitors of This Step

After the repeated process of polypeptide chain elongation, ribosomes come to the end of the cistron where chain termination takes place. The termination signal is one of three triplets, UAA, UAG or UGA (Stretton and Brenner, 1965; Brenner and Beckwith, 1965; Nirenberg et al., 1965; Brenner et al., 1967). Convincing evidence that these are indeed termination signals has come from the elucidation of nucleotide sequences of the terminal region of the coat cistron of RNA phage. The RNA sequence corresponding to the carboxyl end of the coat cistron has been determined as follows: 5' AUC·UAC·UAA·UAG·CCG···AUG·UCG·3' (Nichols, 1970). As is clear from this sequence, there are two nonsense codons UAA, UAG, side by side at the end of the coat protein cistron. This is a built-in safety device to assure correct termination of the polypeptide chain. If most of the cistrons end with two nonsense codons, as in this case, one can explain why bacterial strains carrying a suppressor mutation of a nonsense mutant can grow almost as normally as wild type. Mutants carrying suppressor mutations have aminoacyl tRNA which reads *one* of these nonsense codons as sense. Thus, one might expect that these mutants may have difficulties in terminating polypeptide chain elongation at the end of the cistron. Since each suppressor mutant carries tRNA reading only one of these three nonsense codons, regular termination will not be hampered if there are two nonsense codons side by side, as in the case of RNA phage. However, a recent finding with Qβ phage suggest that not all cistrons end with two nonsense codons (Moore et al., 1971). The RNA of phage Qβ directs synthesis of coat proteins, assembly protein, RNA replicase and a fourth protein known as A_1. The first eight amino acids of the NH_2 terminal end of A_1 are the same as those of the coat protein. However, A_1 contains tryptophan, histidine and methionine, which are not in the Qβ coat protein. A_1 is never synthesized by mutant Qβ phage with an amber mutation in the coat-protein cistron. These data suggest that protein A_1 may consist of coat protein which has additional peptides attached at the COOH end. Thus, A_1 may be produced by the failure of ribosomes to read the termination signal properly. Since UAG and UAA are efficient termination signals while UGA is not, it has been postulated that UGA may be the termination codon of Qβ phage coat protein. There are ten triplet codons between the coat cistron and the RNA polymerase cistron of phage R17. The role of these nucleotides is not understood, but polynucleotides which are in between two cistrons may be read occasionally, as shown in this case, depending on the nature of the termination region. The exact mechanism through which these termination codons accomplish the release of completed chains remains obscure. It has been established, however, that instead of tRNA, proteins would respond to these termination codons (Capecchi, 1967; Scolnick et al., 1968).

For the assay of polypeptide chain release at the termination codon, three methods are available. The first method makes use of R17 phage which carries an amber mutation (UAG) in the coat cistron region. In the absence of suppressor tRNA, an oligopeptide corresponding to the first six amino acids from the NH_2 terminal end of the coat protein is synthesized because of the presence of UAG at the seventh triplet codon in this cistron. Thus, RNA from this mutant phage directs the synthesis of a fragment of coat protein (fMet·Ala·Ser·Asn·Phe·Thr). From studies on the requirement for the release of this hexapeptide from ribosomes, protein factors involved in the termination process were first isolated. The second assay method for termination is less natural but much easier than the first one. In this system, a complex of fMet-tRNA, ribosome and AUG triplet was prepared and release of formyl methionine from this complex was followed in the presence of the releasing factor and one of the termination triplets. One can visualize this system as peptide synthesis directed by the shortest messenger RNA in which the initiation codon is followed by the termination codon. In this system two protein factors, R_1 or R_2, are required for the release of formylmethionine, depending on the termination triplet used. R_1 and R_2 have been recently renamed by workers in this field as RF_1 and RF_2 (Releasing factor 1 and 2). RF_1 is required with UAA or UAG, and RF_2 is required with UAA or UGA. In addition to these, a third factor, S, (recently renamed as RF_3) was found to stimulate the release reaction [MILMAN et al., 1969; GOLDSTEIN et al., 1970 (2)]. The mode of action of RF_1 and RF_2 is not clear. They stimulate the binding of the termination triplet to the ribosome (SCOLNICK and CASKEY, 1969) but this may not be the actual function of these factors in the physiological termination process. When the ribosomes have completed a polypeptide chain, they should have the completed polypeptidyl tRNA at the acceptor site and unesterified tRNA on the donor site. By translocation the completed polypeptidyl tRNA moves to the donor site, thereby bringing the termination codon to the acceptor site. Thus, *before* RF_1 or RF_2 functions, the termination codon is already at the acceptor site. Therefore there is no need for these factors to "stimulate" the interaction between the termination codon and the ribosome. As for the function of RF_3, it has two opposite effects on the complex of ribosomes, release factor, and termination codon. In the absence of GTP or GDP, RF_3 stimulated formation of the RF^1·UAA·ribosome complex, while in the presence of GTP or GDP, RF_3 stimulated the dissociation of this complex (GOLDSTEIN and CASKEY, 1970). It is probable that this latter function of RF_3 is physiologically significant. Together with other factors which are described in Section (D-2), RF_3 probably functions to dissociate the ribosome from messenger RNA. At any rate, the ribosomal termination complex consisting of the termination codon, RF_1 or RF_2, RF_3 and peptidyl tRNA must be formed before the release of the completed peptidyl group. The third method of assay for peptidyl termination takes advantage of the fact that the presence of 20% alcohol eliminates the need for terminator codons (TOMPKINS et al., 1970). Thus, both RF- and RF_2 promote the release of formylmethionine from ribosomes without the terminator codon in a reaction mixture containing ethanol and either an f^3 H Met-tRNA·AUG·ribosome complex or ribosomes and f^3 H Met-tRNA. For this codon-independent release of formylmethionine, neither the termination codon, nor the initiation codon (AUG) is necessary. Ethanol stimulates the formation of a complex of release factor and ribosome. It also promotes the formation of the complex consisting of fMet-tRNA, ribosome and initiation factor. Both ribosomal subunits were necessary for this codon-independent release of formylmethionine. It should therefore be pointed out that the ribosomal requirements for codon-directed and codon-independent release of methionine are identical, and differ from those for the formation of fMet-puromycin or formylmethionyl ester, which can take place in the absence of 30 S subunits. This suggests that the release factor may be bound at the acceptor site, which encompasses both 30 S and 50 S ribosomal subunits.

Tetracycline, streptomycin, sparsomycin and chloramphenicol inhibited release of formylmethionine in the system described above. We can understand the reasons why these antibiotics are inhibitory to this system from their known mode of action. Thus, tetracycline probably binds to the acceptor site of the ribosome and prevents the binding of RF_1 and RF_2 which would be expected to bind to the acceptor site. Similarly, due to the possible distortion of the acceptor site by SM, RF_1 and RF_2 would have difficulties in binding to that site. In addition, SM and TC appear to inhibit the interaction of termination codons with ribosomes (TOMPKINS et al., 1970). However, as mentioned above, this effect may not be important in the physiological release of the completed chain because the termination codon is already on the ribosome at the time release takes place. Inhibition of release by CM and sparsomycin suggests that peptidyl transferase is involved in chain termination (VOGEL et al., 1969). In addition to these antibiotics, EM, amicetin and LM were also inhibitory. Except for EM, they are all peptidyl transferase inhibitors. Although EM inhibits translocation, it binds very close to or on the peptidyl transferase. The effect of these peptidyl transferase inhibitors on termination can be studied independently of the termination codon recognition in the presence of alcohol (see above). In an attempt to correlate the peptidyl transferase activity with the release of formylmethionine by RF_1 or RF_2, various concentrations of antibiotics were added to both systems. A close parallel between these two reactions was observed when sparsomycin, CM, LM, amicetin, hydroxypuromycin and acetyl puromycin were added. Although the effective concentrations of the different antibiotics were widely different, each antibiotic inhibited the two reactions over the same concentration range and to about the same extent for each concentration tested. Inhibition of the termination step by chloramphenicol, sparsomycin, gougerotin and tetracycline has also been observed in the termination step studied with the amber mutant of R17 phage (CAPECCHI and KLEIN, 1969).

The notion that peptidyl transferase is involved in some steps of the termination process was further strengthened by the use of ribosomes in which the peptidyl transferase was reversibly inactivated (VOGEL et al., 1969). Thus, 50 S ribosomes lost their ability to catalyze peptide bond synthesis when depleted of NH_4^+ and K^+ ions. Activity was restored if, in the presence of Mg^{++}, either NH_4^+ or K^+ was added, and ribosomes were incubated at 30 °C. The ribosomal activity to release formylmethionine in the model termination assay closely followed the peptidyl transferase activity.

Polypeptide chain termination in eukaryotes appears to be essentially the same as in the prokaryotic system. Genetic studies with yeast gave evidence that UAA and UAG are the termination codons (HAWTHORNE, 1969). A termination release factor has been isolated from rabbit reticulocytes [GOLDSTEIN et al., 1970 (2)]. In this system advantage was taken of the fact that mammalian Met-$tRNA^f$ (methionyl tRNA specific for initiation) binds to reticulocyte ribosomes without added oligo- or polyribonucleotide mRNA templates. To facilitate the binding of Met-$tRNA^f$ to the donor site, the Met-tRNA had been formylated by an *E. coli* formylase. The release of formylmethionine from this fMet-tRNA^f was used as an assay method for the release in the presence of various triplets. Polymers such as poly UA (1:3), and poly UG (1:2) stimulated release of formylmethionine from the complex, whereas a number of other randomly ordered polyribonucleotides had no detectable effect. Poly UG (1:2) was 30% as active as poly UA (1:3) in directing formylmethionine release from *E. coli* fMet-tRNA^f AUG·ribosome complexes with either *E. coli* RF_1 or RF_2. Since *E. coli* RF_1 and RF_2 recognize the terminator codons UAA, UGA and UAG, but not UGU, etc., poly UG's stimulatory effect was regarded as ambiguous codon recognition. Thus, it was concluded that UAA is the termination codon in this mammalian system. The reticulocyte releasing factor had no detectable activity on the *E. coli* ribosome-termination complex. The mammalian termination system was also sensitive to inhibitors of peptidyl transferase such as sparsomycin and gougerotin, indicating that this enzyme is involved in the eukaryote termination system as well as in the prokaryote system. There has been a suggestion that cyclic AMP may be involved in the mammalian termination system (KHAIRALLAH et al., 1967), but this has not been pursued.

2. Release of Ribosomes from Messenger RNA and Inhibitors of This Step

The fate of ribosomes after the release of the completed protein is a controversial issue. One school of thought believes that ribosomes are released as subunits which have a tendency to reassociate to form inactive 70 S ribosomes. The concept that ribosomes are dissociated as subunits, and 70 S ribosomes as such may not be present in cells had been presented (MANGIAROTTI and SCHLESSINGER, 1966; SCHLESSINGER et al., 1967; KAEMPFER et al., 1968; PHILLIPS et al., 1969). The initiation factor IF_3 keeps the released subunits apart so that they can be engaged in protein synthesis. In this view, IF_3 would play a role as a regulator of protein synthesis. When protein synthesis is actively taking place, the ribosomes will be released as subunits which immediately engage in a new round of protein synthesis. On the other hand, when the environmental situation does not call for rapid protein synthesis, the amount of subunits released exceeds the amount of IF_3 and these will accumulate as 70 S ribosomes (KAEMPFER, 1971). The 70 S ribosomes thus formed *slowly* dissociate into subunits which may again engage in protein synthesis.

The other school of thought suggests that ribosomes run off mRNA as 70 S ribosomes which are then rapidly dissociated into subunits by IF_3 (KOHLER et al., 1968; RON et al., 1968; ALGRANATI et al., 1969; KELLEY and SCHAECHTER, 1969). Data have been presented which show that the subunits of ribosomes observed after sucrose density gradient centrifugation are partially due to the hydrostatic pressure-induced dissociation of ribosomes which occurs during sedimentation under a high gravitational force (INFANTE and BAIERLEIN, 1971; SUBRAMANIAN and DAVIS, 1971). If the ribosomes are fixed with glutaraldehyde, pressure-induced dissociation is prevented. According to this view, ribosomal subunits which had been observed in the extract may have been an artifact of sedimentation analysis. Furthermore, 70 S ribosomes exchange their subunits very quickly under certain conditions. This is in sharp contrast to the concept of the other school of thought which maintains that 70 S ribosomes would undergo a slow dissociation.

Studies on the release of ribosomes from mRNA revealed that two heat-stable factors and GTP are required (HIRASHIMA and KAJI, 1972). In this experiment, polysomes were isolated from growing *E. coli* and separated from soluble factors. The nascent polypeptides of these polysomes were released by puromycin. Under the experimental conditions, most of the peptidyl tRNA is apparently located at the donor site, and therefore almost all the nascent peptides were released from the ribosome by puromycin treatment. The polysomes thus formed consist of ribosomes, mRNA and unesterified tRNA at the donor site. This complex is very similar to the ribosomes which have just released the completed chain at the termination signal.

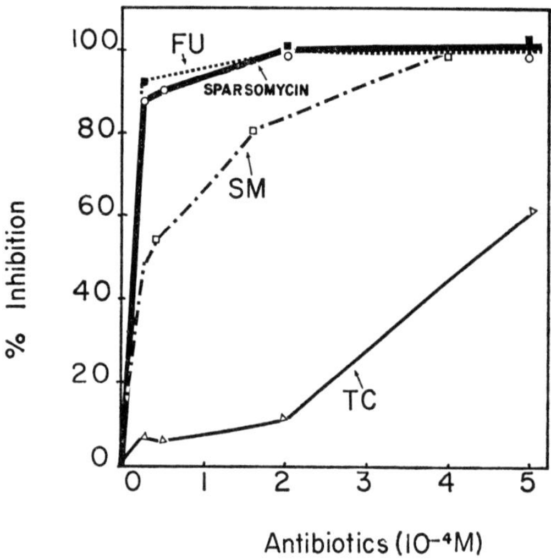

Fig. 3. Effect of antibiotics on the release of ribosomes from mRNA (HIRASHIMA and KAJI, 1972). The reaction mixture contained polysomes isolated from growing *E. coli*, EFG and a heat-stable protein factor (ribosome releasing factor), puromycin, GTP and various antibiotics. The mixture was incubated at 37 .C for 15 min and monosome formed was measured by sedimentation analysis

Release of ribosomes from these polysomes is regarded as a process similar to the release of ribosomes at the termination step. Examination of the factors involved in this system revealed that EFG and an additional heat-stable factor are required. All the ribosomes released were in the form of 70 S ribosomes and no subunits were observed. At first glance, this result appears to support the concept that ribosomes are released as 70 S ribosomes. However, it is still possible that they are released as subunits which rapidly associate into 70 S ribosomes. The 70 S ribosomes released from the messenger RNA in this experiment could undergo dissociation in the presence of IF_3. The role of EFG in the release of ribosomes from mRNA is somewhat mysterious because unesterified tRNA on the ribosome is already at the donor site, there being no need for translocation. As shown in Fig. 3, the process of run-off studied in this system was sensitive to fusidic acid, sparsomycin and streptomycin. One can readily understand the reasons for the inhibition by these antibiotics. Sparsomycin inhibits the puromycin reaction of nascent peptide chains on the ribosomes while fusidic acid and streptomycin inhibit the action of EFG. With the use of these antibiotics, it was possible to demonstrate that these two factors, EFG and the additional heat-stable factor (tentatively named as ribosome-releasing factor or RR factor) had to be present simultaneously. In the experiment illustrated in Table 8, the run-off of ribosomes was carried out in two steps. In the first step, polysomes were incubated with EFG and GTP but not with the RR factor. In the second step, polysomes were then incubated with RR in the presence of fusidic acid. As can be seen from this table, prior treatment of polysomes with EFG did not reverse the inhibitory effect of fusidic acid (expt. 1). On the other hand, as one would expect, the inhibitory effect of sparsomycin was reduced greatly if the polyomes were pre-treated with puromycin (expts. 2, 3). This indicates that the action of sparsomycin is to stop the reaction of ribosome-bound peptidyl tRNA with puromycin. Once

Table 8. Simultaneous requirement for EFG, GTP and RR for run-off ribosomes from mRNA. (Hirashima and Kaji, 1972)

Expt.	Additions to the first step	Additions to the second step	% inhibition by antibiotics (FA or SPM)
1	EFG, GTP, puromycin	RR, FA	100
2	EFG, GTP, SPM	RR	92
3	EFG, GTP, puromycin	RR, SPM	25
4	EFG, GTP, RR	Puromycin, FA	84
5	EFG, GTP, RR	Puromycin, SPM	99.4
6	EFG, GTP, RR	Puromycin, G, FA	96.8
7	EFG, RR, puromycin	GTP, FA	99.5

puromycin has performed its function, the addition of sparsomycin has very little effect. Similarly, pre-treatment of polysomes with RR, EFG and GTP in the absence of puromycin caused no release of ribosomes. The addition of fusidic acid in the second step again strongly inhibited the release or ribosomes from mRNA (expt. 4). This indicates that RR and EFG have to be present simultaneously for the release of ribosomes from messenger RNA. The 70 S ribosomes produced in this reaction had no mRNA or tRNA on them indicating that RR and EFG remove tRNA as well as mRNA from ribosomes. This is an essential step for the subsequent dissociation of the 70 S ribosomes into subunits because the presence of tRNA prevents 70 S ribosomes from dissociating (Schlessinger et al., 1967). In fact, the 70 S ribosomes produced by RR and EFG were readily dissociated by IF_3 into their subunits, while 70 S ribosomes with tRNA on them under similar conditions were not (Hirashima and Kaji, 1972; Ishitsuka and Kaji, 1972).

V. Epilogue

In this article an attempt has been made to give an overall picture of the protein-synthesizing mechanism in relation to the action of various inhibitors. The overall scheme of protein synthesis is summarized in Fig. 4. The points of action of various antibiotics are also included. As can be seen from the diagram, most of our present understanding of the mechanism of protein synthesis was derived from the discovery of the partial reactions involved in protein synthesis, such as the initiation process, binding of tRNA to ribosomes, peptide bond formation, translocation, release of completed chains, and release of ribosomes at the end of a cistron. In the past, elucidation of the mode of action of antibiotics had to wait for the discoveries of

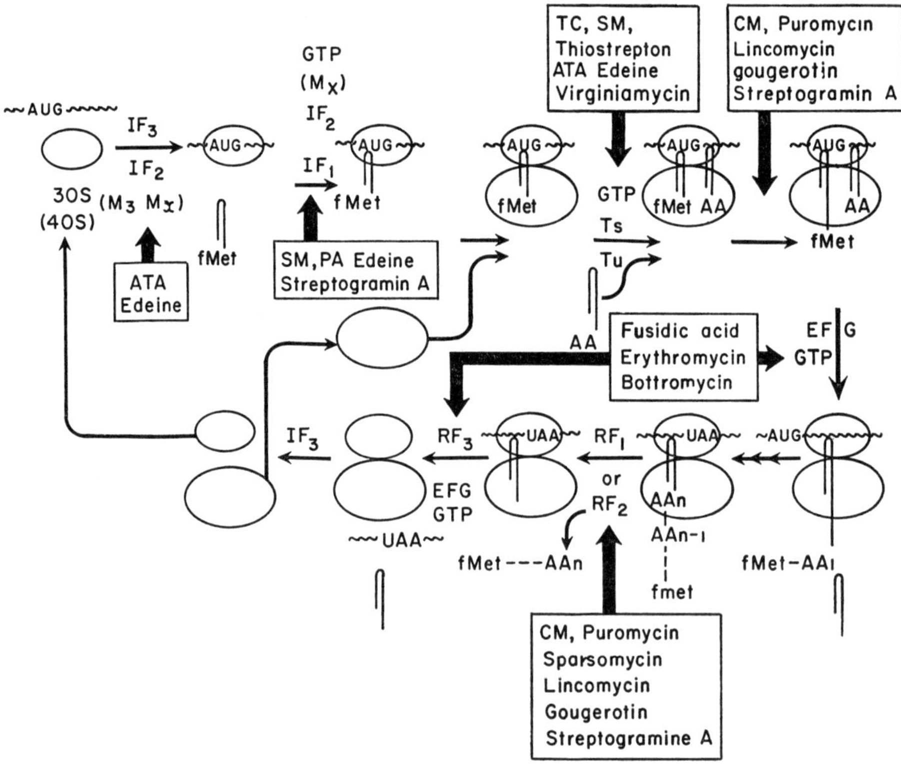

Fig. 4. Overall scheme for protein biosynthesis and mode of action of various inhibitors

these partial reactions. Now that the major steps in protein synthesis are understood. and numerous inhibitors have been described, the time has come when the use of antibiotics is almost a necessity in elucidating the finer details of the mechanism of protein synthesis. For example, the suggestion that the translocation step may consist of two discrete steps only came after this reaction was studied with antibiotics, such as erythromycin. I believe that careful, discrete application of inhibitors, together with analysis of various factors involved in protein synthesis, will eventually lead us to a thorough understanding of this most complicated system for macromolecular synthesis.

References

ALGRANATI, I. D., GONZALEZ, N. S., BADE, E. G.: Physiological role of 70 S ribosomes in bacteria. Proc. nat. Acad. Sci. (Wash.) **62**, 574 (1969).

ANAND, N., DAVIS, B. D.: Effect of streptomycin on Escherichia coli. Nature (Lond.) **185**, 22 (1960).

ANDERSON, B., HODGKIN, D. C., VISWAMITRA, M. A.: The structure of thiostrepton. Nature (Lond.) **225**, 233 (1970).

ANDERSON, J. S., BRETSCHER, M. S., CLARK, B. F. S., MARCKER, K. A.: A GTP requirement for binding initiator tRNA to ribosomes. Nature (Lond.) **215**, 490 (1967).

ALLEN, D. W., ZAMECNIK, P. C.: The effect of puromycin on rabbit reticulocyte ribosomes. Biochim. biophys. Acta (Amst.) **55**, 865 (1962).

APIRION, D.: Three genes that affect *Escherichia coli* ribosomes. J. molec. Biol. **30**, 255 (1967).

APIRION, D., SCHLESSINGER, D., PHILLIPS, S., SYPHERD, P.: *Escherichia coli:* reversion from streptomycin dependence, a mutation in a specific 30 S ribosomal protein. J. molec. Biol. **43**, 327 (1969).

ARLINGHAUS, R. J., SHAEFFER, J., SCHWEET, R.: Mechanism of peptide bond formation in polypeptide synthesis. Proc. nat. Acad. Sci. (Wash.) **51**, 1291 (1964).

BALIGA, B. S., COHEN, S. A., MUNRO, H. N.: Effect of cycloheximide on the reaction of puromacin with polysome-bound peptidyl tRNA. FEBS Letters **8**, 249 (1970).

BERISSI, H., GRONER, Y., REVEL, M.: Effect of a purified initiation factor F3 (B) on the selection of ribosomal binding sites on phage MS2 RNA. Nature (Lond.) New Biol. **234**, 44 (1971).

BERMAN, M. L., MONIER, R.: Influence of the 30 S ribosomal subunit on the peptidyl transferase activity of the 50 S ribosomal subunit from *E. coli.* Biochimie **53**, 233 (1971).

BHUYAN, B. K.: Pactamycin, an antibiotic that inhibits protein synthesis. Biochem. Pharmacol. **16**, 1411 (1967).

BIRGE, E. A., KURLAND, C. G.: Altered ribosomal protein in streptomycin dependent Escherichia coli. Science **166**, 1282 (1969).

BISSEL, D. M.: Formation of an altered enzyme by Escherichia coli in the presence of neomycin. J. molec. Biol. **14**, 619 (1965).

BODLEY, J. W., DAVIE, E. W.: A study of the mechanism of ambiguous amino acid coding by poly U: the nature of the products. J. molec. Biol. **18**, 344 (1966).

BODLEY, J. W., LIN, L.: Interaction of *E. coli* and G-factor with the 50 S ribosomal subunit. Nature (Lond.) **227**, 60 (1970).

BODLEY, J. W., ZIEVE, F. J., LIN, L., ZIEVE, S. T.: Formation of the ribosome-G-factor-GDP complex in the presence of fusidic acid. Biochem. biophys. Res. Commun. **37**, 437 (1969).

BODLEY, J. W., ZIEVE, F. J., LIN, L., ZIEVE, S. T.: (1) Studies on translocation III. Conditions necessary for the formation and detection of a stable ribosome-G-factor-GDP complex in the presence of fusidic acid. J. biol. Chem. **245**, 5656 (1970).

BODLEY, J. W., ZIEVE, F. J., LIN, L.: (2) Studies on translocation IV. The hydrolysis of a single round of GTP in the presence of fusidic acid. J. biol. Chem. **245**, 5662 (1970).

BODLEY, J. W., LIN, L., HIGHLAND, J. H.: (3) Studies on translocation VI. Thiostrepton prevents the formation of a ribosome-G-factor-guanosine triphosphate complex. Biochem. biophys. Res. Commun. **41**, 1406 (1970).

BOEDTKER, H.: The reaction of ribonucleic acid with formaldehyde. I. Optical absorbance studies. Biochemistry **6**, 2718 (1967).

BRECKENRIDGE, L., GORINI, L.: The dominance of streptomycin sensitivity re-examined. Proc. nat. Acad. Sci. (Wash.) **62**, 979 (1969).

BRENNER, S., BECKWITH, J. R.: Ochre mutants, a new class of suppressible nonsense mutants. J. molec. Biol. **13**, 629 (1965).

BRENNER, S., BARNETT, L., KATZ, E. R., CRICK, F. H. C.: UGA: a third nonsense triplet in the genetic code. Nature (Lond.) **213**, 449 (1967).

BRETSCHER, M. S.: Translocation in protein synthesis: a hybrid structure model. Nature (Lond.) **218**, 675 (1968).

BRETSCHER, M. S., MARCKER, K. A.: Polypeptidyl-s RNA and aminoacyl-s RNA binding sites on ribosomes. Nature (Lond.) **211**, 380 (1966).

Brot, N., Redfield, B., Weissbach, H.: Studies on the reaction of the aminoacyl-tRNA-Tu-GTP complex with ribosomal subunits. Biochem. biophys. Res. Commun. **41**, 1388 (1970).

Brot, N., Spears, C. L., Weissbach, H.: The formation of a complex containing ribosomes, transfer factor G and a guanosine nucleotide. Biochem. biophys. Res. Commun. **34**, 843 (1969).

Brot, N., Spears, C. L., Weissbach, H.: The interaction of transfer factor G, ribosomes and guanosine nucleotides in the presence of fusidic acid. Arch. Biochem. Biophys. **143**, 286 (1971).

Brown, J. C., Smith, A. E.: Initiator codons in eukaryotes. Nature (Lond.) **226**, 610 (1970).

Brownstein, B. L., Lewandowski, L. J.: A mutation supressing streptomycin dependence. I. An effect on ribosome formation. J. molec. Biol. **25**, 99 (1967).

Busiello, E., Digirolamo, M., Felicetti, L.: Role of mammalian ribosomal subunits and elongation factors in poly U-directed protein synthesis. Biochim. biophys. Acta (Amst.) **228**, 289 (1971).

Cameron, H. J., Julian, G. R.: The effect of chloramphenicol on the polysome formation of starved stringent Escherichia coli. Biochim. biophys. Acta. (Amst.) **169**, 373 (1968).

Capecchi, M. R.: Polypeptidyl chain termination *in vitro*. Isolation of a release factor. Proc. nat. Acad. Sci. (Wash.) **58**, 1144 (1967).

Capecchi, M. R., Klein, H. A.: Characterization of three proteins involved in polypeptide chain termination. Cold Spr. Harb. Symp. quant. Biol. **34**, 469 (1969).

Cardani, C. D., Ghiringhelli, R., Mondelli, R., Pavan, M., Quilico, A.: Properties biologiques et composition chimique de la pederine. Ann. Soc. Entomol. Franc. **1**, 813 (1965).

Casjens, S. R., Morris, A. J.: The selective inhibition of protein assembly by gougerotin. Biochim. biophys. Acta (Amst.) **108**, 677 (1965).

Cerna, J., Jonak, J., Rychlik, I.: (2) Effects of macrolide antibiotics on the ribosomal peptidyl transferase in cell-free systems derived from *E. coli B* and erythromycin-resistant mutant of E. coli B. Biochim. biophys. Acta (Amst.) **240**, 109 (1971).

Cerna, J., Lichtenhaler, F. W., Rychlik, I.: (1) The effect of gougerotin analogues on ribosomal peptidyl transferase. FEBS Letters **14**, 45 (1971).

Cerna, J., Rychlik, I., Pulkrabek, P.: The effect of antibiotics on the coded binding of peptidyl tRNA to the ribosome and on the transfer o[5] the peptidyl residue to puromycin. Europ. J. Biochem. **9**, 27 (1969).

Cerna, J., Rychlik, I., Zemlicka, J., Chladek, S.: Substrate specificity of ribosomal peptidyl transferase II 2'(3')-O-aminoacyl nucleosides as acceptors of the peptide chain in the fragment reaction. Biochim. biophys. Acta (Amst.) **204**, 203 (1970).

Chae, Y-B., Mazumder, R., Ochoa, S.: Polypeptide chain initiation in *E. coli*. Studies on the function of initiation factor F_1. Proc. nat. Acad. Sci. (Wash.) **63**, 828 (1969).

Chang, F. N., Flaks, J. G.: Topography of the *E. coli* 30 S ribosomal subunit and streptomycin binding. Proc. nat. Acad. Sci. (Wash.) **67**, 1321 (1970).

Chang, F. N., Weisblum, B.: The specific of lincomycin binding to ribosomes. Biochemistry **6**, 836 (1967).

Chen, Chong-Maw, Ofengand, J.: Inactivation of the Tu-GTP recognition site in aminoacyl-tRNA by chemical modification of the tRNA. Biochem. biophys. Res. Commun. **41**, 190 (1970).

Chuang, D. M., Simpson, M. V.: A translocation-associated ribosomal conformational change detected by hydrogen exchange and sedimentation velocity. Proc. nat. Acad. Sci. (Wash.) **68**, 1474 (1971).

Clark, B. F. C., Marcker, K. A.: The role of N-formylmethionyl sRNA in protein biosynthesis. J. molec. Biol. **17**, 394 (1966).

Clark, J. M., Chang, A. Y.: Inhibitors of the transfer of amino acids from aminoacyl soluble ribonucleic acid to proteins. J. biol. Chem. **240**, 4734 (1965).

Cocito, C.: Metabolism of macromolecules in bacteria treated with virginiamycin. J. gen. Microbiol. **57**, 179 (1969).

Cocito, C., Kaji, A.: Virginiamycin M: a specific inhibitor of the acceptor site of ribosomes. Biochimie **53**, 763 (1971).

Cohen, B. B.: Two fractions required for cell-free protein synthesis with components from rabbit reticulocytes. Biochem. J. **115**, 523 (1969).

Cohen, L. B., Goldberg, I. H.: Inhibition of peptidyl-sRNA binding to ribosomes by pactamycin. Biochem. biophys. Res. Commun. **29**, 617 (1967).

Cohen, L. B., Goldberg, I., Herner, A.: Inhibition by pactamycin of the initiation of protein synthesis. Effect on the 30 S ribosomal subunit. Biochemistry **8**, 1327 (1969).

Colombo, B., Felicetti, L., Baglioni, C.: Inhibition of protein synthesis in reticulocytes by antibiotics — Effects on polysomes. Biochim. biophys. Acta (Amst.) **119**, 109 (1966).

Connamacher, R. H., Mandel, H. G.: Studies on the intracellular localization of tetracycline in bacteria. Biochim. biophys. Acta (Amst.) **166**, 475 (1968).

Cooper, D., Banthorpe, D., Wilkie, D.: Modified ribosomes conferring resistance to cycloheximide in mutants of *saccharomyces cerevisiae*. J. molec. Biol. **26**, 347 (1967).

Cooper, D., Gordon, J.: Effect of aminoacyl transfer RNA on competition between guanosine 5'-triphosphate and guanosine 5'-diphosphate for binding to a polypeptide chain elongation factor from Escherichia coli. Biochemistry **8**, 4289 (1969).

Culp, W. J., McKeehan, W. L., Hardesty, B.: Deacylated $tRNA^{Phe}$ binding to a reticulocyte ribosomal site for the initiation of polyphenylalanine synthesis. Proc. nat. Acad. Sci. (Wash.) **63**, 1431 (1969).

Culp, W. J., Morrisey, J., Hardesty, B.: Initiator tRNA for the synthesis of globin peptides. Biochem. biophys. Res. Commun. **40**, 777 (1970).

Cundliffe, E.: Antibiotics and polyribosomes. II. Some effects of lincomycin, spiramycin and streptomycin A in vivo. Biochemistry **8**, 2063 (1969).

Cundliffe, E.: The mode of action of thiostrepton *in vivo*. Biochem. biophys. Res. Commun. **44**, 912 (1971).

Cundliffe, E.: Polyribosomes and ribosomal subunits of bacterial protoplasts. Biochem. biophys. Res. Commun. **33**, 247 (1968).

Cundliffe, E., McQuillen, K.: Bacterial protein synthesis: the effects of antibiotics. J. molec. Biol. **30**, 137 (1967).

Das, H. K., Goldstein, A., Kanner, L. C.: Inhibition by chloramphenicol of the growth of nascent protein chains in *Escherichia coli*. Molec. Pharmacol. **2**, 158 (1966).

Davies, J., Gilbert, W., Gorini, L.: Streptomycin, suppression and the code. Proc. nat. Acad. Sci. (Wash.) **51**, 883 (1964).

Davies, J., Jones, D. S., Khorana, H. G.: A further study of misreading of codons induced by streptomycin and neomycin using ribonucleotides in alternating sequence as templates. J. molec. Biol. **18**, 48 (1966).

Davies, B. D.: Action of aminoglycoside antibiotics. Asian med. J. **11**, 78 (1968).

Day, L. E.: Tetracycline inhibition of cell-free protein synthesis I. Binding of tetracycline to components of the system. J. Bact. **91**, 1917 (1966).

Dintzis, H. M., L.: Assembly of the peptide chains of hemoglobin. Proc. nat. Acad. Sci. (Wash.) **47**, 247 (1961).

Dresden, M. H., Hoagland, M. B.: Polyribosomes of *Escherichia coli*. Breakdown during glucose starvation. J. biol. Chem. **242**, 1065 (1967).

Dube, S. K., Rudland, P. S.: Control of translation by T4 phage: Altered binding of disfavoured messenger. Nature (Lond.) **226**, 820 (1970).

Dubnoff, J. S., Maitra, U.: Protein factors involved in polypeptide chain initiation in *E. coli*. Cold Spr. Harb. Symp. quant. Biol. **34**, 301 (1969).

Ellis, R. J., Hartley, M. R.: Sites of synthesis of chloroplast proteins. Nature (Lond.) New Biol. **233**, 193 (1971).

Ennis, H. L.: Synergistin: a synergistic antibiotic complex which selectively inhibits protein synthesis. Proc. Int. Cong. Chemotherapy, 6th, Tokyo **2**, 489 (1970).

Ennis, H. L.: Interaction of vernamycin A with *Escherichia coli* ribosomes. Biochemistry **10**, 1265 (1971).

Erbe, R. W., Nau, M. M., Leder, P.: Translation and translocation of defined RNA messengers. J. molec. Biol. **39**, 441 (1969).

Fahnstock, S., Neumann, H., Shashoua, V., Rich, A.: Ribosome-catalyzed ester formation. Biochemistry **9**, 2477 (1970).

Fernandez-Munoz, R., Monro, R. E., Torres Pinedo, R., Vazquez, D.: Substrates and antibiotic binding sites at the peptidyl transferase centre of Escherichia coli ribosomes. Studies on the chloramphenicol, lincomycin and erythromycin sites. Europ. J. Biochem. **23**, 185 (1971).

Flessel, C. P.: Chloramphenicol protects polyribosomes. Biochem. biophys. Res. Commun. **32**, 438 (1968).

Galasinski, W., Moldave, K.: Purification of aminoacyl transferase II (translocation factor) from rat liver. J. biol. Chem. **244**, 6527 (1969).

Ghosh, H. P., Khorana, H. G.: Studies on polynucleotides LXXXIV. On the role of ribosomal subunits in protein synthesis. Proc. nat. Acad. Sci. (Wash.) **58**, 2455 (1967).

Gilbert, W.: Polypeptide synthesis in *Escherichia coli*. II. The polypeptide chain and sRNA. J. molec. Biol. **6**, 389 (1963).

Gill, D. M., Pappenheimer, A. M., Jr., Baseman, J. B.: (2) Studies on transferase II using diphtheria toxin. Cold Spr. Harb. Symp. quant. Biol. **34**, 595 (1969).

Gill, D. M., Pappenheimer, A. M., Jr., Brown, R., Kurnick, J. T.: (1) Studies on the mode of action of diphtheria toxin. VII. Toxin-stimulated hydrolysis of nicotinomide adenine dinucleotide in mammalian cell extracts. J. exp. Med. **129**, 1 (1969).

Godchaux, W. III., Adamson, S. D., Herbert, E.: Effect of cycloheximide on polyribosome function in reticulocytes. J. molec. Biol. **27**, 57 (1967).

Goldberg, I. H., Mitsugi, K.: (1) Inhibition by sparsomycin and other antibiotics of the puromycin induced release of polypeptide from ribosomes. Biochemistry **6**, 383 (1967).

Goldberg, I. H., Mitsugi, K.: (2) Sparsomycin inhibition of polypeptide synthesis promoted by synthetic and natural polynucleotides. Biochemistry **6**, 372 (1967).

Goldstein, J. L., Beaudet, A. L., Caskey, C. T.: Peptide chain termination with mammalian release factor. Proc. nat. Acad. Sci. (Wash.) **67**, 99 (1970).

Goldstein, J. L., Caskey, C. T.: Peptide chain termination: effect of proteins on ribosomal binding of release factors. Proc. nat. Acad. Sci. (Wash.) **67**, 537 (1970).

Goldstein, J. L., Milman, G., Scolnick, E., Caskey, T.: Peptide chain termination, VI. Purification and site of action of *s*. Proc. nat. Acad. Sci. (Wash.) **65**, 430 (1970).

Goor, R. S., Maxwell, E. S.: The diphtheria toxin-dependent adenosine diphosphate ribosylation of rat liver aminoacyl transferase II. J. biol. Chem. **245**, 616 (1970).

Gordon, J.: Hydrolysis of guanosine 5'-triphosphate associated with binding of aminoacyl transfer ribonucleic acid to ribosomes. J. biol. Chem. **244**, 5680 (1969).

Gordon, J.: Regulation of the *in vivo* synthesis of the polypeptide chain. Biochemistry **9**, 912 (1970).

Gorini, L.: In formational suppression. Ann. Rev. Genet. **4**, 107 (1970).

Gorini, L., Kataja, E.: Phenotypic repair by streptomycin of defective genotypes in *E. coli*. Proc. nat. Acad. Sci. (Wash.) **51**, 487 (1964).

Gottesman, M. E.: Reactions of ribosome-bound peptidyl transfer ribonucleic acid with aminoacyl transfer ribonucleic acid or puromycin. J. biol. Chem. **242**, 5564 (1967).

Greenshpan, H., Revel, M.: Initiator protein dependent binding of messenger RNA to ribosomes. Nature (Lond.) **224**, 331 (1969).

Grollman, A. P., Stewart, M. L.: Inhibition of the attachment of messenger ribonucleic acid to ribosomes. Proc. nat. Acad. Sci. (Wash.) **61**, 719 (1968).

Groner, Y., Herzberg, M., Revel, M.: Translation initiation factor C(f2): selective inactivation of its f-Met-tRNA binding activity which does not affect messenger RNA binding to the 30 S ribosome. FEBS Letters **6**, 315 (1970).

Grunberg-Manago, M., Robinowitz, J. C., Dondon, J., Lelong, J. C., Gros, F.: Different classes of initiation factor F3 and their dissociation activity. FEBS Letters **19**, 193 (1971).

Gualerzi, C., Pon, C. L., Kaji, A.: Initiation factor dependent release of aminoacyl-tRNA's from complexes of 30 S ribosomal subunits, synthetic polynucleotide, and aminoacyl tRNA. Biochem. biophys. Res. Commun. **45**, 1312 (1971).

Gurgo, C., Apirion, D., Schlessinger, D.: Effects of chloramphenicol and fusidic acid on polyribosome metabolism in *Escherichia coli*. FEBS Letters **3**, 34 (1969).

Gurgo, C., Apirion, D., Schlessinger, D.: (1) Polyribosome metabolism in *Escherichia coli* treated with chloramphenicol, neomycin, spectinomycin or tetracycline. J. molec. Biol. **45**, 205 (1969).

GUTHRIE, C., NOMURA, M.: Initiation of protein synthesis: a critical test of the 30 S subunit model. Nature (Lond.) **219**, 232 (1968).
HACHMANN, J., MILLER, D. L., WEISSBACH, H.: Purification of factor Ts: studies on the formation and stability of nucleotide complexes containing transfer factor Tu. Arch. Biochem. Biophys. **147**, 457 (1971).
HAWTHORNE, D. C.: Identification of nonsense codons in yeast. J. molec. Biol. **43**, 71 (1969).
HEINTZ, R., MCALLISTER, H., ARLINGHAUS, R., SCHWEET, R.: Formation and function of the active ribosome complex. Cold Spr. Harb. Symp. quant. Biol. **31**, 633 (1966).
HELSER, T. L., DAVIES, J. E., DAHLBERG, J. E.: Mechanism of kasugamycin resistance in *E. coli*. Nature (Lond.) New Biol. **235**, 6 (1972).
HERSHEY, J. W. B., DEWEY, K. F., THACH, R. E.: Purification and properties of initiation factor f-1. Nature (Lond.) **222**, 944 (1969).
HERZBERG, M., LELONG, J. C., REVEL, M.: Purification of initiator C from *Escherichia coli:* a protein which binds messenger RNA and initiator tRNA to the 30 S ribosome. J. molec. Biol. **44**, 297 (1969).
HETTINGER, T. P., CRAIG, L. C.: Edeine. IV. Structures of the antibiotic peptides edeines A_1 and B_1. Biochemistry **9**, 1224 (1970).
HIEROWSKI, M., KURYLO-BOROWSKA, Z.: On the mode of action of edeine. I. Effect of edeine on the synthesis of polyphenylalanine in a cell-free system. Biochim. biophys. Acta (Amst.) **95**, 578 (1965).
HIRASHIMA, A., KAJI, A.: Release of ribosomes from messenger RNA—a process dependent on G-factor and an additional factor. J. molec. Biol. (1972) (in press).
HOEKSEMA, H.: Chemical studies on lincomycin I. The structure of lincomycin. J. Amer. chem. Soc. **86**, 4223 (1964).
HOERZ, W., MCCARTY, K. S.: Evidence for a proposed initiation complex for protein synthesis in reticulocyte polyribosome profiles. Proc. nat. Acad. Sci. (Wash.) 63 1206 (1969).
HONJO, T., NISHIZUKA, Y., HAYAISHI, O., KATO, I.: Diphtheria toxin-dependent adenosine diphosphate ribosylation of aminoacyl transferase II and inhibition of protein synthesis. J. biol. Chem. **243**, 3553 (1968).
HOSOKAWA, K., FUJIMURA, R. K., NOMURA, M.: Reconstitution of functionally active ribosomes from inactive subparticles and proteins. Proc. nat. Acad. Sci. (Wash.) **55**, 198 (1966).
HSU, W-T., WEISS, S. B.: Selective translation of T4 template RNA by ribosomes from T4-infected *Escherichia coli*. Proc. nat. Acad. Sci. (Wash.) **64**, 345 (1969).
HURWITZ, C., BRAUN, C. B.: Measurement of binding of chloramphenicol by intact cells. J. Bact. **93**, 1671 (1967).
IBUKI, F., MOLDAVE, K.: The effect of guanosine triphosphate, other nucleotides and aminoacyl transfer RNA on the activity of transferase I and on its binding to ribosomes. J. biol. Chem. **243**, 44 (1968).
IGARASHI, K., ISHITSUKA, H., KAJI, A.: Comparative studies on the mechanism of action of lincomycin, streptomycin and erythromycin. Biochem. biophys. Res. Commun. **37**, 499 (1969).
IGARASHI, K., KAJI, A.: On the nature of two ribosomal sites for specific sRNA binding. Proc. nat. Acad. Sci. (Wash.) **58**, 1971 (1967).
IGARASHI, K., KAJI, A.: Evidence for one functional phenylalanyl tRNA binding site on the 30 s ribosomal subunit. Proc. nat. Acad. Sci. (Wash.) **62**, 498 (1969).
IGARASHI, K., KAJI, A.: Relationship between sites 1,2 and acceptor, donor sites for the binding of aminoacyl tRNA to ribosomes. Europ. J. Biochem. **14**, 41 (1970).
IGARASHI, K., TANAKA, S., KAJI, A.: On the aminoacyl-tRNA binding site of the 30 s ribosomal subunit and its relation to the chain initiation site of the ribosome. Biochim. biophys. Acta (Amst.) **228**, 728 (1971).
INFANTE, A., BAIERLEIN, R.: Pressure-induced dissociation of sedimenting ribosomes: effect on sedimentation patterns. Proc. nat. Acad. Sci. (Wash.) **68**, 1780 (1971).
IRVIN, J. D., JULIAN, G. R.: The distribution of ^{14}C-proline peptides synthesized *in vitro* directed by polycytidylic acid; the effect of chloramphenicol. FEBS Letters **8**, 129 (1970).
ISHITSUKA, H., KAJI, A.: Prevention of ribosomal donor site from being occupied by aminoacyl tRNA during polypeptide synthesis. FEBS Letters (1972) (in press).

Ishitsuka, H., Kaji, A.: Release of tRNA by breakdown of messenger RNA. Biochim. biophys. Acta (Amst.) (1972) (in press).

Ishitsuka, H., Kuriki, Y., Kaji, A.: Release of transfer ribonucleic acid from ribosomes. A G-factor and guanosine triphosphate-dependent reaction. J. biol. Chem. **245**, 3346 (1970).

Iwasaki, K., Mizumoto, K., Tanaka, M., Kaziro, Y.: Studies on the *in vitro* system for protein synthesis from mammalian organs. Seikagaku **43**, 722 (1971).

Iwasaki, K., Sabol, S. L., Wahba, A. J., Ochoa, S.: Translation of the genetic message VII. Role of initiation factors in formation of the chain initiation complex with *Escherichia coli* ribosomes. Arch. Biochem. Biophys. **125**, 542 (1968).

Jacobs-Lorena, M., Brega, A., Baglioni, C.: Inhibition of protein synthesis in reticulocytes by antibiotics. V. Mechanism of action of pederine, an inhibitor of initiation and elongation. Biochem. biophys. Acta (Amst.) **240**, 263 (1971).

Jayaraman, J., Goldberg, I. H.: Localization of sparsomycin action to the peptide-bond forming step. Biochemistry **7**, 418 (1968).

Jerez, C., Sandoval, A., Allende, J. E., Henes, C., Ofengand, J.: Specificity of the interaction of aminoacyl RNA with a protein-guanosine triphosphate complex from wheat embryo. Biochemistry **8**, 3006 (1969).

Julian, G. R.: ^{14}C lysine peptides synthesized in an *in vitro Escherichia coli* system in the presence of chloramphenicol. J. molec. Biol. **12**, 9 (1965).

Kaempfer, R.: Control of single ribosome formation by an initiation factor for protein synthesis. Proc. nat. Acad. Sci. (Wash.) **68**, 2458 (1971).

Kaempfer, R., Messelson, M., Raskas, H. J.: Cyclic dissociation into stable subunits and reformation of ribosomes during bacterial growth. J. molec. Biol. **31**, 277 (1968).

Kaji, A.: Partial proteolytic digestion of yeast hexokinase and its relation to multiple forms of the enzyme. Arch. Biochem. Biophys. **112**, 54 (1965).

Kaji, A., Kaji, H.: Specific interaction of soluble RNA with polyribonucleic acid induced polysomes. Biochem. biophys. Res. Commun. **13**, 186 (1963).

Kaji, H.: Intraribosomal environment of the nascent peptide chain. Int. Rev. Cytol. **29**, 169 (1970).

Kaji, H.: Genetic code and streptomycin: their relation to ribosomal subunits. Biochim. biophys. Acta (Amst.) **134**, 134 (1967).

Kaji, H., Kaji, A.: Specific binding of sRNA with the template-ribosome complex. Proc. nat. Acad. Sci. (Wash.) **52**, 1541 (1964).

Kaji, H., Kaji, A.: Specific binding of sRNA to ribosomes: effect of streptomycin. Proc. nat. Acad. Sci. (Wash.) **54**, 213 (1965).

Kaji, H., Suzuka, I., Kaji, A.: Binding of specific ribonucleic acid to ribosomes. Binding of soluble ribonucleic acid to template — 30 s subunits complex. J. biol. Chem. **241**, 1251 (1966).

Kaji, H., Tanaka, Y.: Binding of dihydrostreptomycin to ribosomal subunits. J. molec. Biol. **32**, 221 (1968).

Kan, Y. W., Golini, F., Thach, R. E.: A new protein synthesis factor from *Escherichia coli*. Proc. nat. Acad. Sci. (Wash.) **67**, 1137 (1970).

Kaziro, Y., Inoue, N., Kuriki, Y., Mizumoto, K., Tanaka, M., Kawakita, M.: Purification and properties of factor G. Cold Spr. Harb. Symp. quant. Biol. **34**, 385 (1969).

Kelley, W. S., Schaechter, M.: Magnesium ion-dependent dissociation of polysomes and free 70 s ribosomes in *Bacillus megaterium*. J. molec. Biol. **42**, 599 (1969).

Khairallah, E., Pitot, H. C.: 3', 5', cyclic AMP and the release of polysome-bound proteins *in vitro*. Biochem. biophys. Res. Commun. **29**, 269 (1967).

Kinoshita, T., Kawano, G., Tanaka, N.: Association of fusidic acid sensitivity with G-factor in a protein-synthesizing system. Biochem. biophys. Res. Commun. **33**, 769 (1968).

Kinoshita, T., Tanaka, N.: On the site of action of bottromycin A_2. J. Antibiot. **23**, 311 (1970).

Kinoshita, T., Liou, Y-F., Tanaka, N.: Inhibition by thiopeptin of ribosomal functions associated with T and G factors. Biochem. biophys. Res. Commun. **44**, 859 (1971).

KIRSCHMANN, C., DAVIS, B. D.: Phenotypic suppression in *E. coli* by chloramphenicol and other reversible inhibitors of the ribosome. J. Bact. **98**, 152 (1969).

KNIPPENBERG, P. H., VAN, RAVENSWAAY CLAASEN, J. C., GRIJM-VOS, M., VELDSTRA, H., BOSON, L.: Stimulation and inhibition of polypeptide synthesis by streptomycin in ribosomal systems of *Escherichia coli*, programmed with various messengers. Biochim. biophys. Acta (Amst.) **95**, 461 (1965).

KOHLER, R. E., RON, F. Z., DAVIS, B. D.: Significance of the free 70 s ribosomes in *Escherichia coli* extracts. J. molec. Biol. **36**, 71 (1968).

KOLAKOFSKY, D., DEWEY, K. F., HERSHEY, J. W. B., THACH, R. E.: Guanosine 5'-triphosphatase activity of initiation factor f_2. Proc. nat. Acad. Sci. (Wash.) **61**, 1066 (1968).

KREIDER, G., BROWNSTEIN, B.: A mutation suppressing streptomycin dependence. II. An altered protein on the 30 s ribosomal subunit. J. molec. Biol. **61**, 135 (1971).

KREMBEL, J., APIRION, D.: Changes in ribosomal proteins associated with mutants in a locus that affects Escherichia coli ribosomes. J. molec. Biol. **33**, 363 (1968).

KRISKO, I., GORDON, J., LIPMANN, F.: Studies on the interchangeability of one of the mammalian and bacterial supernatant factors in protein biosynthesis. J. biol. Chem. **244**, 6117 (1969).

KUCAN, Z., LIPMANN, F.: Differences in chloramphenicol sensitivity of cell-free amino acid polymerization systems. J. biol. Chem. **239**, 516 (1964).

KUECHLER, E.: Role of GTP in the positioning of formylmethionyl tRNAf on the *E. coli* ribosome. Nature (Lond.) New Biology **234**, 216 (1971).

KUNTZEL, H.: Specificity of mitochondrial and cytoplasmic ribosomes from *neurospora crassa.* FEBS Letters **4**, 140 (1969).

KURYLO-BOROWSKA, Z., HIEROWSKI, M.: On the mode of action of edeine. II. Studies on the binding of edeine to Escherichia coli ribosomes. Biochim. biophys. Acta (Amst.) **95**, 590 (1965).

LAI, C. J., WEISBLUM, B.: Altered methylation of ribosomal RNA in an erythromycin-resistant strain of Staphylococcus aureus. Proc. nat. Acad. Sci. (Wash.) **68**, 856 (1971).

LEE-HUANG, S., OCHOA, S.: Messenger discriminating species of initiation factor F3. Nature (Lond.) New Biology **234**, 236 (1971).

LEIS, J. P., KELLER, E. B.: Protein chain initiation by methionyl-tRNA. Biochem. biophys. Res. Commun. **40**, 416 (1970).

LELONG, J. C., COUSIN, M. A., GROS, D., GRUNBERG-MANAGO, M., GROS, F.: Streptomycin induced release of fmet-tRNA from the ribosomal initiation complex. Biochem. biophys. Res. Commun. **42**, 530 (1971).

LELONG, J. C., GRUNBERG-MANAGO, M., DONDON, J., GROS, D., GROS, F.: Interaction between guanosine derivatives and factors involved in the initiation of protein synthesis. Nature (Lond.) **226**, 505 (1970).

LENGYEL, P., SOLL, D.: Mechanism of protein biosynthesis. Bact. Rev. **33**, 264 (1969).

LENNETTE, E., APIRION, D.: The level of fmet-tRNA on ribosomes from streptomycin treated cells. Biochem. biophys. Res. Commun. **41**, 804 (1970).

LEVINTHAL, C., FAN, D. P., HIGA, A., ZIMMERMANN, R. A.: The decay and protection of messenger RNA in bacteria. Cold Spr. Harb. Symp. quant. Biol. **28**, 183 (1963).

LEWANDOWSKI, L. J., BROWNSTEIN, B. L.: An altered pattern of ribosome synthesis in a mutant of *E. coli*. Biochem. biophys. Res. Commun. **25**, 554 (1966).

LIN, Y. C., KINOSHITA, T., TANAKA, N.: Mechanism of protein synthesis inhibition by bottromycin A_2: Studies with puromycin. J. Antibiot. Ser. A (Tokyo) **21**, 471 (1968).

LOCKWOOD, A. H., CHAKRABORTY, P. R., MAITRA, U.: A complex between initiation factor IF2, guanosine triphosphate, and fmet-tRNAf: an intermediate in initiation complex formation. Proc. nat. Acad. Sci. (Wash.) **68**, 3122 (1971).

LOCKWOOD, A. H., MAITRA, U.: Function of GTP in polypeptide chain initiation in *E. coli.* Fed. Proc. (1972) (in press).

LODISH, H. F.: Bacteriophage f2 RNA: control of translation and gene order. Nature (Lond.) **220**, 345 (1968).

LODISH, H. F.: Specificity in bacterial protein synthesis: role of initiation factors and ribosomal subunits. Nature (Lond.) **226**, 705 (1970).

Lodish, H. F., Robertson, H. D.: Regulation of *in vitro* translation of bacteriophage f2 RNA. Cold Spr. Harb. Symp. quant. Biol. **34**, 655 (1969).

Lucas-Lenard, J., Haenni, A-L.: Release of transfer RNA during peptide chain elongation. Proc. nat. Acad. Sci. (Wash.) **63**, 93 (1969).

Lucas-Lenard, J., Lipmann, F.: Protein biosynthesis. Ann. Rev. Biochem. **40**, 409 (1971).

Lucas-Lenard, J., Lipmann, F.: Initiation of polyphenylalanine synthesis by N-acetyl-phenylalanyl sRNA. Proc. nat. Acad. Sci. (Wash.) **57**, 1050 (1967).

Lucas-Lenard, J., Lipmann, F.: Separation of three microbial amino acid polymerization factors. Proc. nat. Acad. Sci. (Wash.) **55**, 1562 (1966).

Lucas-Lenard, J., Tao, P., Haenni, A.-L.: Further studies on bacterial polypeptide elongation. Cold Spr. Harb. Symp. quant. Biol. **34**, 455 (1969).

Luzzatto, L., Apirion, D., Schlessinger, D.: Mechanism of action of streptomycin in *E. coli:* interruption of the ribosome cycle at the initiation of protein synthesis. Proc. nat. Acad. Sci. (Wash.) **60**, 873 (1968).

Luzzatto, L., Apirion, D., Schlessinger, D.: Polyribosome depletion and blockage of the ribosome cycle by streptomycin in Escherichia coli. J. molec. Biol. **42**, 315 (1969).

MacDonald, J. S., Goldberg, I. H.: An effect of pactamycin on the initiation of protein synthesis in reticulocytes. Biochem. biophys. Res. Commun. **41**, 1 (1970).

McKeehan, W., Hardesty, B.: (1) The mechanism of cycloheximide inhibition of protein synthesis in rabbit reticulocytes. Biochem. biophys. Res. Commun. **36**, 625 (1969).

McKeehan, W., Hardesty, B.: (2) Purification and partial characterization of the aminoacyl transfer ribonucleic acid binding enzyme from rabbit reticulocytes. J. biol. Chem. **244**, 4330 (1969).

Maden, B. E. H., Traut, R. R., Monro, R. E.: Ribosome-catalyzed peptidyl transfer: the polyphenylalanine system. J. molec. Biol. **35**, 333 (1968).

Malkin, M., Lipmann, F.: Fusidic acid: inhibition of factor T2 in reticulocyte protein synthesis. Science **164**, 71 (1969).

Mangiarotti, G., Schlessinger, D.: Polyribosome metabolism in *E. coli*. I. Extraction of polyribosomes and ribosomal subunits from fragile, growing *E. coli*. J. molec. Biol. **20**, 123 (1966).

Mao, J. C.-H., Robishaw, E. E.: Effects of macrolides on peptide-bond formation and translocation. Biochemistry **10**, 2054 (1971).

Marcus, A., Bewley, J. D., Weeks, D. P.: Aurintricarboxylic acid and initiation factors of wheat embryo. Science **167**, 1735 (1970).

Matsua, S., Shiratori, O., Katagiri, K.: Antitumor activity of showdomycin. J. Antibiot. Ser. A (Tokyo) **17**, 234 (1964).

Matthaei, H., Amelunxen, F., Eckert, K., Heller, G.: The mechanism of protein biosynthesis I. The binding of messenger-RNA and aminoacyl sRNA to ribosomes. Ber. Bunsengesellschaft **68**, 735 (1964).

Maxwell, I. H.: Studies of the binding of tetracycline to ribosomes *in vitro*. Molec. Pharmacol. **4**, 25 (1968).

Maxwell, R. E., Nickel, V. S. The antibacterial activity of the isomers of chloramphenicol. Antibiot. and Chemother. **4**, 289 (1954).

Mazumder, R., Chae, Y.-B., Ochoa, S.: Polypeptide chain initiation in *E. coli:* sulfhydryl groups and the function of initiation factor F2. Proc. nat. Acad. Sci. (Wash.) **63**, 98 (1969).

Miller, D. L., Hachmann, J., Weissbach, H.: The reaction of the sulfhydryl groups on the elongation factors Tu and Ts. Arch. Biochem. Biophys. **144**, 115 (1971).

Miller, D. L., Weissbach, H.: An interaction between the transfer factors required for protein synthesis. Arch. Biochem. Biophys. **132**, 146 (1969).

Miller, D. L., Weissbach, H.: Studies on the purification and properties of factor Tu from E. coli. Arch. Biochem. Biophys. **141**, 26 (1970).

Miller, M. J., Zasloff, M., Ochoa, S.: Association of polypeptide initiation factors with 30 S ribosomal subunits. FEBS Letters **3**, 50 (1969).

Milman, G., Goldstein, J., Scolnick, E., Caskey, T.: Peptide chain termination III. Stimulation of *in vitro* termination. Proc. nat. Acad. Sci. (Wash.) **63**, 183 (1969).

Miskin, R., Zamir, A., Elson, D.: Inactivation and reactivation of ribosomal subunits s: the peptidyl transferase activity of the 50 S subunit of *E. coli*. J. molec. Biol. **54**, 355 (1970).

MIZUSHIMA, S., NOMURA, M.: Assembly mapping of 30 S ribosomal proteins from *E. coli*. Nature (Lond.) **226**, 1214 (1970).

MODOLELL, J.: Int. Cong. Biochem. 8th (Switz.) (1970).

MODOLELL, J., CABRER, B., PARMEGGIANI, A., VAZQUEZ, D.: (1) Inhibition by siomycin and thiostrepton of both aminoacyl-tRNA and factor G binding to ribosomes. Proc. nat. Acad. Sci. (Wash.) **68**, 1796 (1971).

MODOLELL, J., DAVIS, B.: Rapid inhibition of polypeptide chain extension by streptomycin. Proc. nat. Acad. Sci. (Wash.) **61**, 1279 (1968).

MODOLELL, J., DAVIS, B. D.: A unitary mechanism for the several effects of streptomycin on the ribosome. Cold Spr. Harb. Symp. quant. Biol. **34**, 113 (1969).

MODOLELL, J., DAVIS, B. D.: Breakdown by streptomycin of initiation complexes formed on ribosomes of *E. coli*. Proc. nat. Acad. Sci. (Wash.) **67**, 1148 (1970).

MODOLELL, J., VAZQUEZ, D., MONRO, R. E.: (2) Ribosomes, G-factor and siomycin. Nature (Lond.) New Biol. **230**, 109 (1971).

MONRO, R. E.: Catalysis of peptide bond formation by 50 S ribosomal subunits from *Escherichia coli*. J. molec. Biol. **26**, 147 (1967).

MONRO, R. E., CELMA, M. L., VAZQUEZ, D.: Action of sparsomycin on ribosome-catalyzed peptidyl transfer. Nature (Lond.) **222**, 356 (1969).

MONRO, R. E., MARCKER, K. A.: Ribosome-catalyzed reaction of puromycin with a formyl-methionine-containing oligonucleotide. J. molec. Biol. **25**, 347 (1967).

MONRO, R. E., CERNA, J., MARCKER, K. A.: Ribosome-catalyzed peptidyl transfer: substrate specificity at the P-site. Proc. nat. Acad. Sci. (Wash.) **61**, 1042 (1968).

MONRO, R. E., VAZQUEZ, D.: Ribosome-catalyzed peptidyl transfer: effects of some inhibitors of protein synthesis. J. molec. Biol. **28**, 161 (1967).

MOORE, C. H., FARRON, F., BOHNERT, D., WEISSMANN, C.: Possible origin of a minor virus specific protein (a_1) in Qβ particles. Nature (Lond.) New Biol. **234**, 204 (1971).

MORSE, D.: Polarity induced by chloramphenicol and relief by suA. J. molec. Biol. **55**, 113 (1971).

NAKAMOTO, T.: The initial phase in the polyuridylic acid-directed polymerization of phenylalanine. J. biol. Chem. **242**, 4534 (1967).

NATHANS, D.: Puromycin inhibition of protein synthesis: incorporation of puromycin into peptide chains. Proc. nat. Acad. Sci. (Wash.) **51**, 585 (1964).

NATHANS, D., NEIDLE, P.: Structural requirements for puromycin inhibition of protein synthesis. Nature (Lond.) **197**, 1076 (1963).

NATHANS, D., VON EHRENSTEIN, G., MONRO, R., LIPMANN, F.: Protein synthesis from aminoacyl soluble ribonucleic acid. Fed. Proc. **21**, 127 (1962).

NETH, R., MONRO, R. E., HELLER, G., BATTANER, E., VAZQUEZ, D.: Catalysis of peptidyl transfer by human tonsil ribosomes and effects of some antibiotics. FEBS Letters **6**, 198 (1970).

NICHOLS, J. L.: Nucleotide sequence from the polypeptide chain termination region of the coat protein cistron in bacteriophage R17 RNA. Nature (Lond.) **225**, 147 (1970).

NIRENBERG, W. M., LEDER, P.: RNA codewords and protein synthesis. Science **145**, 1399 (1964).

NIRENBERG, W. M., LEDER, P., BERNFIELD, M., BRIMACOMBE, R., TRUPIN, J., ROTTMAN, F., O'NEAL, C.: RNA codewords and protein synthesis. VII. On the general nature of the RNA code. Proc. nat. Acad. Sci. (Wash.) **53**, 1161 (1965).

NIRENBERG, M. W., MATTHAEI, J. H.: The dependence of cell-free protein synthesis upon naturally occurring or synthetic polyribonucleotides. Proc. nat. Acad. Sci. (Wash.) **47**, 1588 (1961).

NISHIMURA, H., MAYAMA, M., KOMATSU, Y., KATO, H., SHIMAOKA, N., TANAKA, Y.: Showdomycin: a new antibiotic from a *Streptomyces* SP. J. Antibiot. Ser. A (Tokyo) **17**, 148 (1964).

NISHIZUKA, Y., LIPMANN, F.: (1) Comparison of guanosine triphosphate split and polypeptide synthesis with a purified *E. coli* system. Proc. nat. Acad. Sci. (Wash.) **55**, 212 (1966).

NISHIZUKA, Y., LIPMANN, F.: (2) The interrelationship between GTP and amino acid polymerization. Arch. Biochem. Biophys. **116**, 344 (1966).

NOMURA, M., LOWRY, C.: Phage F2 RNA-directed binding of formylmethionyl-tRNA to ribosomes and the role of 30 S ribosomal subunits in initiation of protein synthesis. Proc. nat. Acad. Sci. (Wash.) **58**, 946 (1967).

NOMURA, M., LOWRY, C. V., GUTHRIE, C.: The initiation of protein synthesis: joining of the 50 S ribosomal subunit to the initiation complex. Proc. nat. Acad. Sci. (Wash.) **58**, 1487 (1967).

NOMURA, M., MIZUSHIMA, S., OZAKI, M., TRAUB, P., LOWRY, C. V.: Structure and function of ribosomes and their molecular components. Cold Spr. Harb. Symp. quant. Biol. **34**, 49 (1969).

OBRIG, T. G., CULP, W. J., MCKEEHAN, W. L., HARDESTY, B.: (1) The mechanism by which cycloheximide and related glutarimide antibiotics inhibit peptide synthesis on reticulocyte ribosomes. J. biol. Chem. **246**, 174 (1971).

OBRIG, T. G., IRVIN, J., CULP, W., HARDESTY, B.: (2) Inhibition of peptide initiation on reticulocyte ribosomes by edeine. Europ. J. Biochem. **21**, 31 (1971).

OHTA, T., SARKAR, S., THACH, R. E.: The role of guanosine 5'-triphosphate in the initiation of peptide synthesis. III. Binding of formylmethionyl-tRNA to ribosomes. Proc. nat. Acad. Sci. (Wash.) **58**, 1638 (1967).

OKURA, A., KINOSHITA, T., TANAKA, N.: Complex formation of fusidic acid with G-factor, ribosome and guanosine nucleotide. Biochem. biophys. Res. Commun. **41**, 1545 (1970).

OKUYAMA, A., MACHIYAMA, M., KINOSHITA, T., TANAKA, N.: Inhibition by kasugamycin of initiation complex formation on 30 S ribosomes. Biochem. biophys. Res. Commun. **43**, 196 (1971).

OLD, D., GORINI, L.: Amino acid changes provoked by streptomycin in a polypeptide synthesized *in vitro*. Science **150**, 1290 (1965).

OLEINICK, N. L., CORCORAN, J. W.: Two types of binding of erythromycin to sibosomes from antibiotic sensitive and resistant *Bacillus subtilis* **168**. J. biol. Chem. **244**, 727 (1969).

OLEINICK, N. L., WILHELM, J. M., CORCORAN, J. W.: Nonidentity of the site of action of erythromycin A and chloramphenicol on *Bacillus subtilis* ribosomes. Biochim. biophys. Acta (Amst.) **155**, 290 (1968).

ONO, Y., SKOULTCHI, A., KLEIN, A., LENGYEL, P.: Peptide chain elongation: discrimination against the initiator transfer RNA by microbial amino acid polymerization factors. Nature (Lond.) **220**, 1304 (1968).

ONO, Y., SKOULTCHI, A., WATERSON, J., LENGYEL, P.: (1) Stoichiometry of aminoacyl-transfer RNA binding and GTP during chain elongation and translocation. Nature (Lond.) **223**, 697 (1969).

ONO, Y., SKOULTCHI, A., WATERSON, J., LENGYEL, P.: (2) Peptide chain elongation: GTP cleavage catalyzed by factors binding aminoacyl-transfer RNA to the ribosome. Nature (Lond.) **222**, 645 (1969).

OTAKA, E., TERAOKA, H., TAMAKI, M., TANAKA, K., OSAWA, S.: Ribosomes from erythromycin resistant mutants of *E. coli* Q13. J. molec. Biol. **48**, 499 (1970).

OZAKI, M., MIZUSHIMA, S., NOMURA, M.: Identification and functional characterization of the protein controlled by the streptomycin resistant locus in *E. coli*. Nature (Lond.) **222**, 333 (1969).

PANET, A., DEGROOT, N., LAPIDOT, Y.: Substrate specificity of *E. coli* peptidyl transferase. Europ. J. Biochem. **15**, 222 (1970).

PARENTI-ROSINA, R., EISENSTADT, A., EISENSTADT, J. M.: Isolation of protein initiation factors from 30 S ribosomal subunits. Nature (Lond.) **221**, 363 (1969).

PARMEGGIANI, A.: Crystalline transfer factors from *E. coli*. Biochem. biophys. Res. Commun. **30**, 613 (1968).

PARMEGGIANI, A., GOTTSCHALK, E. M.: (1) Isolation and some properties of the amino acid polymerization factors from *Escherichia coli*. Cold Spr. Harb. Symp. quant. Biol. **34**, 377 (1969).

PARMEGGIANI, A., GOTTSCHALK, E. M.: (2) Properties of the crystalline amino acid polymerization factors from *Escherichia coli:* binding of G to ribosomes. Biochem. biophys. Res. Commun. **35**, 861 (1969).

PAVAN, M., BO, G.: Pederin, toxic principle obtained in the crystalline state from the beetle *Paederus fuscipes curt*. Physiol. comp. ('s-Grav.) **3**, 307 (1953).

PERANI, A., PARISI, B., DECARLI, L., CIFERRI, O.: Incorporation of amino acids by a cell-free system prepared from human cells cultured *in vitro*. Biochim. biophys. Acta (Amst.) **161**, 223 (1968).

PESTKA, S.: The action of streptomycin on protein synthesis in vitro. Bull. N. Y. Acad. Med. **43**, 126 (1967).

PESTKA, S.: Inhibitors of ribosome functions. Ann. Rev. Microbiol. **25**, 487 (1971).

PESTKA, S.: Studies on the formation of transfer ribonucleic acid-ribosome complexes. V. On the function of a soluble transfer factor in protein synthesis. Proc. nat. Acad. Sci. (Wash.) **61**, 726 (1968).

PESTKA, S.: (1) Studies on the formation of transfer ribonucleic acid-ribosome complexes. VIII. Survey of the effect of antibiotics on N-acetyl-phenylalanyl-puromycin formation: possible mechanism of chloramphenicol action. Arch. Biochem. Biophys. **136**, 80 (1970).

PESTKA, S.: (2) Studies on the formation of transfer ribonucleic acid-ribosome complexes. IX. Effect of antibiotics on translocation and peptide bond formation. Arch. Biochem. Biophys. **136**, 89 (1970).

PESTKA, S.: (1) Studies on the formation of transfer ribonucleic acid-ribosone complexes. X. Phenylalanyl-oligonucleotide binding to ribosomes and the mechanism of chloramphenicol action. Biochem. biophys. Res. Commun. **36**, 589 (1969).

PESTKA, S.: (2) Studies on the formation of transfer ribonucleic acid-ribosome complexes. XI. Antibiotic effects on phenylalanyl-oligonucleotide binding to ribosomes. Proc. nat. Acad. Sci. (Wash.) **64**, 709 (1969).

PESTKA, S.: Translocation, aminoacyl-oligonucleotides, and antibiotic action. Cold Spr. Harb. Symp. quant. Biol. **34**, 395 (1969).

PESTKA, S., HISHIZAWA, T., LESSARD, J. L.: Studies on the formation of transfer ribonucleic acid-ribosome complexes. XIII. Aminoacyl oligonucleotide binding to ribosomes: characteristics and requirements. J. biol. Chem. **245**, 6208 (1970).

PESTKA, S., MARSHALL, R., NIRENBERG, M.: RNA codewords and protein synthesis. V. Effect of streptomycin on the formation of ribosome-sRNA complexes. Proc. nat. Acad. Sci. (Wash.) **53**, 639 (1965).

PESTKA, S., NIRENBERG, M. W.: Regulatory mechanisms and protein synthesis. X. Codon recognition on 30 S ribosomes. J. molec. Biol. **21**, 145 (1966).

PHILLIPS, L. A., HOTHAM-IGLEWSKI, B., FRANKLIN, R. M.: Ribosomes of *Escherichia coli*. I. Effzcts of monovalent cations on the distribution of polysomes, ribosomes and ribosomal subunits. J. molec. Biol. **40**, 279 (1969).

PLOTZ, P. H., DAVIS, B. D.: Absence of chloramphenicol-insensitive phase of streptomycin action. J. Bact. **83**, 802 (1962).

PRICHARD, P. M., GILBERT, J. M., SHAFRITZ, D. A., ANDERSON, W. F.: Factors for the initiation of haemoglobin synthesis by rabbit reticulocyte ribosomes. Nature (Lond.) **226**, 511 (1970).

RAEBURN, S., GOOR, R. S., SCHNEIDER, J. A., MAXWELL, E. S.: Interaction of aminoacyl transferase II and guanosine triphosphate: inhibition by diphtheria toxin and nicotinamide adenine dinucleotide. Proc. nat. Acad. Sci. (Wash.) **61**, 1428 (1968).

RAO, P., MOLDAVE, K.: The binding of aminoacyl sRNA and GTP to transferase I. Biochem. biophys. Res. Commun. **28**, 909 (1967).

RAO, P., MOLDAVE, K.: Interaction of polypeptide chain elongation factors with rat liver ribosomal subunits. J. molec. Biol. **46**, 447 (1969).

RAO, S. S., GROLLMAN, A. P.: Cycloheximide resistance in yeast: a property of the 60 S ribosomal subunit. Biochem. biophys. Res. Commun. **29**, 696 (1967).

RAVEL, J. M., SHOREY, R. L., GARNER, C. W., DAWKINS, R. C., SHIVE, W.: The role of an aminoacyl tRNA-GTP-protein complex in polypeptide synthesis. Cold Spr. Harb. Symp. quant. Biol. **34**, 321 (1969).

RAVEL, J. M., SHOREY, R. L., SHIVE, W.: The composition of the active intermediate in the transfer of aminoacyl-RNA to ribosomes. Biochem. biophys. Res. Commun. **32**, 9 (1968).

RAVEL, J. M., SHOREY, R. L., SHIVE, W.: Relationship between peptidyl transferase activity and interaction of ribosomes with phenylalanyl transfer ribonucleic acid-guanosine 5'-triphosphate TIu complex. Biochemistry **9**, 5028 (1970).

Revel, M., Greenshpan, H., Herzberg, M.: Specificity in the binding of *E. coli* ribosomes to natural messenger RNA. Europ. J. Biochem. **16**, 117 (1970).

Revel, M., Herzberg, M., Becarevic, A., Gros, F.: (1) Role of a protein factor in the functional binding of ribosomes to natural messenger RNA. J. molec. Biol. **33**, 231 (1968).

Revel, M., Herzberg, M., Greenshpan, H.: Initiator protein dependent binding of messenger RNA to the ribosome. Cold Spr. Harb. Symp. quant. Biol. **34**, 261 (1969).

Revel, M., Lelong, J. C., Brawerman, G., Gros, F.: (2) Function of three protein factors and ribosomal subunits in the initiation of protein synthesis in *E. coli*. Nature (Lond.) **219**, 1016 (1968).

Richter, D.: Formation of a ternary complex between yeast aminoacyl-tRNA binding factor, GTP, and aminoacyl-tRNA. Biochem. biophys. Res. Commun. **38**, 864 (1970).

Richter, D., Lin, L., Bodley, J. W.: Studies on translocation IX: the pattern of action of antibiotic translocation inhibitors in eukaryotic and prokaryotic systems. Arch. Biochem. Biophys. **147**, 186 (1971).

Roberts, W. K., Coleman, W. H.: Polyuridylic acid binding by protein from Ehrlich ascites cell ribosomes and its inhibition by Aurintricarboxylic acid. Biochemistry **10**, 4304 (1971).

Ron, E. Z., Kohler, R. E., Davis, B. D.: Magnesium ion dependence of free and polysomal ribosomes from *Escherichia coli*. J. molec. Biol. **36**, 83 (1968).

Rosset, R., Gorini, L.: A ribosomal ambiguity mutation. J. molec. Biol. **39**, 95 (1969).

Roufa, D. J., Doctor, B. P., Leder, P.: The two site model of ribosomal function: a test using degenerate serine codons in bacteriophage f2 mRNA. Biochem. biophys. Res. Commun. **39**, 231 (1970).

Rychlik, I., Cerna, C., Stanislav, C., Pulkrapek, P., Zemlicka, J.: Substrate specificity of ribosomal peptidyl transferase. The effect of the nature of the amino acid side chain on the acceptor activity of 2'(3')-O-amino acyladenosines. Europ. J. Biochem. **16**, 136 (1970).

Rychlik, I., Chladek, S., Zemlicka, J.: Release of peptide chains from the polylysyl-tRNA ribosome complex by cytidylyl-(3'-5')-2'(3')-O-glycyladenosine. Biochim. biophys. Acta (Amst.) **138**, 640 (1967).

Sabol, S., Ochoa, S.: Ribosomal binding of labeled initiation factor F3. Nature (Lond.) New Biol. **234**, 233 (1971).

Salas, M., Hille, M. B., Last, J. A., Wahba, A. J., Ochoa, S.: (1) Translation of the genetic message, II. Effect of initiation factors on the binding of formylmethionyl tRNA to ribosomes. Proc. nat. Acad. Sci. (Wash.) **57**, 387 (1967).

Salas, M., Miller, M. J., Wahba, A. J., Ochoa, S.: (2) Translation of the genetic message. V. Effect of Mg^{++} an formylation of methionine in protein synthesis. Proc. nat. Acad. Sci. (Wash.) **57**, 1865 (1967).

Sarkar, S., Thach, R. E.: Inhibition of formylmethionyl transfer RNA binding to ribosomes by tetracycline. Proc. nat. Acad. Sci. (Wash.) **60**, 1479 (1968).

Scaife, J., Beckwith, J. R.: Mutational alteration of the maximal level of lac operon expression. Cold Spr. Harb. Symp. quant. Biol. **31**, 403 (1966).

Schedl, P., Singer, R. E., Conway, T. W.: A factor required for the translation of bacteriophage f2 RNA in extracts of T4-infected cells. Biochem. biophys. Res. Commun. **38**, 631 (1970).

Schlessinger, D., Mangiarotti, G., Apirion, D.: The formation and stabilization of 30 S and 50 S ribosome couples in *Escherichia coli*. Proc. nat. Acad. Sci. (Wash.) **58**, 1782 (1967).

Schlessinger, D., Gurgo, C., Luzzatto, L., Apirion, D.: Polyribosome metabolism in growing and non-growing *E. coli*. Cold Spr. Harb. Symp. quant. Biol. **34**, 231 (1969).

Schneir, M., Moldave, K.: The isolation and biological activity of multiple forms of aminoacyl transferase I of rat liver. Biochim. biophys. Acta (Amst.) **166**, 58 (1968).

Schreier, M. H., Noll, H.: Conformational changes in ribosomes during protein synthesis. Proc. nat. Acad. Sci. (Wash.) **68**, 805 (1971).

Scolnick, E. M., Caskey, C. T.: Peptide chain termination. V. The role of release factors in mRNA terminator codon recognition. Proc. nat. Acad. Sci. (Wash.) **64**, 1235 (1969).

Scolnick, E. M., Tompkins, R., Caskey, T., Nirenberg, M.: Release factors differing in specificity for terminator codons. Proc. nat. Acad. Sci. (Wash.) **61**, 768 (1968).

SHAFRITZ, D. A., ANDERSON, W. F.: Factor dependent binding of methionyl-tRNAs to reticulocyte ribosomes. Nature (Lond.) **227**, 918 (1970).

SHAFRITZ, D. A., LAYCOCK, D. G., CRYSTAL, R. G., ANDERSON, W. F.: Requirement for GTP in the initiation process on reticulocyte ribosomes and ribosomal subunits. Proc. nat. Acad. Sci. (Wash.) **68**, 2246 (1971).

SHAFRITZ, D. A., PRICHARD, P. M., GILBERT, J. M., ANDERSON, W. F.: Separation of two factors, M_1 and M_2, required for poly U dependent polypeptide synthesis by rabbit reticulocyte ribosomes at low magnesium ion concentration. Biochem. biophys. Res. Commun. **38**, 721 (1970).

SHERMAN, M. I., SIMPSON, M. V.: The role of ribosomal conformation in protein biosynthesis: the streptomycin-ribosome interaction. Proc. nat. Acad. Sci. (Wash.) **64**, 1388 (1969).

SHOREY, R. L., RAVEL, J. M., GARNER, C. W., SHIVE, W.: Formation and properties of the aminoacyl transfer ribonucleic acid-guanosine triphosphate-protein complex. An intermediate in the binding of aminoacyl transfer ribonucleic acid. J. biol. Chem. **244**, 4555 (1969).

SIEGELMAN, F. L., APIRION, D.: Inhibition of polyuridylic acid-directed protein synthesis by aurintricarboxylate in extracts of Escherichia coli. J. Bact. **105**, 451 (1971).

SISLER, J., MOLDAVE, K.: Studies on the binding of phenylalanyl transfer RNA to rat-liver ribosomes. Biochim. biophys. Acta (Amst.) **195**, 123 (1969).

SKOGERSON, L., ROUFA, D., LEDER, P.: Characterization of the initial peptide of Qβ RNA polymerase and control of its synthesis. Proc. nat. Acad. Sci. (Wash.) **68**, 276 (1971).

SKOULTCHI, A., ONO, Y., WATERSON, J., LENGYEL, P.: Peptide chain elongation. Cold Spr. Harb. Symp. quant. Biol. **34**, 437 (1969).

SKOULTCHI, A., ONO, Y., MOON, H. M., LENGYEL, P.: On three complementary amino acid polymerization factors from *Bacillus stearothermophilus* separation of a complex containing two of the factors, guanosine-5'-triphosphate and aminoacyl transfer RNA. Proc. nat. Acad. Sci. (Wash.) **60**, 675 (1968).

SKOULTCHI, A., ONO, Y., WATERSON, J., LENGYEL, P.: Peptide chain elongation: indications for the binding of an amino acid polymerization factor, guanosine 5'-triphosphate-aminoacyl transfer ribonucleic acid complex to the messenger ribosome complex. Biochemistry **9**, 508 (1970).

SLECHTA, L.: Mode of action of sparsomycin in *Escherichia coli*. Antimicrobial Agent Chemotherapy **326** (1965).

SMITH, A. E., MARCKER, K. A.: Cytoplasmic methionine transfer RNA's from eukaryotes. Nature (Lond.) **226**, 607 (1970).

SPARLING, P. F., MODOLELL, J., TAKEDA, Y., DAVIS, B. D.: Ribosomes from *Escherichia coli* merodiploids heterozygous for resistance to streptomycin and to spectinomycin. J. molec. Biol. **37**, 407 (1968).

SPIRIN, A. S.: How does the ribosome work? A hypothesis based on the two subunit construction of the ribosome. Curr. modern Biol. **2**, 115 (1968).

SPYRIDES, G. J., LIPMANN, F.: The effect of univalent cations on the binding of sRNA to the template-ribosome complex. Proc. nat. Acad. Sci. (Wash.) **51**, 1220 (1964).

STAEHELIN, T., MAGLOTTI, D., MONRO, R. E.: On the catalytic center of peptidyl transfer: a part of the 50 S ribosome structure. Cold Spr. Harb. Symp. quant. Biol. **34**, 39 (1969).

STEITZ, J. A.: Polypeptide chain initiation: nucleotide sequences of the three ribosomal binding sites in bacteriophage R17 RNA. Nature (Lond.) **224**, 957 (1969).

STEITZ, J. A., DUBE, S. K., RUDLAND, P. S.: Control of translation by T4 phage: altered ribosome binding at R17 initiation sites. Nature (Lond.) **226**, 824 (1970).

STERN, J. L., COHEN, S. S.: Lethality and the stimulation of RNA synthesis by streptomycin. Proc. nat. Acad. Sci. (Wash.) **51**, 859 (1964).

STEWART, M. L., GROLLMAN, A. P., HUANG, M.-T.: (1) Aurintricarboxylic acid: inhibitor of initiation of protein synthesis. Proc. nat. Acad. Sci. (Wash.) **68**, 97 (1971).

STEWART-BLAIR, M. L., YANOWITZ, I. S., GOLDBERG, I. H.: Inhibition of synthesis of new globin chains in reticulocyte lysates by pactamycin. Biochemistry **10**, 4198 (1971).

STRETTON, A. O. W., BRENNER, S.: Molecular consequences of the amber mutation and its supression. J. molec. Biol. **12**, 456 (1965).

Suarez, G., Nathans, D.: Inhibition of aminoacyl-sRNA binding to ribosomes by tetracycline. Biochem. biophys. Res. Commun. **18**, 743 (1965).

Subramanian, A. R., Davis, B. D.: Rapid exchange of subunits between free ribosomes in extracts of Escherichia coli. Proc. nat. Acad. Sci. (Wash.) **68**, 2453 (1971).

Subramanian, A. R., Davis, B. D., Beller, R. J.: The ribosome dissociation factor and the ribosome-polysome cycle. Cold Spr. Harb. Symp. quant. Biol. **34**, 223 (1969).

Suzuka, I., Kaji, A.: Studies on diphenylalanine synthesis. Biochim. biophys. Acta (Amst.) **149**, 540 (1967).

Suzuka, I., Kaji, H., Kaji, A.: Binding of specific sRNA to 30 S ribosomal subunits: effect of 50 S ribosomal subunits. Proc. nat. Acad. Sci. (Wash.) **55**, 1483 (1966).

Suzuka, I., Kaji, H., Kaji, A.: Binding of specific sRNA to 30 S-subunits. Comparison with the binding to 70 S ribosomes. Biochem. biophys. Res. Commun. **21**, 187 (1965).

Swan, D., Matthaei, H.: Further evidence for the four-site model of ribosomal function: bacteriophage fr mRNA coded binding of aa-tRNA. FEBS Letters **17**, 215 (1971).

Symons, R. H., Harris, R. J., Clarke, L. P., Wheldrake, J. F., Elliott, W. H.: Structural requirements for inhibition of polyphenylalanine synthesis by aminoacyl and nucleotidyl analogues of puromycin. Biochim. biophys. Acta (Amst.) **179**, 248 (1969).

Szer, W., Kurylo-Borowska, Z.: Effect of edeine on aminoacyl-tRNA binding to ribosomes and its relationship to ribosomal binding sites. Biochim. biophys. Acta (Amst.). **224**, 477 (1970).

Takanami, M., Okamoto, T.: Interaction of ribosomes and synthetic polyribonucleotides. J. molec. Biol. **7**, 323 (1963).

Tanaka, K., Teraoka, H., Tamaki, M.: (1) Peptidyl puromycin synthesis: effect of several antibiotics which act on 50 S ribosomal subunits. FEBS Letters **13**, 65 (1971).

Tanaka, K., Watanabe, S., Teraoka, H., Tamaki, M.: (1) Effect of siomycin on protein synthesizing activity of *Escherichia coli* ribosomes. Biochem. biophys. Res. Commun. **39**, 1189 (1970).

Tanaka, K., Watanabe, S., Tamaki, M.: (2) Mode of action of siomycin. J. Antibiot. Ser. A (Tokyo) **23**, 13 (1970).

Tanaka, N., Kawano, G., Kinoshita, T.: (1) Chromosomal location of a fusidic acid resistant marker in *Escherichia coli*. Biochem. biophys. Res. Commun. **42**, 564 (1971).

Tanaka, N., Lin, Y.-C., Okuyama, A.: (2) Studies on translocation of f-Met-tRNA and peptidyl tRNA with antibiotics. Biochem. biophys. Res. Commun. **44**, 477 (1971).

Tanaka, N., Masukawa, H., Umezawa, H.: Structural basis of kanamycin for miscoding activity. Biochem. biophys. Res. Commun. **26**, 544 (1967).

Tanaka, S., Igarashi, K., Kaji, A.: Further studies on the action of puromycin and tetracycline. J. biol. Chem. (1972) (in press).

Tanaka, S., Kaji, A.: Does translocase (G-factor) require the presence of unesterified tRNA on the donor site for its action? The effect of fusidic acid. Biochem. biophys. Res. Commun. **46**, 136 (1972).

Tanaka, Y., Kaji, H.: The role of ribosomal protein for the binding of dihydrostreptomycin to ribosomes. Biochem. biophys. Res. Commun. **32**, 313 (1968).

Teraoka, H.: Reversal of the inhibitory action of chloramphenicol on the ribosomal peptidyl transfer reaction by erythromycin. Biochim. biophys. Acta (Amst.) **213**, 535 (1970).

Teraoka, H., Tanaka, K., Tamaki, M.: The comparative study on the effects of chloramphenicol, erythromycin and lincomycin on polylysine synthesis in an *Escherichia coli* cell-free system. Biochim. biophys. Acta (Amst.) **174**, 776 (1969).

Thach, S. S., Thach, R. E.: (2) Translocation of messenger RNA and "accomodation" of fMet-tRNA. Proc. nat. Acad. Sci. (Wash.) **68**, 1791 (1971).

Thach, S. S., Thach, R. E.: (1) One molecule of guanosine triphosphate is present in each 30 S initiation complex. Nature (Lond.) New Biol. **229**, 219 (1971).

Tocchini-Valentini, G. P., Felicetti, L., Rinaldi, G. M.: Mutants of *Escherichia coli* blocked in protein synthesis: mutants with an altered G-factor. Cold Spr. Harb. Symp. quant. Biol. **34**, 463 (1969).

Tompkins, R. K., Scolnick, E. M., Caskey, C. T.: Peptide chain termination. VII. The ribosomal and release factor requirements for peptide release. Proc. nat. Acad. Sci. (Wash.) **65**, 702 (1970).

TRAUB, P., NOMURA, M.: Structure and function of *E. coli* ribosomes. V. Reconstruction of functionally active 30 S ribosomal particles from RNA and proteins. Proc. nat. Acad. Sci. (Wash.) **59**, 777 (1968).

VANDERHAEGHE, H., VAN DYCK, P., PARMENTIER, G., DESOMER, P.: Isolation and properties of the components of staphylomycin. Antibiot. and Chemother. **7**, 606 (1957).

VAZQUEZ, D.: Antibiotics which affect protein synthesis: the uptake of ^{14}C-chloramphenicol by bacteria. Biochem. biophys. Res. Commun. **12**, 409 (1963).

VAZQUEZ, D.: (2) Binding of CM to ribosomes: the effect of a number of antibiotics. Biochim. biophys. Acta (Amst.) **114**, 277 (1966).

VAZQUEZ, D.: (3) Antibiotics affecting chloramphenicol uptake by bacteria. Their effect on amino acid incorporation in a cell-free system. Biochim. biophys. Acta (Amst.) **114**, 289 (1966).

VAZQUEZ, D.: (1) Studies on the mode of action of the streptomycin antibiotics. J. gen. Microbiol. **42**, 93 (1966).

VAZQUEZ, D., BATTANE, E., NETH, R. R., HELLER, G., MONRO, R. E.: The function of 80 S ribosomal subunits and effects of some antibiotics. Cold Spr. Harb. Symp. quant. Biol. **34**, 369 (1969).

VAZQUEZ, D., MONRO, R. E.: Effects of some inhibitors of protein synthesis on the binding of aminoacyl tRNA to ribosomal subunits. Biochim. biophys. Acta (Amst.) **142**, 155 (1967).

VERMEER, C., TALENS, J., BLOEMSMA-JONKMAN, F., BOSCH, L.: Studies on the ribosomal initiation factor F3. Multiplicity of F3 and DF activities. FEBS Letters **19**, 201 (1971).

VOGEL, Z., VOGEL, T., ELSON, D.: The effect of erythromycin on peptide bond formation and the termination reaction. FEBS Letters **15**, 249 (1971).

VOGEL, Z., VOGEL, T., ZAMIR, A., ELSON, D.: Ribosome activation and the binding of dihydrostreptomycin: effect of polynucleotides and temperature on activation. J. molec. Biol. **54**, 379 (1970).

VOGEL, Z., ZAMIR, A., ELSON, D.: The possible involvement of peptidyl transferase in the termination step of protein synthesis. Biochemistry **8**, 5161 (1969).

WAHBA, J. A., CHAE, Y.-B., IWASAKI, K., MAZUMDER, R., MILLER, M. J., SABOL, S., SILLERO, M. A. G.: Initiation of protein synthesis in *Escherichia coli*. I. Purification and properties of the initiation factors. Cold Spr. Harb. Symp. quant. Biol. **34**, 285 (1969).

WALLER, J.-P., ERDOS, T., LEMOINE, F., GUTTMANN, S., SANDRIN, E.: Inhibition of protein synthesis by aminoacyl 3'(2')-adenosine. Biochim. biophys. Acta (Amst.) **119**, 566 (1966).

WARNER, J. R., RICH, A.: The number of soluble RNA molecules on reticulocyte polyribosomes. Proc. nat. Acad. Sci. (Wash.) **51**, 1134 (1964).

WATANABE, S., TANAKA, K.: Effect of siomycin on the acceptor site of *Escherichia coli* ribosomes. Biochem. biophys. Res. Commun. **45**, 728 (1971).

WATERSON, J., BEAUD, G., LENGYEL, P.: The S_1 factor in peptide chain elongation. Nature (Lond.) **227**, 34 (1970).

WEBER, M. J., DEMOSS, J. A.: The inhibition by chloramphenicol of nascent protein formation in *E. coli*. Proc. nat. Acad. Sci. (Wash.) **55**, 1224 (1966).

WEBER, M. J., DEMOSS, J. A.: Inhibition of the peptide bond syntheizing cycle by chloramphenicol. J. Bact. **97**, 1099 (1969).

WEBSTER, R. E., ZINDER, N. D.: Fate of the message-ribosome complex upon translation of termination signals. J. molec. Biol. **42**, 425 (1969).

WEISBLUM, B., DAVIES, J.: Antibiotic inhibitors of the bacterial ribosome. Bact. Rev. **32**, 493 (1968).

WEISBLUM, B., DEMOHN, V.: Inhibition by thiostrepton of the formation of a ribosome-bound guanine nucleotide complex. FEBS Letters **11**, 149 (1970).

WEISSBACH, H., BROT, N.: Inhibition of transfer factor Ts by aurintricarboxylic acid. Biochem. biophys. Res. Commun. **39**, 1194 (1970).

WEISSBACH, H., KURYLO-BOROWSKA, Z., SZER, W.: (1) Effect of edeine on aminoacyl tRNA binding to ribosomes. Arch. Biochem. Biophys. **146**, 356 (1971).

WEISSBACH, H., REDFIELD, B., BROT, N.: (2) Further studies on the role of factors Ts and Tu in protein synthesis. Arch. Biochem. Biophys. **144**, 224 (1971).

WEISSBACH, H., REDFIELD, B., BROT, N.: (3) Aminoacyl tRNA-Tu-GTP interaction with ribosomes. Arch. Biochem. Biophys. **145**, 676 (1971).

WEISSBACH, H., REDFIELD, B., HACHMAN, J.: Studies on the role of factor Ts in aminoacyl-tRNA binding to ribosomes. Arch. Biochem. Biophys. **141**, 384 (1970).

WETTSTEIN, F. O., NOLL, H.: Binding of transfer ribonucleic acid to ribosomes engaged in protein synthesis: number and properties of ribosomal binding sites. J. molec. Biol. **11**, 35 (1965).

WETTSTEIN, F. O., NOLL, H., PENMAN, S.: Effect of cycloheximide on ribosomal aggregates engaged in protein synthesis in vitro. Biochim. biophys. Acta (Amst.) **87**, 525 (1964).

WHITE, J. P., CANTOR, C. R.: Role of magnesium in the binding of tetracycline to *Escherichia coli* ribosomes. J. molec. Biol. **58**, 397 (1971).

WILEY, P. F., MACKELLAR, F. A.: The structure of sparsomycin. J. Amer. chem. Soc. **92**, 417 (1970).

WILHELM, J. M., CORCORAN, J. W.: Antibiotic glycosides. VI. Definition of the 50 S ribosomal subunit of *Bacillus subtilis* 168 as a major determinant of sensitivity to erythromycin A. Biochemistry **6**, 2578 (1967).

WITTMANN, H. G., STOEFFLER, G., HINDENNACH, I., KURLAND, C. G., RANDALL-HAZALBAUER, L., BIRGE, E. A., NOMURA, M., KALTSCHMIDT, E., MIZUSHIMA, S., TRAUT, R. R, BICKLE, T. A.: Correlation of 30 S ribosomal proteins of *E. coli* isolated in different laboratories. Molec. gen. Genet. **111**, 327 (1971).

WOLFE, A. D., HAHN, F. E.: Erythromycin mode of action. Science **143**, 1445 (1964).

WOLFE, A. D., HAHN, F. E.: Mode of action of chloramphenicol. IX. Effects of chloramphenicol upon a ribosomal amino acid polymerization system and its binding to bacterial ribosome. Biochim. biophys. Acta (Amst.) **95**, 146 (1965).

YAMAGUCHI, H., TANAKA, N.: Site of action of mikamycins A and B in polypeptide synthesizing systems. J. Biochem. **61**, 18 (1967).

ZIMMERMANN, R. A., GARVIN, R. T., GORINI, L.: (1) Alteration of a 30 S ribosomal protein accompanying the *ram* mutation in *E. coli*. Proc. nat. Acad. Sci. (Wash.) **68**, 2263 (1971).

ZIMMERMANN, R. A., ROSSET, R., GORINI, L.: The nature of phenotypic masking exhibited by drug dependent *str* A mutants of *E. coli*. Pers. communication (1972).

Structural Features of Immunoglobulin Light Chains[1]

ALLEN B. EDMUNDSON, MARIANNE SCHIFFER, KATHRYN R. ELY, and MICAL K. WOOD

I. Introduction

Immunoglobulins are proteins with known antibody activity or with structural features closely resembling those of antibodies. The basic multi-chain structure of immunoglobulins consists of two light and two heavy chains linked by interchain disulfide bonds (PORTER, 1959, 1969; COHEN and MILSTEIN, 1967; HABER, 1968; EDELMAN and GALL, 1969; MILSTEIN and PINK, 1970). In the most common class of immunoglobulins, IgG[2], the protein is a monomer with a molecular weight of 145,000 to 160,000. Within a single molecule the two light chains (MW = 22,000 to 23,000) are identical, as are the heavy chains (MW = 50,000 to 55,000). In IgM proteins, the molecules are pentamers stabilized by disulfide bonds; and in IgA immunoglobulins, the molecules form dimers or higher aggregates. The heavy chains differ in molecular weight and chemical properties, but the light chains are similar in representatives of different classes of immunoglobulins. Free light chains excreted into the urine in patients with multiple myeloma are called Bence-Jones proteins (EDELMAN and GALLY, 1962), the presence of which is pathognomonic of the disease. These light chains are monoclonal and can be isolated in large quantities for chemical and physical studies. In the present article we shall concentrate on the structural features of the light chains, with emphasis on aspects important to consider in the crystallographic study of a Bence-Jones protein.

There are two principal antigenic types of light chains, K and L (= $\varkappa$ and λ chains). The antigenic specificity of antibodies is determined by the amino acid sequences of the constituent chains (KOSHLAND and ENGLBERGER, 1963; HABER, 1964; WHITNEY and TANFORD, 1965), and the two types of light chains can be distinguished by their sequences. Within each type, the sequences of the carboxyl ("constant") halves of the polypeptide chains are relatively invariant, while the amino halves (V) are variable both in sequence and in length. HILSCHMANN and CRAIG (1965) provided direct evidence for this generalization with partial sequences of two human $\varkappa$ chains, and it rapidly proved valid for other human (TITANI et al., 1965) and murine (GRAY et al., 1967) $\varkappa$ chains, human λ chains [PUTNAM et al., 1967 (1, 2); MILSTEIN et al., 1968; PONSTINGL et al., 1967, 1968], and light chains prepared from mixtures of functional γ-globulins (MILSTEIN, 1965).

In the "constant" (C) halves of human $\varkappa$ chains there is one position (191) in which isopolar substitution occurs. This position is associated with the antigenic locus Inv: in Inv a$^+$ proteins, leucine occupies this position, and valine is present in

[1] This work was supported by the U. S. Atomic Energy Commission.

[2] See Nomenclature for Human Immunoglobulins, Bull. Wld. Hlth. Org. **30**, 447 (1964).

Inv b+ proteins (MILSTEIN, 1966; BAGLIONI et al., 1966). The C halves of λ chains are slightly more variable. An arginine-lysine interchange associated with the Oz antigenic locus (EIN and FAHEY, 1967; APPELLA and EIN, 1967) is found in a region comparable to that of the Inv factor (i.e., in locations varying from 191 to 194 in different proteins). Serine-154, occurring in most λ chain sequences, is replaced by glycine in the Kern protein (PONSTINGL et al., 1967), and two isopolar substitutions are observed in the Mz protein [MILSTEIN, 1967 (1, 2)] in positions 145 (Val for Ala) and 173 (Asn for Lys).

In the "variable" (V) regions the sequences of either κ chains or λ chains may differ as much as 50% (see PUTNAM, 1969; also Refs. listed in Footnote 3). If chemical homologies are assumed, however, the sequences of chains in both major classes can be divided into subgroups[4], within which members do not vary by more than about 25% [NIALL and EDMAN, 1967; SMITHIES, 1967; HOOD et al., 1967, 1970; HOOD and TALMAGE, 1970; MILSTEIN, 1967 (1), 1969; HILSCHMANN et al., 1969]. The chain length is usually maintained within a subgroup, but may change from subgroup to subgroup. For example, κ chains of subgroups I and III are shorter than those of group II by four to six residues with the gap occurring around residue 30. In two λ chain groups there is a characteristic gap of three residues beginning at position 28.

Despite these differences it is apparent that the similarity in chain length of the V and C parts has been conserved during evolution. The positions of the two *intra*chain disulfide bonds have also been conserved. Each pair of half-cystine residues connects a loop of approximately 60 residues, which are found in homologous positions in the V and C halves (see Refs. in Footnote 3). The *inter*chain half-cystine residue involved in linking light and heavy chains is the C-terminus in κ chains and the penultimate residue in λ chains (MILSTEIN, 1965; HILSCHMANN and CRAIG, 1965; TITANI et al., 1965).

The switch from V to C parts occurs in homologous positions (i.e., residues 107 to 110) in the κ and λ chains. The structures are such that the protein can be cleaved into V and C parts by proteolysis in the switch regions (DEUTSCH, 1963; CIOLI and BAGLIONI, 1966; SOLOMON and MC LAUGHLIN, 1969; SOLOMON et al., 1970; KARLSSON et al., 1969; BJÖRK et al., 1971; SCHRAMM, 1971). Optical rotatory dispersion (ORD) and circular dichroism (CD) studies show that the conformational changes accompanying this cleavage are minimal (BJÖRK et al., 1971). Hydrodynamic measurements (KARLSSON et al., 1969) and calculations of "average hydrophobicities" [WELSCHER, 1969 (1)] support the hypothesis that both V and C parts have compact globular structures. It is of particular interest that the V parts can be crystallized in some cases (SOLOMON et al., 1970; SCHRAMM, 1971).

Factors affecting the crystallization of light chains and their fragments, as well as the properties that are important in the determination of the three-dimensional structures of these molecules, will be discussed in the following sections. The 6 Å electron density map of a λ-type Bence-Jones (Mcg) dimer will also be described.

---

[3] For general reviews of the subject, see Gamma Globulins, Proc. Nobel Symp., vol. 3 (1967); Cold Spr. Harb. Symp. quant. Biol. **32** (1967); FEBS Symp. **15** (1969).

[4] The subgroup classifications were proposed at the Conference on Nomenclature for Animal Immunoglobulins, Prague, 1969. See J. biol. Chem. **245**, 3033 (1970).

II. Association of Light Chains

Bence-Jones proteins and light chains of human IgG proteins exist chiefly as dimers at neutral pH in aqueous salt solutions (GALLY and EDELMAN, 1964; BERGGÅRD and PETERSON, 1969). In solvents like acetic or propionic acid, some but not all dimers dissociate to monomers. MILSTEIN (1965) showed that the stable dimers are composed of two polypeptide chains covalently linked by an interchain disulfide bond homologous with that between light and heavy chains. The second participant in the corresponding disulfide bond in a "monomer" is a half-cystine residue (MILSTEIN, 1965). The conversion of a covalently linked dimer to a monomer may occur in the urine by a disulfide interchange reaction with free cystine. The two forms, which are not strictly monomers and dimers in the chemical usage of $2M \rightleftarrows M_2$, can be represented as M-S-S-Cys and M-S-S-M.

In comparing "covalent" and "non-covalent" dimers of a $\varkappa$-type Bence-Jones protein, SEON et al. (1969) found no significant differences in the results of ORD and CD measurements. However, one tyrosine residue in the non-covalent dimer was iodinated more rapidly than its counterpart in the covalently linked dimer. SEON et al. interpreted this result as an indication of a conformational difference between the two dimers. The two complexes also behaved differently in the ultracentrifuge.

III. Thermal Behavior of Bence-Jones Proteins

The distinct thermal properties of the Bence-Jones proteins provide the basis for one of the earliest clinical tests for a specific disease (BENCE-JONES, 1848). As acidified urine from a patient with multiple myeloma is heated to 50 to 60°, the Bence-Jones protein precipitates. At ~100° the protein dissolves, but precipitates again as the urine is cooled. NEET and PUTNAM (1966) and WELSCHER [1969 (1)] suggested that the thermal behavior of Bence-Jones proteins is strongly influenced by hydrophobic interactions. This suggestion is supported by the properties of such interactions, which are endothermic and enhanced with increasing temperature to about 60° (NÉMETHY and SCHERAGA, 1962). Above this temperature the process is exothermic.

After examining purified samples of the variable and constant halves of a human $\varkappa$ chain, SOLOMON and MC LAUGHLIN (1969) reported that the V part alone was responsible for the distinct thermal properties of the *intact* Bence-Jones protein. However, the "average hydrophobicities" are not significantly different in the V and C parts (see Section IV. 1.), and it is not clear why the C halves do not contribute more substantially to the thermal behavior. A comparison of the three-dimensional structures of the amino and carboxyl halves of the Bence-Jones protein under crystallographic investigation may eventually provide a partial explanation for the differences in thermal properties.

IV. Distribution of Polar and Apolar Residues in Amino Acid Sequences of Light Chains

In proteins like the immunoglobulins, for which the three-dimensional structures are just beginning to be determined, correlation of local structures with association properties is not possible, but an overview of structural features can be informative.

As a generalization, associating proteins have apolar residues which are not accommodated in the interior of the molecule and therefore are available for interchain hydrophobic interactions (Fisher, 1964; Bigelow, 1967). Association resulting from a high proportion of apolar residues is more likely to occur if the fractional charge is low. Both conditions are met in the light chain, and association is favored [Welscher, 1969 (1)].

1. "Average Hydrophobicities" and Fractional Charge

Using procedures developed by Tanford (1962) and Bigelow (1967), Welscher [1969 (1)] calculated the "average hydrophobicities"[5] ($H\phi_{av}$) and the "fractional charges"[6] (FCh) of light chains for which the *sequences* were known. The $H\phi_{av}$ values were taken as a measure of the contribution of hydrophobic interactions to the stability of the globular conformation. Mean values, which were almost identical in the $\varkappa$ and λ chains (1000 *vs* 1010 cal/residue), were well above the lower limit (~900 cal/residue) set for a stable globular conformation by comparison with proteins of known structure. When considered separately, the V and C parts met the same criteria for globular folding. Values calculated from amino acid *compositions* of light chains from various animals also clustered around the value of 1000 cal/residue.

The fractional charges, with mean values of 0.20 and 0.18 units/residue in *human* $\varkappa$ and λ chains were similar in the light chains from eleven species. Values for variable parts (0.18 and 0.16 units/residue for $\varkappa$V and λV) were slightly lower than those of the constant parts (0.22 and 0.19 for $\varkappa$C and λC). These values were higher than those (0.10 to 0.13) for subtilisin and chymotrypsinogen A, but lower than those for hemoglobin chains (0.25 to 0.27), cytochrome c from different species (0.31 to 0.35), and sperm whale myoglobin (0.36).

2. Apolar Residues and Globular Folding

In addition to their assumed influence on association and thermal properties, apolar residues are believed to be important in the stabilization of three-dimensional structures of individual polypeptide chains. In examining the frequency and distribution of such residues, Welscher [1969 (2)] found common patterns in the sequences of 15 $\varkappa$ and λ chains. The V and C halves each contained about 55 apolar residues (of a total of 107 to 110). Of these, 30 sites in the V parts and 33 positions in the C regions were occupied by apolar residues in all light chains examined. However, the distribution of apolar sites was significantly different in the V and C parts. On the assumption that globular folding is largely stabilized by intrachain hydrophobic interactions, the V parts in the 15 light chains should have similar folding. The same suggestion

[5] Free enthalpy change (ΔF) when an apolar side-chain goes from an aqueous phase into an environment of lower dielectric constant, such as the non-aqueous interior of a globular protein molecule. The movement results in a destabilization of the unfolded state of the polypeptide chain. Values of hydrophobicity (in cal per mole of side-chain) depend on the structure of the side-chain. In decreasing order in the range of 3000 to 440 cal/mole, tryptophan, isoleucine, tyrosine, phenylalanine, proline, leucine, valine, lysine, methionine, half-cystine, alanine, arginine, and threonine were assumed to contribute to the hydrophobicities of light chains. Average hydrophobicity is defined as the Σ hydrophobicities/total no. of residues.

[6] Mole fraction of acidic and basic residues in the proteins.

can be made for the C parts, but the V and C parts probably do not have the same structures.

Such suggestions are based on proteins like the myoglobins and hemoglobins, in which one type of globular structure is found in animals as diverse as mammals (KENDREW, 1962; PERUTZ et al., 1965; WATSON, 1969), lamprey (HENDRICKSON et al., 1971), a marine annelid (PADLAN and LOVE, 1968), and *Chironomus* larvae (HUBER et al., 1971). The structures of the globins are stabilized by hydrophobic interactions involving residues in about 30 internal sites (KENDREW, 1962; PERUTZ et al., 1965; WATSON, 1969). There are only about 6 of 150 sites that are occupied by the same residues throughout the phylogenetic tree (BRAUNITZER, 1966), but isopolar substitution is consistently observed in the 30 apolar positions.

3. Prediction of Helices

Detailed comparisons have to be made with caution, because the globins are ~75% helical, and the light chains are not. In the globin helices the conserved apolar sites tend to lie along "hydrophobic arcs" (SCHIFFER and EDMUNDSON, 1967). Similar arrangements of apolar residues (i.e., in n, $n \pm 3$, $n \pm 4$ positions) are found with much lower frequency in sequences of light chains. Predictions based on the "helical wheels" method (SCHIFFER and EDMUNDSON, 1967), alone [WELSCHER, 1969 (2)] or in combination with a nearest neighbor analysis of tripeptide sequences (WU and KABAT, 1971) indicate a low degree of helicity (10 to 15%) in the light chains. The predictions are consistent with the results of ORD and CD measurements (HAMAGUCHI and MIGITA, 1964; JIRGENSONS et al., 1966). Moreover, there is no evidence for long helices in the electron density map, although short segments of one to two turns might not be conspicuous at the present resolution of 6 Å (see Section VI. 5.).

4. Graphical Procedures to Illustrate the Distribution of Polar and Apolar Residues

The distribution of polar and apolar residues can be examined in more detail with graphical procedures such as those shown in Figs. 1 and 2 (EDMUNDSON et al., unpublished results). In Fig. 1 the ratios of the summed molal volumes of the polar (V_p) and apolar (V_a) residues for nonapeptide segments of the Ag $\varkappa$ chain (TITANI et al., 1965; PUTNAM et al., 1967 (2)] and the Bo λ chain [PUTNAM et al., 1967 (1)] are plotted against the residue numbers. Plots of the ratios of the numbers of polar (N_p) and apolar (N_a) residues for the same segments are presented in Fig. 2.

To construct the graphs the sequences were first divided into polar and apolar groups of amino acid residues We included Lys, His, Arg, Asp, Asn, Glu, Gln, Ser, and Thr in the polar group and Pro, Gly, Ala, Cys, Val, Ile, Leu, Met, Tyr, Phe, and Trp in the apolar group. These assignments are open to question in specific cases, especially if they are correlated with hydrophobicity and the tendency to be "inside" or "outside" in a protein for which the three-dimensional structure is unknown. For example, both tyrosine and tryptophan have polar groups on their aromatic rings, but act *predominantly* as hydrophobic residues in the myoglobins and hemoglobins (WATSON, 1969). Two of the three Tyr and both Trp residues in the mammalian globins occupy internal positions. Using the reactivity of tyrosine side-chains to iodination as a reflection of exposure to solvent, SEON et al. (1971) concluded that

Tyr-49, -91, and -173 were accessible to solvent in a Bence-Jones $\varkappa$ chain (Col). Five Tyr residues (-36, -86, -140, -186, and -192) were less reactive, and one residue (-87) was considered to be "buried".

Since both local interactions and interactions between residues from different segments of the chain are important in the maintenance of the conformation of a protein, attention was given to the choice of the length of segments used for the calculations. Sequences of 5, 7, 9, 11, and 13 residues were examined. Selection of a

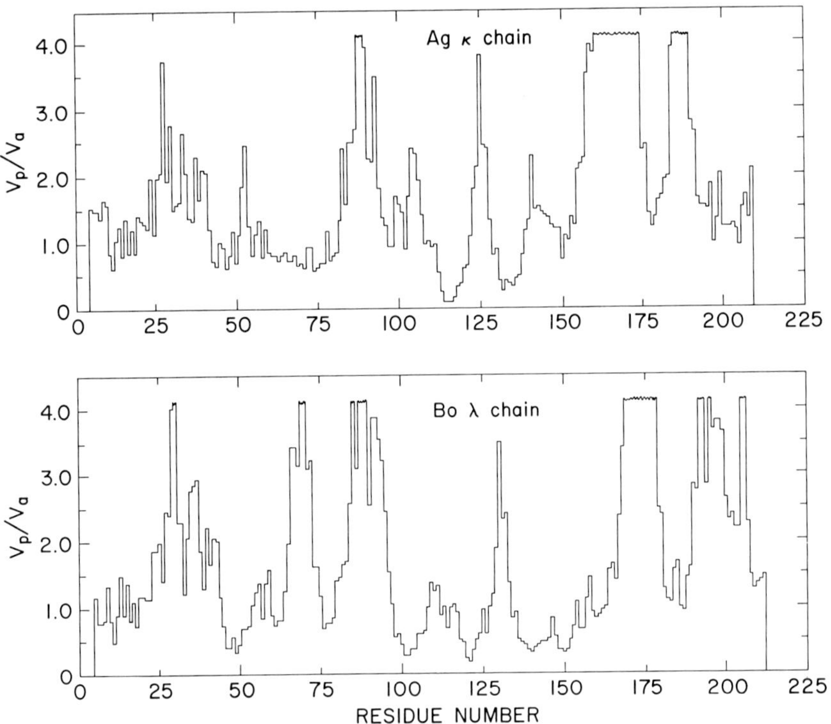

Fig. 1. Plots of V_p/V_a ratios in all possible nonapeptide segments of the Ag $\varkappa$ chain and the Bo λ chain. Each ratio is plotted at the midpoint of the segment (i.e., at the fifth residue). V_p and V_a represent the combined molal volumes of polar and apolar residues in each segment

nonapeptide sequence offers a reasonable compromise between the wider fluctuations associated with the shorter segments and the damping effects noted with longer sequences.

A polypeptide chain of nine residues can vary in length from about 13.5 Å in an α-helix (1.5 Å per residue) to about 31 Å in a fully extended chain ($\sim$3.5 Å per residue). The conformations of the nonapeptide segments within an intact protein can be determined by X-ray diffraction, but it should be mentioned that the conformations of free nonapeptides with identical sequences might be (and probably will be) different.

The ratios of V_p/V_a were calculated for all possible nonapeptides in a light chain. For example, the first sequence included residues 1 to 9, the second 2 to 10, etc. Each ratio was plotted against the residue number at the midpoint of the segment: e.g., the value for residues 1 to 9 was listed at residue 5. The points for the sequence of the entire protein were connected as in a bar graph. High values and "peaks" on the plots indicate the preponderance of polar side-chains, while valleys between peaks reflect the influence of apolar residues.

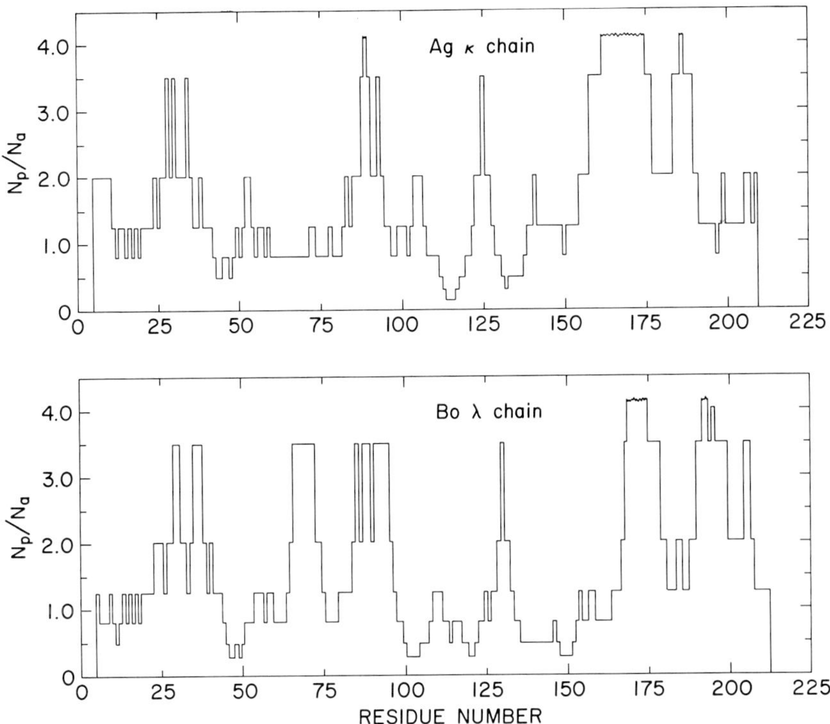

Fig. 2. Plots of N_p/N_a ratios in nonapeptide segments of the Ag κ chain and the Bo λ chain. N_p and N_a correspond to the combined numbers of polar and apolar residues in each segment

The graphs can be conveniently compared in halves. Because of variations in chain length of Bence-Jones proteins, the graphs may have to be shifted along the abscissa to maximize the correspondence. To align the carboxyl halves, for example, the Ag graph should be moved the equivalent of six residues to bring the peak centered around residue 125 into the position occupied by the corresponding peak around residue 131 in the Bo plot. When this change is made, the graphs for the C-terminal halves of the two classes of human Bence-Jones proteins are quite similar except for two regions: (1) the segments around residue 200 and (2) the peak around residue 140 in the Ag plot which is not present in the appropriate segment of the Bo graph.

The most conspicuous features of the graphs for the carboxyl halves are the two large peaks reflecting the disproportionately high polar fractions beyond residue 150. The initial segments of the C parts (i.e., the third quarters of the polypeptide chains) have the highest proportion of apolar residues in the proteins, a distribution resulting in fewer peaks and lower average values in the V_p/V_a plots.

As expected, the graphs differ more widely in the amino halves. For example, the sizable peak centered around residue 70 in the Bo plot is absent in the Ag graph. Moreover, the peaks encompassing residues 80 to 95 are considerably lower in some examples not illustrated, such as the Roy $\varkappa$ chain [Hilschmann and Craig, 1965; Hilschmann, 1967 (1)] and the X λ chain (Milstein et al., 1968). General features of the graphs for $\varkappa$ and λ chains are similar in the first 25 residues, which are widely used in the studies of genetic variations and in the identification of subgroups to which the proteins belong.

The plots of the ratios (N_p/N_a) of the *numbers* of polar and apolar residues approximate the V_p/V_a graphs in most of the nonapeptide sequences (compare Figs. 1 and 2) because the average molal volumes of the two types of residues are almost the same in the light chains. The results of the N_p/N_a calculations are step values ranging from 0 ($N_p/N_a = 0/9$) to ∞ ($N_p/N_a = 9/0$).

5. Application of Fisher's Equations to Light Chains

We used the V_p/V_a ratios in Fisher's equations (1964, 1965, 1967) to set up a model for the *limiting* case in which *all* polar residues are external, and apolar residues are internal. Application of these equations seemed to be of particular interest in the light chains because Fisher (1964, 1967) correlated the proportions of polar and apolar residues in other proteins with the tendency to associate. For example, the results of applying the equations indicate correctly that myoglobin should be a monomer in solution, while the α and β chains of hemoglobin, as well as insulin and glucagon, should associate.

In the equations the ratio of V_p/V_a for a chain is used to define the quantity p. The value of p is compared with p_s, which is equal to p if the protein is a spherical molecule:

$$p_s = \frac{r^3}{(r-d)^3} - 1; \text{ where } r = 3\sqrt{\frac{3}{4\pi}(V_p + V_a)}, \text{the radius of a sphere of volume } (V_p + V_a),$$

and d = thickness of the external shell of polar residues; d is taken to be 4 Å.

For most globular proteins with molecular weights up to about 50,000, p/p_s values larger than about 0.85 are characteristic of monomeric molecules. For example, sperm whale myoglobin has a p/p_s value of 0.90, while the α and β chains of human hemoglobin, porcine insulin, and porcine glucagon have ratios of 0.69, 0.66, 0.46, and 0.54, respectively.

The p/p_s ratios for examples of $\varkappa$ and λ chains are presented in Table 1. The values varied from 1.05 to 1.37, reflecting the diversity in the overall compositions of the proteins. These high values imply that the number of polar residues is more than adequate to cover an apolar center if a light chain is a spherical or nearly spherical molecule. In addition, the light chains should be capable of forming stable monomers in solution.

As mentioned previously, however, the predominant form of light chains in aqueous, neutral salt solutions is the dimer, which is probably not spherical in solution. In an ultracentrifugal study of a dimer, for example, HOLASEK et al. (1963) obtained a value of 1.19 for the frictional ratio f/f_0. This ratio was coupled with the hydration value given by small-angle X-ray scattering measurements (HOLASEK et al., 1963) to calculate an axial ratio of 1:2.5 for an ellipsoid of revolution. The parallel analysis of the X-ray scattering curves indicated particle axes of 21.0, 48.3, and 74.8 Å (axial ratios = 1:2.3:3.6). Comparison of the two methods cannot be exact because frictional ratios are not determined for triaxial particles, but an ellipsoid of revolution derived from the X-ray data had an axial ratio of 1:2.8.

Table 1. Values of p/p_s for Bence-Jones proteins. The values were calculated with an IBM 1131A computer, used in conjunction with FISHER's equations (1964, 1965, 1967). After computation of ratios for the entire proteins, the sequences were divided into halves and quarters, and the calculations were repeated. Because the carboxyl halves of molecules from each class and species are relatively invariant, the results for these segments are given for only one representative of each group. References for the sequences are listed at the bottom of the table

Protein	Entire Molecule	Amino Half	Carboxyl Half	First Quarter	Second Quarter	Third Quarter	Fourth Quarter
Human ϰ							
Ag	1.31	0.86	0.91	0.54	0.55	0.37	0.94
Eu	1.26	0.80		0.46	0.56		
Cum	1.24	0.78		0.43	0.59		
Ti	1.23	0.76		0.46	0.55		
Roy	1.18	0.70		0.45	0.44		
Mil	1.14	0.66		0.38	0.46		
Murine ϰ							
41	1.37	0.85	1.02	0.49	0.60	0.34	1.33
70	1.16	0.62		0.33	0.46		
Human λ							
Bo	1.22	0.82	0.82	0.52	0.52	0.28	1 00
Hul	1.20	0.79		0.44	0.57		
Ha	1.17	0.76		0.57	0.41		
Sh	1.17	0.77		0.46	0.51		
X	1.15	0.73		0.54	0.40		
New	1.09	0.66		0.46	0.38		
Kern	1.05	0 62		0.41	0.38		

Human ϰ *Ag:* TITANI et al. (1965); PUTNAM et al. [1967 (2)]; *Eu:* CUNNINGHAM et al. (1968); *Cum:* HILSCHMANN [1967 (2)]; *Ti:* SUTER et al. (1969); *Roy:* HILSCHMANN and CRAIG (1965); HILSCHMANN [1967 (1)]; *Mil:* DREYER et al. (1967).
Murine ϰ *41* and *70:* GRAY et al. (1967).
Human λ *Bo:* PUTNAM et al. [1967 (1)]; *Hul:* EDMUNDSON et. al. (1968); *Ha:* PUTNAM et al. [1967 (1)]; *Sh:* WIKLER et al. (1967); *X:* MILSTEIN et al. (1968); *New:* LANGER et al. (1968); *Kern:* PONSTINGL et al. (1967).

As a detailed method of analysis, the use of FISHER's equations can be criticized because the distinction between the "inside" and "outside" of molecules is not always clear and because side-chains accessible to solvent are not all polar. However, the discrepancies between theory and observation are far greater than expected in the light chains. We think that such large discrepancies can occur only if the light chains are composed of independently folded regions. Under this assumption treatment of an entire polypeptide chain as one globular unit is inappropriate.

Accordingly, we looked for variations in the Fisher index in different parts of the molecules. The p/p_s values obtained when the V and C halves were treated as independent molecules are given in Table 1. The results can be used to classify the fragments most likely to associate or remain monomeric by Fisher's criteria. For example, the amino halves of the Mil, murine 70, New, and Kern proteins are expected to form higher complexes, while the V parts of Ag, murine 41, and Bo proteins are more likely to form monomers. In this regard it is of interest to calculate the p/p_s ratios for isolated V or C fragments. The V fragment obtained from the Au $\varkappa$ chain (Schramm, 1971) exists as a dimer in solution at neutral pH, and its p/p_s ratio is 0.59.

The isolation of stable fragments significantly smaller than half a light chain has not been described. The smallest spherical monomer with an apolar center and a polar cover would have a molecular weight of 5,000 to 6,000, or approximately one-fourth the size of a light chain. When the sequences of light chains are divided into quarters, the calculated p/p_s ratios indicate that only the final quarters could possibly form molecules of this type (see Table 1). It is the high proportion of polar residues which accounts for this result (see Figs. 1 and 2). The contrast with the low values for the third quarters underlines the conclusion that the residues available for inter- or intra-chain hydrophobic interactions in the C parts are distributed unevenly, with the concentrations heaviest before position ~160.

6. Evolutionary Models for the Light Chains

On the basis of internal homologies in the sequences of immunoglobulins, Singer and Doolittle (1966) and Hill et al. (1966) proposed that a primitive light chain gene was half the size of modern genes. This proposal is consistent with the isolation of relatively stable, globular half-molecules which crystallize. The stability and globular folding of these fragments can be predicted with the aid of simple calculations of "average hydrophobicities" and Fisher indices. Surprisingly, the latter calculations also suggest that a primordial molecule only one-fourth as large as the present light chains could also have existed if its composition were similar to those of the last quarters of the proteins listed in Table 1.

V. Criteria of Purity in the Bence-Jones Protein Used in the Crystallographic Study

1. Molecular Species

Calibrated Sephadex columns can be used to separate the M-S-S-M (MW 46,000) and M-S-S-Cys (MW 23,000) forms of the Bence-Jones proteins and to determine the relative amounts of the two forms in the samples (Edmundson et al., 1969). The range of diversity in samples of different proteins is illustrated in Fig. 3. The Dil $\varkappa$-type protein and the Gos λ-type protein were obtained as mixtures of M-S-S-M (peak 2 in

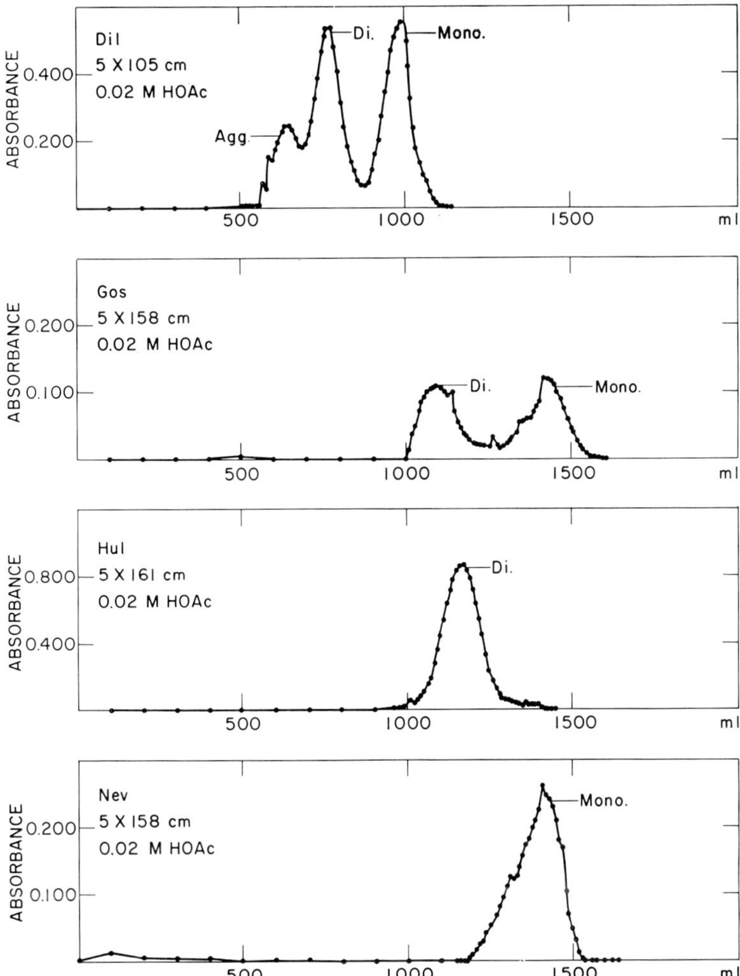

Fig. 3. Behavior of Bence-Jones proteins on Sephadex G-75 at 4°, with 0.02 M acetic acid as eluent. The absorbance at 280 nm is plotted against the effluent volume. Dimensions of columns are given in the appropriate panels. Fractions of 10 ml were collected. The Dil $\varkappa$ chain and the Gos λ chain were equal mixtures of M-S-S-M (MW = 46,000: peak 2 in the Dil tracing) and M-S-S-Cys (MW = 23,000: Dil peak 3) molecules; the Hul λ chain consisted solely of M-S-S-M molecules; and the Nev λ chain was composed of M-S-S-Cys molecules. Component 1 of the Dil protein corresponds to a conglomerate of the type $(\text{M-S-S-M})_a$ $(\text{M-S-S-Cys})_b$, which is dissociated in 0.2 M acetic acid

the Dil elution profile) and M-S-S-Cys (peak 3) molecules; the Hul λ chain was isolated exclusively as M-S-S-M forms; and the Nev λ chain was composed of M-S-S-Cys molecules. In most cases the M-S-S-M and M-S-S-Cys forms did not associate under the mildly acidic conditions (pH 3.2) used in the fractionation, but the Dil protein was an exception (see peak 1 in Fig. 3). A conglomerate of the mixed type $(\text{M-S-S-M})_a$ $(\text{M-S-S-Cys})_b$ was detected in both 0.02 M acetic acid and 0.5 M NaCl. The conglomerate was dissociated in 0.2 M acetic acid (pH 2.7).

It was assumed that homogeneity with regard to the molecular species was an important factor to consider in crystallization of the Bence-Jones proteins. This assumption was supported by the properties of the Mcg λ chain used in the crystallographic study to be described. The behavior of this protein on a Sephadex column is shown in Fig. 4 (SCHIFFER et al., 1970). The column was calibrated with the Dil and Hul proteins. Crystalline samples of Mcg protein consisted only of M-S-S-M molecules. The chromatographic results were consistent with the value of 45,500 obtained

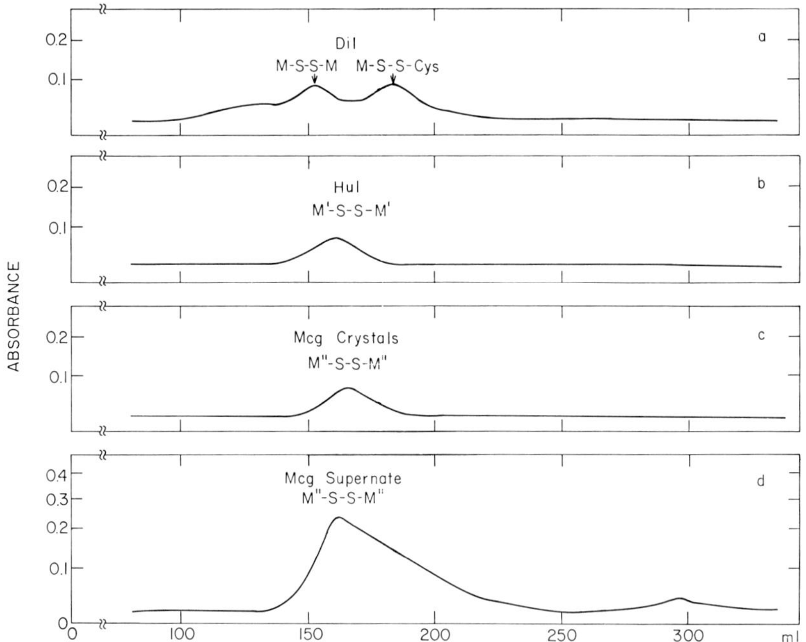

Fig. 4. Behavior of a crystalline sample of a Bence-Jones λ chain (Mcg) on a 2 × 101 cm column of Sephadex G-100, equilibrated at 4° with 0.5 N NaCl. The column was calibrated with the Dil ϰ chain and the Hul λ chain (see Fig. 3). The two samples of the Mcg protein were dissolved crystals (see panel c) and the supernate from the suspension of crystals (see panel d). Impurities in the supernate (see second band in panel d) were not found in the crystals even when samples three times as large were placed on the column

for the molecular weight of the twice-crystallized protein by the sedimentation equilibrium method (H. F. DEUTSCH, personal communication).

2. Classification and Purity

The principal antigenic class of a Bence-Jones protein is usually identified by reaction with antisera to ϰ and λ chains. Because of the characteristic C-terminal sequences of the ϰ and λ chains, carboxypeptidase A can also be used to classify the proteins and simultaneously to assess their purity (EDMUNDSON et al., 1969). The C-terminal half-cystine residue in ϰ chains acts as a barrier to the action of carboxypeptidase A and no amino acids are released. Serine is the only amino acid liberated from pure λ chains, which have a C-terminal sequence of Cys-Ser. The presence of other amino acids indicates contaminants, which can generally be removed by a single passage through an appropriate Sephadex column, or in the case of the Mcg protein, by recrystallization.

While the supernatant over the Mcg crystals (see Fig. 4) contained impurities from which relatively large quantities of amino acids (i.e., up to 50% of the serine value) were released by carboxypeptidase A, the twice-crystallized protein and the column-purified sample yielded 1.9 to 2.0 residues of serine per M-S-S-M molecule and contaminants ranging to only ~10% of the serine value (SCHIFFER et al., 1970). The reactions with rabbit antisera verified that the Mcg protein was a λ chain.

3. Bence-Jones Glycoproteins

Some Bence-Jones proteins (e.g., the Hul and Nev λ chains) are glycoproteins (EDMUNDSON et al., 1968, 1970; MELCHERS, 1969; MELCHERS and KNOPF, 1967; ABEL et al., 1968; SOX and HOOD, 1970), but the carbohydrate moieties tend to be heterogeneous. Because of this heterogeneity, Bence-Jones proteins devoid of carbohydrate seem more likely to crystallize. The glycoprotein linkages are believed to be acyl-glucosamine bonds with asparaginyl residues, and analysis of amino sugars are sufficient to indicate the presence of carbohydrate. Such an analysis of the Mcg protein showed that glucosamine and galactosamine were absent (EDMUNDSON et al., 1971).

VI. Crystallography of the Mcg Bence-Jones Protein

1. Properties of the Crystals

H. F. DEUTSCH first crystallized the Mcg protein by dialysis against distilled water at 4°. Crystals sufficiently large for diffraction (i.e., $0.3 \times 2.0 \times 0.5$ mm $= a \times b \times c$) were prepared at 20° by dialysis against water with a microdiffusion technique (ZEPPEZAUER et al., 1968). The crystals were colorless and bladed, and were bounded by {100} and {001} faces (SCHIFFER et al., 1970). A photograph of one of these crystals is shown in Fig. 5.

The *mmm* symmetry of the three-dimensional diffraction pattern and the systematic extinctions observed on precession photographs (i.e., for $h00$, $h = 2n + 1$; and for $0k0$, $k = 2n + 1$) indicated an orthorhombic unit cell, with a $P2_12_12$ space group. The unit cell dimensions were $a = 72.6 \pm 0.2$; $b = 81.9 \pm 0.2$; and $c = 71.0 \pm 0.2$ Å. The asymmetric unit was the M-S-S-M dimer, four of which were present in the unit cell. The calculated fractional volume of solvent (V_{solv}) in the crystals was 0.46, a value close to those (~0.43) found most frequently in other protein crystals (MATTHEWS, 1968).

It was difficult to prepare isomorphous derivatives because the crystals of the orthorhombic form tended to dissolve in the heavy-atom solutions. A second crystal form without this disadvantage was obtained in 1.6 to 1.9 M ammonium sulfate, buffered at pH 6.2 with phosphate (EDMUNDSON et al., 1971). The prismatic crystals had 32 symmetry and were elongated along the *c* axis. A photograph of the crystals is presented in Fig. 6.

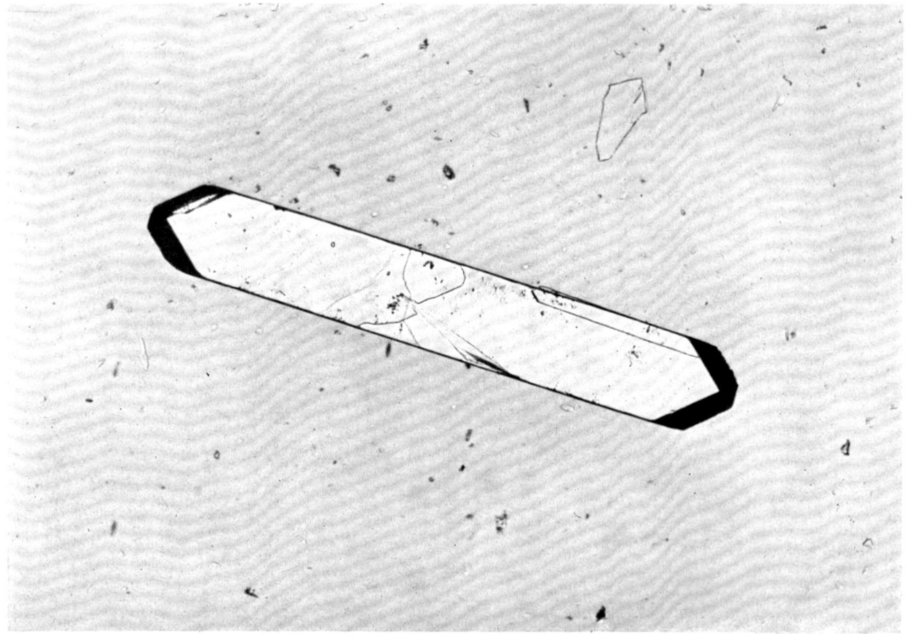

Fig. 5. Photograph of a crystal of the orthorhombic form of the Mcg Bence-Jones protein. The dimensions of this crystal are 2.4 × 0.4 × 0.25 mm (l × w × h). The protein was crystallized by dialysis against water

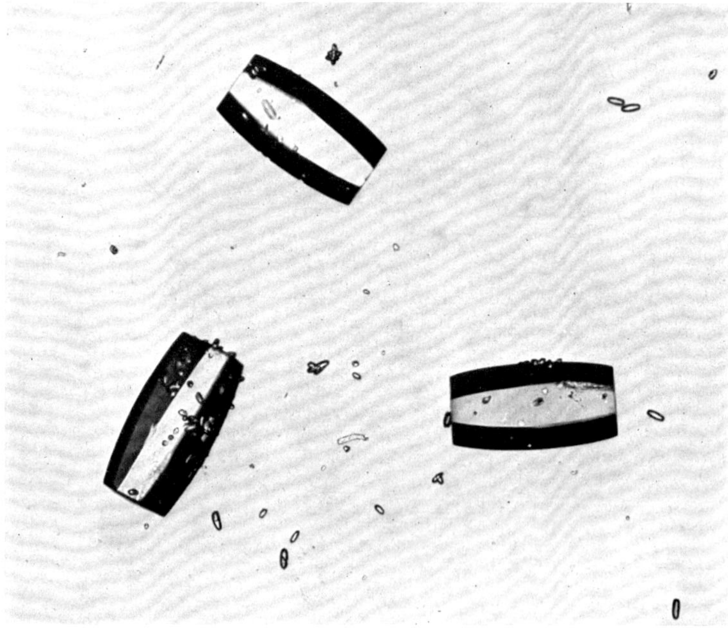

Fig. 6. Photograph of crystals of the trigonal form of the Mcg Bence-Jones protein. The protein was crystallized in 1.9 M ammonium sulfate buffered at pH 6.2 with phosphate

The space group was trigonal $P3_121$ or its enantiomorph $P3_221$, with unit cell dimensions of $a = 72.3 \pm 0.2$ and $c = 185.9 \pm 0.6$ Å. The dimer was the asymmetric unit, with six molecules in the unit cell. The V_{solv} value was calculated to be 0.60.

2. Preparation of Isomorphous Derivatives

Heavy-atom derivatives were prepared by soaking crystals for four to six weeks in solutions of potassium chloroplatinite, methyl mercuric chloride, phenyl mercuric acetate, mersalyl sodium, and *o*-chloromercuriphenol. The results with the platinum derivative were less satisfactory than those for the mercurials, the last two of which were used in the calculation of the 6 Å electron density map.

3. Collection of Data

The long *c* axis (186 Å) in the trigonal form made some reflections difficult to resolve, as illustrated in the precession photograph of the *hh.l* projection (Fig. 7). To improve the resolution, crystals were cut at right angles to *c* to give a length of ~0.5 mm. The fragments were aligned with *c* across the capillary. Data were collected with a Picker diffractometer, controlled by an IBM 1131A computer (MUELLER et al.,

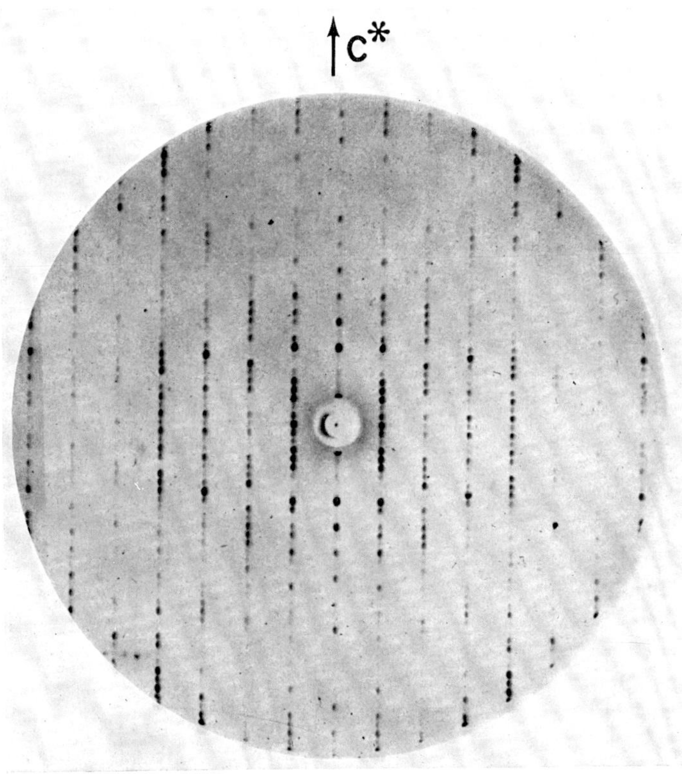

Fig. 7. A 9° precession photograph of the $h\bar{h}.l$ projection for a crystal of the trigonal form of Bence-Jones protein. Note the close spacing of reflections in the c^* direction

1968). Differences in the settings of the ⌀ angles were used to separate the closely spaced reflections. Omega step scans at intervals of 0.03° were employed, and the highest three of six intensity values were summed for each reflection. Data collected by this method have been reproducible to the extent that $\sum|\Delta F|/\sum F = 0.06$.

4. Location of the Heavy-Atom Positions

It is possible to determine all three coordinates of a heavy-atom site by examination of a [110] difference Patterson projection, the only centric projection in the $P3_121$ space group. The difference Patterson projection for the mersalyl derivative is shown

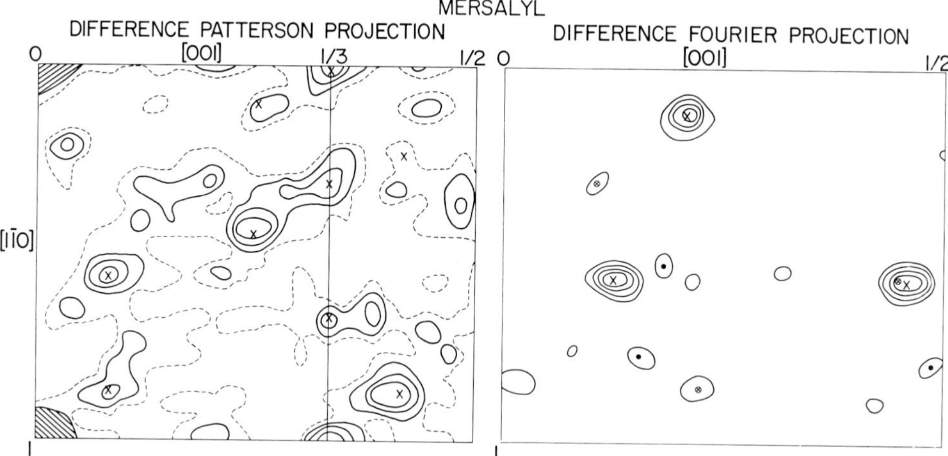

Fig. 8. Difference Patterson and Fourier projections along the [110] axis for the mersalyl derivative of the trigonal form of Bence-Jones protein. Half the unit cell is illustrated. Six double-weight and three single-weight Harker peaks are expected for a single-site derivative. The peaks used to determine the position of the major mersalyl site are indicated by X's. In the second panel, the final difference Fourier projection for half the unit cell shows the positions of the two minor sites (designated by dots and circled X's; for each site three peaks are expected)

in Fig. 8. If one heavy-atom is present in each asymmetric unit, six double-weight and three single-weight Harker peaks are expected in this projection for one-half the unit cell. The peaks used to define the position of the principal site are marked with X's in Fig. 8. The major site occupied by *o*-chloromercuriphenol was located by the same procedure.

Using centric data, the positions, occupancies, and isotropic temperature factors of the two derivatives were refined together by alternate cycles of phase determination and least-squares refinement (Dickerson et al., 1961, 1968). After refinement the improved phases were used to calculate difference Fourier projections, which revealed two minor sites for the mersalyl derivative (see second panel in Fig. 8) and one secondary site for the *o*-chloromercuriphenol derivative. Using all data, the heavy-atom parameters were further refined to the values listed in Table 2. Note that the secondary mersalyl sites are close to the *o*-chloromercuriphenol positions. When examined on the electron density map, all the heavy-atom sites appear to be located on or near the surface of the protein molecule.

Table 2. Final heavy-atom parameters

Derivative	Site	x	y	z	B	A	E	R_K (all)	R_C (centric)	R_f (centric)
Mersalyl	1	0.440	0.872	0.129	56.9	717	394	0.082	0.55	0.19
	2	0.440	0.154	0.107	8.5	120				
	3	0.209	0.452	0.155	5.8	85				
o-chloromercuriphenol	1	0.420	0.142	0.115	9.4	319	398	0.077	0.46	0.23
	2	0.195	0.476	0.156	12.5	240				

B = temperature factor; A = occupancy; E = lack of closure error on an arbitrary scale
Average protein F = 3040 on this scale
Figure of merit = 0.66

$R_K = \frac{\Sigma |F_{PH(obs)} - F_{PH(calc)}|}{\Sigma F_{PH(obs)}}$ $R_C = \frac{\Sigma |F_{H(obs)} - F_{H(calc)}|}{\Sigma F_{H(obs)}}$ $R_f = \frac{\Sigma |F_{PH} - F_P|}{\Sigma F_P}$, where

F = structure factor amplitude; P = protein; H = heavy-atom; PH = protein plus heavy-atom; obs = observed; and calc = calculated.

5. Electron Density Map

The electron density map was calculated with "best phases" (Blow and Crick-1959) and with the observed structure factors weighted by the figure of merit. Superposed sections of the map are illustrated in Fig. 9a and 9b. The sections were com, puted perpendicular to the a^* axis at intervals of $0.04a$, $0.17b$, and $0.008c$. The electron density considered to be part of one asymmetric unit is designated with marker dots and stars.

Sections from $x = 0$ to $x = 0.24$, with y the ordinate and z the abscissa on each section, are presented in Fig. 9a. The most outstanding features in these sections and in the remainder of the map are the large volumes of space occupied by solvent. The boundaries between solvent and protein are clearly defined. These sections also contain regions of continuous electron density between two globular modules which are otherwise separated by a cleft (see junction between marker dots and stars in Fig. 9a).

The module (1) of electron density marked with dots is centered at $x = 0.4$, $y = 0.67$, and $z = 0.17$, while the module (2) designated with stars is centered at $x = 0.36$, $y = 0.5$, and $z = 0.33$ (see Fig. 9b). The volume of module 1 is ~1.5 times that of module 2. Together, they make up an asymmetric unit and therefore represent the Bence-Jones dimer.

It should be noted, however, that the close intermolecular packing of symmetrically related modules (as illustrated in Fig. 9b) prevents their *unequivocal* assignment to an asymmetric unit. The combination of modules 1 and 2 seems the most likely choice because it gives the maximum amount of electron density between modules and

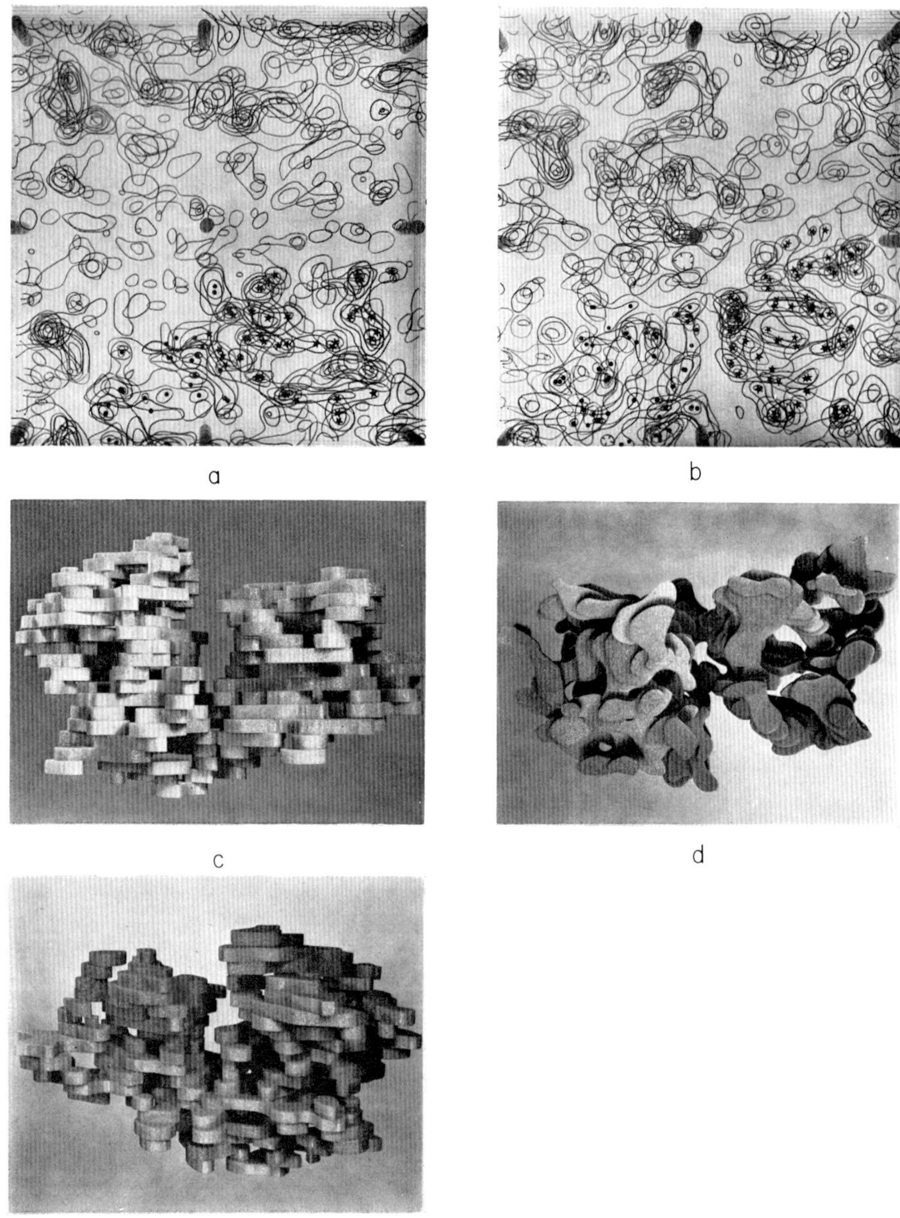

Fig. 9. Photographs of superposed sections of the 6 Å electron density map and the styrofoam model of the asymmetric unit (i.e., the dimer of the Mcg Bence-Jones protein). a Sections normal to the a^* axis, from $x = 0$ to $x = 0.24$, with y the ordinate and z the abscissa in each section. The lowest contour is one-fifth the maximum and zero contours are omitted. In the center of the photograph note the region of low density corresponding to solvent. These superposed sections contain the regions of continuous electron density between the modules which we consider part of one asymmetric unit (see junction between marker dots and stars). b Sections from $x = 0.28$ to $x = 0.52$. The module (1) of electron density marked with dots is centered at $x = 0.4$, $y = 0.67$, and $z = 0.17$, while the module (2) designated with stars is centered at $x = 0.36$, $y = 0.5$, and $z = 0.33$. c Styrofoam model of the asymmetric unit, viewed approximately at right angles to the line joining the centers of modules 1 and 2. Module 1 is on the left. d View of the model in the same orientation as the electron density map above it. e Back view of the model. Module 1 is on the right

results in a particle length (~78 Å) similar to that (~75 Å) found for a Bence-Jones dimer by low-angle X-ray scattering measurements (HOLASEK et al., 1963).

A model of the asymmetric unit was constructed of free-foamed styrofoam to a scale of 1.45 Å to one cm. The photographs of the model are shown in Fig. 9c, 9d, and 9e.

VII. Discussion

The crystallographic results clearly indicate that the Mcg λ chain dimer is composed of two globular regions which are unequal in size and structure. Because of these differences, the regions cannot be monomers. Instead, the electron density map provides strong evidence that the amino- and carboxyl-terminal parts in each light chain fold independently into regions (or "domains") having little affinity for each other, but substantial affinity for homologous parts in a second chain. In the discussion of these results it is important to distinguish the term "module", which we use to describe the electron density for two interacting parts of the dimer, from a "domain", which refers to a structural unit within a single immunoglobulin chain (for a discussion of the domain hypothesis, see EDELMAN and GALL, 1969).

At the present resolution it is not possible to determine which module of electron density contains the two carboxyl domains of the dimer. However, the presence of the *inter*chain disulfide bond between the penultimate residues should facilitate the identification of these domains in higher resolution maps. The amino-terminal domains can then be assigned to the remaining module.

The *intra*chain disulfide loops in the two halves of λ chains are separated by about 48 residues. An amino-terminal segment of 20 to 21 residues and a carboxyl-terminal segment of 19 residues lie outside the loops.

The sensitivity of some light chains to limited proteolysis in the connecting segments is consistent with the hypothesis that the switch regions consist partly of extended chains (KARLSSON et al., 1969). The flexibility conferred on two interacting monomers by such extended chains may help explain the absence of crystallographic and non-crystallographic two-fold axes of symmetry between the Mcg monomers in the crystal. Local two-fold axes within the modules may become apparent at higher resolution.

The serum IgG immunoglobulin from the patient Mcg has also been isolated in a form suitable for X-ray diffraction studies (EDMUNDSON et al., 1971). When isolated from one patient, the Bence-Jones protein and the light chain of the IgG immunoglobulin have the same amino acid sequence and therefore are predisposed toward similar three-dimensional structures. A comparison of the Bence-Jones dimer and the parent IgG molecule should provide direct information about the light chain structure when influenced by interactions with the heavy chain. Our work on the IgG protein parallels the crystallographic study of an IgG "cryoglobulin" by SARMA et al. (1971). As more data become available, the properties of the Mcg λ chain dimer will be compared with the results for a κ chain monomer being examined by EPP et al. (personal communication).

Bence-Jones dimers are similar in size to functional antigen-binding fragments (Fab) released by proteolysis of IgG antibodies or myeloma proteins in the "hinge" regions between the amino and carboxyl halves of the heavy chains (PORTER, 1959, 1969). An Fab fragment is therefore a complex of a light chain and the amino half of a heavy chain, usually linked by an *inter*chain disulfide bond. Poljak et al. (1972) recently described the crystallographic structure of an Fab′ fragment at 6 Å resolution. The similarities between the structures of the Fab′ molecule and Bence-Jones dimer are unmistakable, but the studies will have to be extended to higher resolution before detailed comparisons can be made.

VIII. Summary

A description of the crystallographic structure of a Bence-Jones λ chain dimer at a resolution of 6 Å is given in the present article. Factors affecting the crystallization of this protein and related light chains are discussed, and the properties influencing association of these molecules into dimers are emphasized. Graphical procedures are presented to illustrate the distribution of polar and apolar residues in the amino acid sequences of light chains. Fisher's equations are used to correlate the proportions of polar and apolar residues with the tendency of the light chains and their fragments to associate. The results of the calculations are consistent with the strong crystallographic evidence for the "domain" hypothesis of immunoglobulin structure. The amino- and carboxyl-terminal parts of the Bence-Jones λ chain monomer fold independently into regions ("domains") having little affinity for each other, but substantial affinity for homologous domains in a second molecule.

References

Abel, C. A., Spiegelberg, H. L., Grey, H. M.: The carbohydrate content of fragments and polypeptide chains of human γG-myeloma proteins of different heavy-chain subclasses. Biochemistry **7**, 1271—1278 (1968).

Appella, E., Ein, D.: Two types of lambda polypeptide chains in human immunoglobulins based on an amino acid substitution at position 190. Proc. nat. Acad. Sci. (Wash.) **57**, 1449—1454 (1967).

Baglioni, C., Alescio Zonta, L., Cioli, D., Carbonara, A.: Allelic antigenic factor Inv(a) of the light chains of human immunoglobulins: Chemical basis. Science **152**, 1517—1519 (1966).

Bence-Jones, H.: On a new substance occurring in the urine of a patient with "mollities ossium". Phil. Trans. **138**, 55—62 (1848).

Berggård, I., Peterson, P. A.: Polymeric forms of free normal ϰ and λ chains of human immunoglobulin. J. biol. Chem. **244**, 4299—4307 (1969).

Bigelow, C. C.: On the average hydrophobicity of proteins and the relation between it and protein structure. J. theor. Biol. **16**, 187—211 (1967).

Björk, I., Karlsson, F. A., Berggård, I.: Independent folding of the variable and constant halves of a lambda immunoglobulin light chain. Proc. nat. Acad. Sci. (Wash.) **68**, 1707—1710 (1971).

Blow, D. M., Crick, F. H. C.: The treatment of errors in the isomorphous replacement method. Acta Cryst. **12**, 794—802 (1959).

Braunitzer, G.: Phylogenetic variation in the primary structure of hemoglobins. J. Cell Biol. **67**, Suppl. 1, 1—19 (1966).

Cioli, D., Baglioni, C.: Origin of structural variation in Bence Jones proteins. J. molec. Biol. **15**, 385—388 (1966).

Cohen, S., Milstein, C.: Structure and biological properties of immunoglobulins. Advanc. Immunol. **7**, 1—89 (1967).

Cunningham, B. A., Gottlieb, P. D., Konigsberg, W. H., Edelman, G. M.: The covalent structure of a human γG-immunoglobulin. V. Partial amino acid sequence of the light chain. Biochemistry **7**, 1983—1995 (1968).

Deutsch, H. F.: Crystalline low molecular weight γ-globulin from a human urine. Science **141**, 435—436 (1963).

Dickerson, R. E., Kendrew, J. C., Strandberg, B. E.: The crystal structure of myoglobin: Phase determination to a resolution of 2 Å by the method of isomorphous replacement. Acta Cryst. **14**, 1188—1195 (1961).

Dickerson, R. E., Weinzierl, J. E., Palmer, R. A.: A least-squares refinement method for isomorphous replacement. Acta Cryst. **B 24**, 997—1003 (1968).

Dreyer, W. J., Gray, W. R., Hood, L.: The genetic, molecular, and cellular basis of antibody formation: Some facts and a unifying hypothesis. Cold Spr. Harb. Symp. quant. Biol. **32**, 353—367 (1967).

Edelman, G. M., Gall, W. E.: The antibody problem. Ann. Rev. Biochem. **38**, 415—466 (1969).

Edelman, G. M., Gally, J. A.: The nature of Bence-Jones proteins. J. exp. Med. **116**, 207—227 (1962).

Edmundson, A. B., Ely, K. R., Simonds, N. B., Hutson, N. K., Sheber, F. A., Rossiter, J. L.: Human L type Bence-Jones proteins containing carbohydrate. J. polymer Sci. Part C, **30**, 689—695 (1970).

Edmundson, A. B., Schiffer, M., Wood, M. K., Hardman, K. D., Ely, K. R., Ainsworth, C. F.: Crystallographic studies of an IgG immunoglobulin and the Bence-Jones protein from one patient. Cold Spr. Harb. Symp. quant. Biol. (in press).

Edmundson, A. B., Sheber, F. A., Ely, K. R., Simonds, N. B., Hutson, N. K., Rossiter, J. L.: Characterization of human L type Bence-Jones proteins containing carbohydrate. Arch. Biochem. Biophys. **127**, 725—740 (1968).

Edmundson, A. B., Simonds, N. B., Sheber, F. A., Johnson, T., Bangs, B.: Use of carboxypeptidase A for simultaneous assessment of purity and assignment of human Bence-Jones proteins and light chains to K and L classes. Arch. Biochem. Biophys. **132**, 502—508 (1969).

Ein, D., Fahey, J. L.: Two types of lambda polypeptide chains in human immunoglobulins. Science **156**, 947—948 (1967).

Fisher, H. F.: A limiting law relating the size and shape of protein molecules to their composition. Proc. nat. Acad. Sci. (Wash.) **51**, 1285—1291 (1964).

Fisher, H. F.: An upper limit to the amount of hydration of a protein molecule. A corollary to the "limiting law of protein structure". Biochim. biophys. Acta (Amst.) **109**, 544—550 (1965).

Fisher, H. F.: On the predictability of protein conformation: A limiting law. Abstracts, Symp. I—3, 3, Int. Cong. Biochem. 7th, Tokyo, 29 (1967).

Gally, J. A., Edelman, G. M.: Protein-protein interactions among L polypeptide chains of Bence-Jones proteins and human γ-globulins. J. exp. Med. **119**, 817—836 (1964).

Gray, W. R., Dreyer, W. J., Hood, L.: Mechanism of antibody synthesis: Size differences between mouse kappa chains. Science **155**, 465—467 (1967).

Haber, E.: Recovery of antigenic specificity after denaturation and complete reduction of disulfides in a papain fragment of antibody. Proc. nat. Acad. Sci. (Wash.) **52**, 1099—1106 (1964).

Haber, E.: Immunochemistry. Ann. Rev. Biochem. **37**, 497—520 (1968).

Hamaguchi, K., Migita, S.: Optical rotatory and ultraviolet spectral properties of Bence-Jones proteins. J. Biochem. (Tokyo) **56**, 512—521 (1964).

Hendrickson, W. A., Klock, P. A., Lattman, E. E., Love, W. E., Padlan, E. A.: Cold Spr. Harb. Symp. quant. Biol. (in press).

Hill, R. L., Delaney, R., Fellows, R. E., Jr., Lebovitz, H. E.: The evolutionary origins of the immunoglobulins. Proc. nat. Acad. Sci. (Wash.) **56**, 1762—1769 (1966).

Hilschmann, N.: (1) Die chemische Struktur von zwei Bence-Jones-Proteinen (Roy und Cum.) vom $\varkappa$-Typ. Z. physiol. Chem. **348**, 1077—1080 (1967).

Hilschmann, N.: (2) Die vollständige Aminosäuresequenz des Bence-Jones-Proteins Cum. ($\varkappa$-Typ). Z. physiol. Chem. **348**, 1718—1722 (1967).

Hilschmann, N., Barnikol, H.-U., Hess, M., Langer, B., Ponstingl, H., Steinmetz-Kayne, M., Suter, L., Watanabe, S.: Structural studies on immunoglobulins and their genetic implications for antibody formation. FEBS Symp. **15**, 57—74 (1969).

Hilschmann, N., Craig, L. C.: Amino acid sequence studies with Bence-Jones proteins. Proc. nat. Acad. Sci. (Wash.) **53**, 1403—1409 (1965).

Holasek, A., Kratky, O., Mittelbach, P., Wawra, H.: Small-angle X-ray scattering of Bence-Jones protein. J. molec. Biol. **7**, 321—322 (1963).

Hood, L., Gray, W. R., Sanders, B. G., Dreyer, W. J.: Light chain evolution. Cold Spr. Harb. Symp. quant. Biol. **32**, 133—146 (1967).

Hood, L., Potter, M., McKean, D. J.: Immunoglobulin structure: Amino terminal sequences of kappa chains from genetically similar mice (BALB/c). Science **170**, 1207—1210 (1970).

Hood, L., Talmage, D. W.: Mechanism of antibody diversity: Germ line basis for variability. Science **168**, 325—334 (1970).

Huber, R., Epp, O., Steigemann, W., Formanek, H.: The atomic structure of erythrocruorin in the light of the chemical sequence and its comparison with myoglobin. Europ. J. Biochem. **19**, 42—50 (1971).

Jirgensons, B., Saine, S., Ross, D. L.: The ultraviolet rotatory dispersion and conformation of Bence-Jones proteins. J. biol. Chem. **241**, 2314—2319 (1966).

Karlsson, F. A., Peterson, P. A., Berggård, I.: Properties of halves of immunoglobulin light chains. Proc. nat. Acad. Sci. (Wash.) **64**, 1257—1263 (1969).

Kendrew, J. C.: Side-chain interactions in myoglobin. Brookhaven Symp. Biol. **15**, 216—226 (1962).

Koshland, M. E., Englberger, F. M.: Differences in the amino acid composition of two purified antibodies from the same rabbit. Proc. nat. Acad. Sci. (Wash.) **50**, 61—68 (1963).

Langer, B., Steinmetz-Kayne, M., Hilschmann, N.: Die vollständige Aminosäuresequenz des Bence-Jones-Proteins New (λ-Typ). Subgruppen im variablen Teil bei Immunglobulin-L-Ketten vom λ-Typ. Z. physiol. Chem. **349**, 945—951 (1968).

Matthews, B. W.: Solvent content of protein crystals. J. molec. Biol. **33**, 491—497 (1968).

Melchers, F.: The attachment site of carbohydrate in a mouse immunoglobulin light chain. Biochemistry **8**, 938—947 (1969).

Melchers, F., Knopf, P. M.: Biosynthesis of the carbohydrate portion of immunoglobulin chains: Possible relation to secretion. Cold Spr. Harb. Symp. quant. Biol. **32**, 255—262 (1967).

Milstein, C.: Interchain disulphide bridge in Bence-Jones proteins and in γ-globulin B chains. Nature (Lond.) **205**, 1171—1173 (1965).

Milstein, C.: Variations in amino-acid sequence near the disulphide bridges of Bence-Jones proteins. Nature (Lond.) **209**, 370—373 (1966).

Milstein, C.: (1) Linked groups of residues in immunoglobulin $\varkappa$ chains. Nature (Lond.) **216**, 330—332 (1967).

Milstein, C.: (2) Variations in the C-terminal half of immunoglobulin λ-chains. Biochem. J **104**, 28—30 C (1967).

Milstein, C.: The variability of human immunoglobulin G. FEBS Symp. **15**, 43—56 (1969).

Milstein, C., Clegg, J. B., Jarvis, J. M.: Immunoglobulin λ-chains. Biochem. J. **110**, 631—652 (1968).

Milstein, C., Pink, J. R. L.: Structure and evolution of immunoglobulins. Prog. Biophys. molec. Biol. **21**, 209—263 (1970).

Mueller, M. H., Heaton, L., Amiot, L.: A computer controlled experiment. Res. Develop. **19**, 34—37 (1968).

Neet, K. E., Putnam, F. W.: Characterization of the thermal denaturation of Bence-Jones proteins by ultracentrifugation at elevated temperatures. J. biol. Chem. **241**, 2320—2325 (1966).

Némethy, G., Scheraga, H. A.: Structure of water and hydrophobic bonding in proteins. I. A model for the thermodynamic properties of liquid water. J. chem. Phys. **36**, 3382—3400 (1962).

Niall, H. D., Edman, P.: Two structurally distinct classes of kappa-chains in human immunoglobulins. Nature (Lond.) **216**, 262—263 (1967).

Padlan, E. A., Love, W. E.: Structure of the haemoglobin of the marine annelid worm, *Glycera dibranchiata*, at 5.5 Å resolution. Nature (Lond.) **220**, 376—378 (1968).

PERUTZ, M. F., KENDREW, J. C., WATSON, H. C.: Structure and function of haemoglobin. II. Some relations between polypeptide chain configuration and amino acid sequence. J. molec. Biol. **13**, 669—678 (1965).

POLJAK, R. J., AMZEL, L. M., AVEY, H. P., BECKA, L. N., NISONOFF, A.: Structure of Fab′ New at 6 Å resolution. Nature (Lond.) New Biol. **235**, 137—140 (1972).

PONSTINGL, H., HESS, M., HILSCHMANN, N.: Die vollständige Aminosäure-Sequenz des Bence-Jones-Proteins Kern. Eine neue Untergruppe der Immunglobulin-L-Ketten vom λ-Typ. Z. physiol. Chem. **349**, 867—871 (1968).

PONSTINGL, H., HESS, M., LANGER, B., STEINMETZ-KAYNE, M., HILSCHMANN, N.: Über einen Aminosäureaustausch im konstanten Teil eines Bence-Jones-Proteins vom λ-Typ. Z. physiol. Chem. **348**, 1213—1214 (1967).

PORTER, R. R.: The hydrolysis of rabbit γ-globulin and antibodies with crystalline papain. Biochem. J. **73**, 119—126 (1959).

PORTER, R. R.: Recent studies on the structure of the heavy chain of immunoglobulins. FEBS Symp. **15**, 13—19 (1969).

PUTNAM, F. W.: Structure and variability of immunoglobulin light chains. FEBS Symp. **15**, 21—41 (1969).

PUTNAM, F. W., SHINODA, T., TITANI, K., WIKLER, M.: Immunoglobulin structure: Variation in amino acid sequence and length of human lambda light chains. Science **157**, 1050—1053 (1967).

PUTNAM, F. W., TITANI, K., WIKLER, M., SHINODA, T.: Structure and evolution of kappa and lambda light chains. Cold Spr. Harb. Symp. quant. Biol. **32**, 9—30 (1967).

SARMA, V. R., DAVIES, D. R., LABAW, L. W., SILVERTON, E. W., TERRY, W. D.: Crystal structure of an immunoglobulin molecule by X-ray diffraction and electron microscopy. Cold Spr. Harb. Symp. quant. Biol. **37**, 413—419 (1971).

SCHIFFER, M., EDMUNDSON, A. B.: Use of helical wheels to represent the structures of proteins and to identify segments with helical potential. Biophys. J. **7**, 121—135 (1967).

SCHIFFER, M., HARDMAN, K. D., WOOD, M. K., EDMUNDSON, A. B., HOOK, M. E., ELY, K. R., DEUTSCH, H. F.: A preliminary crystallographic investigation of a human L-type Bence-Jones protein. J. biol. Chem. **245**, 728—730 (1970).

SCHRAMM, H. J.: Die Isolierung und Kristallisation des variablen Fragments eines Bence-Jones-Proteins. Z. physiol. Chem. **352**, 1134—1138 (1971).

SEON, B.-K., ROHOLT, O. A., PRESSMAN, D.: The reactivity of tyrosyl residues in Bence-Jones protein: Differences in rates of iodination in covalent and noncovalent dimers. Biochim. biophys. Acta (Amst.) **194**, 397—405 (1969).

SEON, B.-K., ROHOLT, O. A., PRESSMAN, D.: Topography of Bence-Jones protein. J. biol. Chem. **246**, 887—898 (1971).

SINGER, S. J., DOOLITTLE, R. F.: Antibody active sites and immunoglobulin molecules. Science **153**, 13—25 (1966).

SMITHIES, O.: Antibody variability. Science **157**, 267—273 (1967).

SOLOMON, A., MCLAUGHLIN, C. L.: Bence-Jones proteins and light chains of immunoglobulins. I. Formation and characterization of amino-terminal (variant) and carboxyl-terminal (constant) halves. J. biol. Chem. **244**, 3393—3404 (1969).

SOLOMON, A., MCLAUGHLIN, C. L., WEI, C. H., EINSTEIN, J. R.: Bence-Jones proteins and light chains of immunoglobulins. V. X-ray crystallographic investigation of the amino-terminal half of a ϰ Bence-Jones protein. J. biol. Chem. **245**, 5289—5291 (1970).

SOX, H. C., Jr , HOOD, L : Attachment of carbohydrate to the variable region of myeloma immunoglobulin light chains. Proc. nat. Acad. Sci. (Wash.) **66**, 975—982 (1970).

SUTER, L., BARNIKOL, H.-U., WATANABE, S., HILSCHMANN, N.: Die Primärstruktur einer monoklonalen Immunglobulin-L-Kette der Subgruppe III vom ϰ-Typ (Bence-Jones-Protein Ti). Z. physiol. Chem. **350**, 275—278 (1969).

TANFORD, C.: Contribution of hydrophobic interactions to the stability of the globular conformation of proteins. J. Amer. chem. Soc. **84**, 4240—4247 (1962).

TITANI, K., WHITLEY, E., Jr., AVOGARDO, L., PUTNAM, F. W.: Immunoglobulin structure: Partial amino acid sequence of a Bence-Jones protein. Science **149**, 1090—1093 (1965).

WATSON, H. C.: The stereochemistry of the protein myoglobin. Prog. Stereochem. **4**, 299—333 (1969).

WELSCHER, H. D.: (1) Correlations between amino acid sequence and conformation of immunoglobulin light chains. I. Hydrophobicity and fractional charge. Int. J. Protein Res. **1**, 253—265 (1969).

WELSCHER, H. D.: (2) Correlations between amino acid sequence and conformation of immunoglobulin light chains. II. Sequence comparison and the pattern of nonpolar residues. Int. J. Protein Res. **1**, 267—282 (1969).

WHITNEY, P. L., TANFORD, C.: Recovery of specific activity after complete unfolding and reduction of an antibody fragment. Proc. nat. Acad. Sci. (Wash.) **53**, 524—532 (1965).

WIKLER, M., TITANI, K., SHINODA, T., PUTNAM, F. W.: The complete amino acid sequence of a λ type Bence-Jones protein. J. biol. Chem. **242**, 1668—1670 (1967).

WU, T. T., KABAT, E. A.: An attempt to locate the non-helical and permissively helical sequences of proteins: Application to the variable regions of immunoglobulin light and heavy chains. Proc. nat. Acad. Sci. (Wash.) **68**, 1501—1506 (1971).

ZEPPEZAUER, M., EKLUND, H., ZEPPEZAUER, E. S.: Micro diffusion cells for the growth of single protein crystals by means of equilibrium dialysis. Arch. Biochem. Biophys. **126**, 564—573 (1968).

The Thalassemia Syndromes: Genetically Determined Disorders of the Regulation of Protein Synthesis in Eukaryotic Cells

ALBERT S. BRAVERMAN

I. Introduction

That many familial and congenital human diseases are caused by the deficiency of a specific enzyme or other protein was suspected early in the century (GARROD, 1908), and has been well known for decades. A large group of such diseases has now been identified (STANBURY et al., 1966).

Until surprisingly recently, it was assumed that these enzyme-deficiency states were literally deficiencies of the enzyme protein; thus, PARKER and BEARN (1963) and STANBURY, WYNGAARDEN and FREDRICKSON (1960) postulated that the great majority of these disease states were due to a mutation of putative regulatory genes. However, the elucidation of the structural alteration of sickle hemoglobin (PAULING et al., 1949; INGRAM, 1957) and the subsequent discovery of many other hemoglobinopathies which also involved single amino acid replacements, suggested the possibility that some enzyme deficiency states might also be due to structural mutations which affected the functional activity of the protein. The demonstration, by electrophoretic and kinetic techniques, of the probably structural abnormality of the enzymes isolated from deficient-type cells (see, for example, the human red cell glucose-6-phosphate dehydrogenase deficiencies, BEUTLER, 1969) confirmed this assumption.

These discoveries do not, of course, exclude the existence of regulatory mutations in human or other eukaryotic cells. That is, genetic alterations may occur which lead to the reduced synthesis of a protein whose amino acid sequence is normal. Since little is known concerning the transcription of genetic information in nucleated cells, one cannot *a priori* predict whether such mutations involve operons and their associated regulatory genes as proposed for bacteria by JACOB and MONOD (1961), or cytoplasmic regulatory mechanisms specific for eukaryotes (TOMPKINS et al., 1969). But whatever the mechanism, the identification and study of regulatory mutations in eukaryotic cells is of evident theoretical importance. The following items of knowledge are required for this purpose.

1. Demonstration of the presence of a structurally normal protein. This would involve initially electrophoretic or chromatographic analysis and the study of kinetics of formation but would have to culminate in the determination of the complete amino acid sequence of the protein.

2. In the *absence* of functionally normal protein, the presence of serologically similar protein would have to be ruled out with the use of antibodies to the normal protein.

3. If normal protein is *present* one would need to demonstrate
a) Presence at significantly lower concentrations, or
b) Reduced rates of biosynthesis *in vivo* or *in vitro*, using radioisotopic labelling techniques.

Among eukaryotic mutants known to the author, only delta amino levulinate dehydratase deficiency in mice (Doble and Schimke, 1969), and the human thalassemia syndromes seem to have met most of the criteria outlined above for a regulatory mutation. In the following sections, the clinical and pathophysiological features of the best studied form of thalassemia will be outlined, and present information on the molecular basis of this regulatory gene mutation will be reviewed.

II. The Beta Thalassemias: Clinical Picture and Pathogenesis of Anemia

1. Hemoglobin Nomenclature and Structure

Almost all human and mammalian hemoglobins consist of four proteins chains whose molecular weight is approximately 16,000. One heme group, consisting of a porphyrin molecule and single ferrous iron atom, is ordinarily associated with each chain. The chains are almost invariably of two different types, present in equal quantities.

Normal human adult hemoglobin is one electrophoretic entity, hemoglobin A; its protein (globin) chains are designated α and β so that its molecular formula is $\alpha_2\beta_2$. There is good genetic evidence that the alpha and beta chains are under separate genetic control, and it is probable that their genes are not closely linked (Smith and Torbert, 1958).

About 2.0% of normal adult hemoglobin is designated A_2: its formula is $\alpha_2\delta_2$. The delta chain differs by only eight amino acid residues from the beta chain, and there is good evidence that their genes are closely linked (see Appendix II below).

Human fetal hemoglobin, designated hemoglobin F, can be distinguished from A by its resistance to alkali denaturation or by electrophoresis; its formula is $\alpha_2\gamma_2$, the gamma chain differing from beta chain by 39 alterations in sequence. It is the major hemoglobin of the fetus and neonate.

Mutant forms of all four types of normal human globin chain have been identified; in almost all cases single amino acid substitutions are involved. Mutant forms are identified by a superscript, so that sickle hemoglobin, or hemoglobin S, which involves a beta chain mutation, is written $\alpha_2\beta_2^S$.

2. Severe Beta Thalassemia: Clinical and Hematologic Features

The first form of thalassemia to be recognized as a clinical entity is known as Cooley's Anemia, Mediterranean Anemia, Thalassemia Major, or severe beta thalassemia. This is the best studied of the thalassemia syndromes, and will be our main concern here (Weatherall, 1965, p. 45).

The disease occurs in children of Mediterranean origin and is quite common in certain Italian and Greek populations. Sporadic cases are found throughout the Middle East, and a less severe form is probably not rare in Negroes.

The disease usually has its onset in the second year of life, and by the age of four or five most patients require red cell transfusion monthly to survive. Since there are no physiological mechanisms for iron excretion, iron from transfused blood is stored in these patients' tissues; they are almost invariably dead by their mid-teens, presumably from iron overload.

By the fourth year of life, bone marrow active in the production of the cellular elements of the blood is restricted, in normal individuals, to truncal bones such as the pelvis or vertebrae; in thalassemic patients, as in most individuals with congenital hemolytic anemia, active cellular marrow persists in the bones of the skull and limbs as long as they live. In severe beta thalassemia, which is one of the most severe of congenital anemias, the activity of the skull bone marrow is extremely marked, producing striking deformities of the head and face (the so-called mongoloid faces). Patients are ordinarily extremely undersized, and sexual maturation is delayed or absent.

The liver and spleen are not normally active in erythropoiesis after the first seven months of human fetal life, but permanent extramedullary erythropoiesis in these organs is typical of severe beta thalassemic patients, which is partially responsible for their massive enlargement.

The red cells of thalassemic patients are abnormally small and show a marked reduction in hemoglobin concentration (hypochromia and microcytosis). These changes are not unique to thalassemia, but are typical of anemias in which any phase of hemoglobin synthesis is inhibited, such as iron deficiency. Iron stores are markedly increased in severe beta thalassemia, however.

Examination of the bone marrow in these patients shows a massive increase in erythroid precursors; thus, differentiation from the uncommitted, pluripotential stem cell to the erythroid precursor is presumably normal in these patients.

Severe anemia, in the presence of a hyperactive bone marrow, is often due to the destruction of red cells peripherally (hemolytic anemia) as in sickle cell anemia. That severe beta thalassemia is not an ordinary hemolytic anemia is suggested by the following considerations:

1. Patients with even the most marked reduction of peripheral red cell survival (the usual measure of hemolysis) can, with an active marrow, sustain a red cell count adequate for their survival without transfusion; yet the red cell concentration of thalassemic patients will fall to levels incompatable with life unless they are given blood.

2. Reticulocytes (erythroid cell precursors which have lost their nucleus but retain the polyribosome apparatus) are markedly increased in the circulation if the bone marrow is hyperactive. In severe beta thalassemia, the reticulocyte count in peripheral blood is only modestly increased, if at all, despite the activity of the bone marrow.

3. Although the survival time of thalassemia red cells is reduced, the lifespan is hardly short enough to account for the severity of the anemia (Shahid and Sahli, 1971).

These findings imply that the anemia in severe beta thalassemia is largely due to destruction of a major population of erythroid precursors before they have matured and left the bone marrow (ineffective erythropoiesis). That such destruction does occur in severe beta thalassemia has been demonstrated in the following manner:

If human hemoglobin is tracer-labelled in vivo at the time of its synthesis, almost of the hemoglobin incorporated tracer is recovered in the heme breakdown products stereobilin and bilirubin at about 120 days (the lifespan of the normal red cell) after the tracer is administered. A very small percentage of the tracer is recoverable in these pigments about five days after administration, presumably due to the breakdown of small numbers of erythroid precursors before they leave the bone marrow (the "early labelled peak"; see LONDON et al., 1949). In severe beta thalassemia, however, a marked elevation in the early labelled peak is detected (GRINSTEIN et al., 1960). This supports the hypothesis that in severe beta thalassemia massive destruction of erythroid precursors occurs.

The mechanism of ineffective erythropoiesis in severe beta thalassemia may be more complex than the destruction of differentiated red cell precursors; WICKRAMASINGHE and co-workers (1970) have demonstrated a block to mitosis and DNA synthesis amongst the most mature thalassemic nucleated erythroid cell precursors.

Hemoglobin electrophoresis of lysates from severe beta thalassemic patients reveals no abnormal or mutant hemoglobin; however, between 50 and 100% of hemoglobin from most untransfused severe beta thalassemic patients is normal fetal hemoglobin ($\alpha_2\gamma_2$). Tryptic digest and fingerprint analysis of hemoglobin A from those severe beta thalassemic patients whose lysates do contain it is normal (GUIDOTTI, 1962; BAGLIONI, 1963).

The anemia of severe beta thalassemia cannot, however, be attributed to the high percentage of fetal hemoglobin present in these patients' red cells. Normal adults have less than 1.0% hemoglobin F. but a number of heterozygotes and one homozygote have been reported with Hereditary Persistence of Fetal Hemoglobin (HPFH: WEATHERALL, 1965, p. 134). The homozygote had 100% fetal hemoglobin with no hemoglobin A or A_2, while the heterozygotes had in the vicinity od 25% fetal hemoglobin. All of these individuals are hematologically normal except for minor alterations in red cell morphology.

In severe beta thalassemia and in certain acquired elevations of fetal hemoglobin, such as aplastic anemia, the hemoglobin F is found to be heterogeneously distributed amongst the red cells by histochemical techniques (SHEPARD et al., 1962), while it is distributed homogeneously in HPFH. This contrast suggests that in severe beta thalassemia, as in acquired elevations of fetal hemoglobin, the increased production of hemoglobin F is, in some sense a compensatory or secondary phenomenon, rather than a primary genetic alteration, as in HPFH.

In individuals doubly heterozygous for the beta thalassemia trait (see below) and HPFH, hemoglobin F levels are much greater than those present in simple HPFH heterozygotes, though the double heterozygotes are no more hematologically abnormal than in individuals with thalassemia trait. HUISMAN and co-workers (1971) have shown by analysis of the different types of gamma chain known to exist in fetal hemoglobin, that most of the excess fetal hemoglobin present in double heterozygotes probably arises from the beta thalassemia, rather than the HPFH gene. This data supports the assumption that the elevation of hemoglobin F in severe beta thalassemia is a response to diminished beta chain production.

Severe beta thalassemia may occur in several siblings of the same mating. Both parents of these patients will ordinarily be found to have the *beta thalassemia trait*, which involves a mild to minimal anemia, microcytosis and no symptoms or abnormal physical findings. Starch block electrophoresis reveals, in most cases, an elevated percentage of hemoglobin A_2 ($\alpha_2\delta_2$): fetal hemoglobin levels are normal in most cases, though a population of heterozygotes exists with normal A_2 levels and mild hemoglobin F elevations.

3. Severe Beta Thalassemia: The Defect in Hemoglobin Synthesis

The hypochromia and microcytosis of severe beta thalassemia suggested that the disease involved the suppression of hemoglobin synthesis. The increased production of two globin chains which can substitute for beta chain in the hemoglobin molecule (gamma chain in homozygotes, and delta chain in heterozygotes) pointed to the suppression of beta globin chain synthesis (INGRAM and STRETTEN, 1959). Prior to the biochemical confirmation of this hypothesis, the strongest supporting evidence for it was the phenomenon of sickle-thalassemia (WEATHERALL, 1965, p. 85 et seq.).

Individuals heterozygous for the sickle cell gene are ordinarily asymptomatic and not anemic; because the β^S chain appears to be manufactured at an intrinsically slower rate than the β^A chain; (HEYWOOD et al., 1964; BOYER et al., 1964; BANK et al., 1970) they have about 35% hemoglobin S in their red cells. Untransfused homozygotes of course have close to 100% hemoglobin S. Yet in Negro and Sicilian populations a group of individuals has been observed with between 50 and 100% hemoglobin S. the rest of their hemoglobin being A and/or F. Depending upon their percentage of hemoglobin S, they present as more or less symptomatic sickle cell anemia patients.

Family study reveals that one parent of such patients has sickle cell trait, and the other beta thalassemia trait If the patients themselves have off-spring, the beta thalassemia and S genes invariably segregate independently. The following conclusions (INGRAM and STRETTEN, 1959) were drawn from this experiment in nature:

1. The beta thalassemia trait gene is closely linked to the β^A gene.
2. The beta thalassemia trait gene suppresses beta chain synthesis.
3. The beta thalassemia trait gene is cis-dominent; that is, it suppresses beta chain synthesis by the beta chain gene to which it is linked, and does not suppress the synthetic activity which results from the allelic beta gene.
4. Since in sickle-thalassemia the allelic beta chain gene is β^S, the apparent increase in sickle hemoglobin in this syndrome is simply the normal product of one sickle gene.

This phenomenon, which is known as interaction, and which has also been reported in individuals doubly heterozygous for beta thalassemia trait and other beta chain mutants, such as hemoglobin C (SMITH and KREVANS, 1959), was the most important indirect evidence for the suppression of beta chain synthesis in beta thalassemia In 1965 and 1966, direct proof was obtained [CLEGG et al., 1965; HEYWOOD et al., 1965; BANK and MARKS, 1966 (1)]:

When human reticulocytes are incubated at 37 C. in appropriate medium, there is linear incorporation of radioactive amino acids into the globin chains of hemoglobin. If heme is removed from globin by acid acetone precipitation, alpha, beta and gamma chains may be separated from each other by carboxymethylcellulose chromatography with phosphate buffer in an 8 M urea — mercaptoethanol medium.

Fig. 1 is a chromatogram derived from non-thalassemic reticulocytes and red cells, most of the optical density of course arising from the mature cells. When corrected for differences in extinction, the peaks of alpha and beta chain optical density are equal; that is, the alpha to beta ratio is close to 1.0. The ratio of alpha to beta chain radioactivity is also close to 1.0, indicating that equal quantities of both chains are synthesized by the non-thalassemic reticulocytes. Incubation of non-thalassemic nucleated erythroid cells from bone marrow gives similar results (Braverman and Bank, 1969).

Fig. 2 is a chromatogram of labelled globin from severe beta thalassemic reticulocytes and red cells. Gamma chain is present as well as alpha and beta chain; the ratio

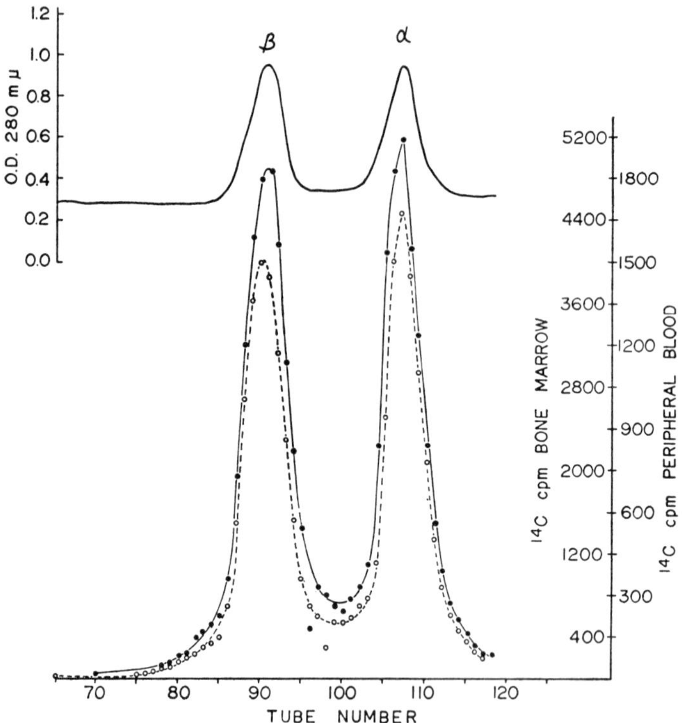

Fig. 1. Elution pattern of from globin chain chromatography of radioactively labelled non-thalassemic peripheral blood (—●—○—) and bone marrow cells (—●—●—). Continuous line represents optical density. (Braverman and Bank, 1969)

of alpha to beta plus gamma optical density (corrected) is also 1.0, reflecting the presence of equal quantities of alpha and beta or gamma chain in any molecule of hemoglobin A or F. Examination of the peaks of globin radioactivity reveals a marked reduction in tracer incorporation into beta chain globin, the alpha to beta synthetic ratio being 15.0. Although the synthesis of gamma chains reduces the imbalance to some degree, the alpha to beta plus gamma synthetic ratio is 7.0, indicating that gamma chain synthesis by no means compensates for the reduction in beta chain synthesis.

This pattern is typical of almost all patients with the severe beta thalassemia syndrome; since Bank and co-workers have shown (1968) that the absolute rate of alpha chain synthesis is normal in severe beta thalassemia, the pattern indicates marked suppression of beta chain synthesis in this syndrome. Most of these patients have alpha to beta synthetic ratios in the vicinity of 10. to 15., indicating a greater than 90% suppression of beta chain synthesis. Occasional patients have ratios as low as 5.0, while a group of patients (many of whom come from the Ferarra region of Italy) have total suppression of beta chain synthesis (BARGELLESI et al., 1967). Bone marrow

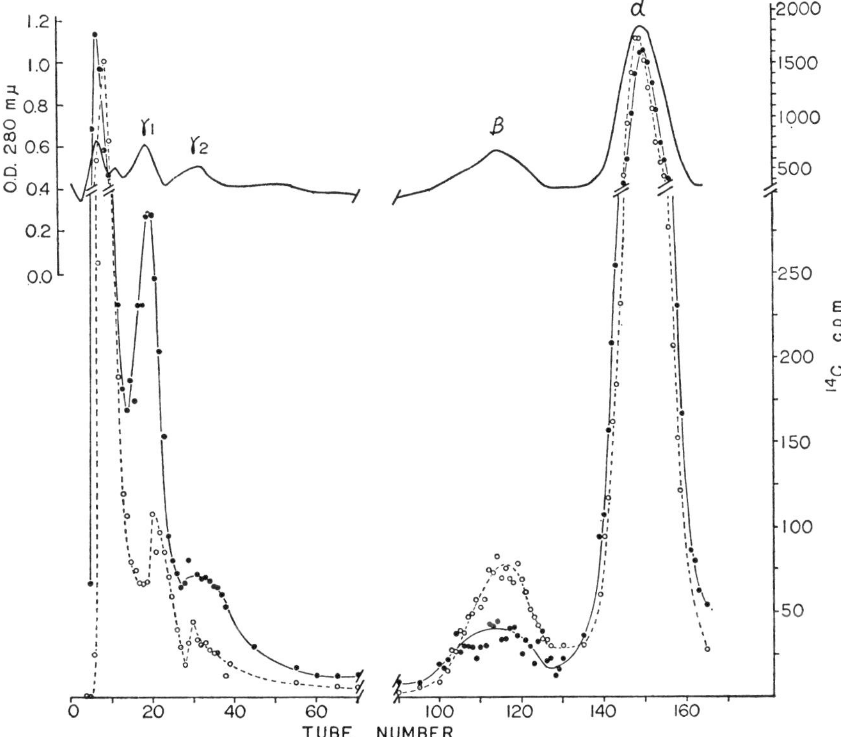

Fig. 2. Elution pattern from globin chain chromatography of radioactivity labelled peripheral blood (●—○) and bone marrow (—○—○—) of a patient with severe beta thalassemia. Continuous line represents optical density. The elution gradient used in these experiments resolves cord blood and thalassemic gamma chain into two peaks; the distinction between these peaks has not been elucidated. (BRAVERMAN and BANK, 1969)

nucleated erythroid cells from severe beta thalassemia patients show a similar order of beta chain suppression, though in almost all cases about 1.5 × more beta chains per alpha chain are synthesized in these cells than in reticulocytes from the same individuals (BRAVERMAN and BANK, 1969).

Similar studies of reticulocytes from individuals with beta thalassemia trait (BANK and MARKS, 1966) reveals a definite, but much less marked suppression of beta chain synthesis; the alpha to beta chain ratios in these individuals are of the order of 2.0, indicating approximately 50% suppression of beta chain synthesis. Interestingly enough, SCHWARTZ has shown (1970) that globin chain synthesis is balanced in the bone marrow erythroid cell precursors of beta thalassemia trait individuals.

Severe beta thalassemia appears to satisfy the criteria for a regulatory (as opposed to structural) mutation outlined above. The following questions are raised by the discovery of suppression of beta chain synthesis in this disease:

1. How does suppression of beta chain synthesis lead to ineffective erythropoiesis and severe anemia in severe beta thalassemia:
2. How is the mechanism of protein synthesis altered in severe beta thalassemia to give rise to diminished beta chain production?

4. The Pathogenesis of Thalassemic Anemia: The Fate of Excess Alpha Chain

We have already noted that, although a marked excess of alpha chain radioactivity is present in severe beta thalassemia reticulocytes, no excess alpha chain optical density can be detected in the same lysates; that is, newly synthesized excess alpha chain is present, but excess alpha chain does not accumulate in these cells.

In the alpha thalassemias, in which alpha chain synthesis has been shown to be suppressed and beta chain synthesis is normal (see Appendix I), excess beta chain optical density does accumulate as the hemoglobin tetramer hemoglobin H, or β_4. The newly synthesized excess alpha chain of severe beta thalassemia is evidently normal during its short lifespan; BANK showed (1968) that alpha chain radioactivity from a fresh beta thalassemia lysate combined with beta chain (hemoglobin H) to form chromatographically normal hemoglobin A. MODELL and co-workers (1969) demonstrated that radioactivity from similar lysates moved in 16,000 and 32,000 peaks by Sephadex chromatography, suggesting that excess alpha chain exists in dimer and (in older lysates) monomer form. Overnight incubation of severe beta thalassemia reticulocytes in which further protein synthesis is blocked after an initial one hour incorporation period, causes an almost 50% decrease in alpha chain radioactivity, though beta chain radioactivity, and alpha chain radioactivity in non-thalassemic reticulocytes do not diminish under these conditions (BANK and O'DONNELL, 1969). That the excess alpha chain is precipitated and partially hydrolyzed in severe beta thalassemia erythroid cells is suggested by the studies of FESSAS and co-workers, who showed that inclusion bodies, which can be demonstrated in severe beta thalassemia erythroid cell precursors and in circulating red cells from splenectomized patients, consist almost entirely of alpha chain peptides (1966).

Evidently then, some feature of alpha chain primary structure prevents the molecule from forming any stable hemoglobin species, and leads to its rapid destruction within the red cell (since excess beta chain can form a relatively stable hemoglobin tetramer under presumably similar conditions).

5. Pathogenesis of Thalassemia Anemia: Alternative Mechanisms

There are three results of the suppression of beta chain synthesis which may, alone or in combination, account for the premature red cell destruction and severe anemia of severe beta thalassemia:

1. Production and rapid destruction of excess alpha chain.
2. Marked reduction of total red cell hemoglobin concentration (hypochromia).

3. Accumulation of iron and other heme precursors within the red cell. Several lines of evidence implicate excess alpha chain in the early destruction of thalassemic erythroid cells:

1. Unstable mutant hemoglobins, which form various types of inclusion bodies within red cells are characteristically associated with hemolysis (RIEDER, R., 1971; RANNEY, 1970).

2. Severe beta thalassemia erythroid cells with the greatest number of alpha peptide inclusion bodies either do not leave the bone marrow or are rapidly destroyed by the spleen (FESSAS et al., 1966).

3. The oldest severe beta thalassemia red cells (i.e. the most dense by gradient centrifugation) have the highest levels of hemoglobin F, implying that those cells which produced the most gamma chain to combine with excess alpha chain have a selective advantage (GABUZDA et al., 1963).

Moreover, severe beta thalassemia reticulocytes appear to manufacture more gamma chain per alpha chain than nucleated erythroid cells from the same patients, suggesting that erythroid precursors with low levels of gamma chain production may never leave the bone marrow (BRAVERMAN and BANK, 1969).

4. There is a strong negative correlation between the survival time of severe beta thalassemia red cells transfused into normal subjects and the quantity of excess alpha chain synthesized by the thalassemic red cell precursors (VIGI et al., 1969).

5. One individual in whom hematologic and family studies confirmed *homozygosity* for *beta* thalassemia trait and *heterozygosity* for *alpha* thalassemia trait (see Appendix I) had an unusually mild beta thalassemia syndrome (KAN and NATHAN, 1970). This data may be interpreted to imply that the suppression of excess alpha chain production by the alpha thalassemia gene ameliorates the hematologic disease.

However, there is evidence that excess alpha chain production may not, in itself, account for the severity of severe beta thalassemia anemia. Another syndrome of homozygous beta thalassemia, which also involves marked hypochromia and high levels of fetal hemoglobin (~50%) is known, and is associated with mild anemia, no transfusion requirement and normal patient survival. When beta chain production is measured in the reticulocytes of these patients with *mild beta thalassemia*, it is found to be as suppressed as it is in patients with typical severe beta thalassemia (BRAVERMAN et al., 1971). If excess alpha chain production alone were responsible for the severity of severe beta thalassemia anemia, these patients should display much less marked suppression of beta chain synthesis in their reticulocytes. The possibility that a disproportionately higher level of beta chain synthesis occurs in the nucleated erythroid cell precursors of patients with the mild syndrome has also been excluded.

Hypochromia per se, or the accumulation of iron and heme precursors within thalassemic red cells may also play a role in their premature destruction. However, in the absence of objective data, we can only speculate on the significance of these factors in thalassemic anemia by considering their effect on red cell lifespan when they occur in disease states which do not involve the accumulation of excess alpha chain.

In only two other conditions is hypochromia comparable to that in severe beta thalassemia commonly observed: iron deficiency anemia and the sideroblastic anemias. In iron deficiency anemia, the red cell lifespan is moderately reduced, due to an intrinsic abnormality of the red cell (LORIA et al., 1967), though it is probably an inhibition of differentiation at the stem cell level which accounts for the severity of the anemia (HERSHKO et al., 1970; HILLMAN and FINCH, 1971). The role of ineffective erythropoiesis (destruction of erythroid precursors) in iron deficiency anemia in humans has not yet been defined.

The sideroblastic anemias are a heterogeneous group of hereditary and acquired anemias characterized by marked hypochromia, iron accumulation within erythroid precursors and red cells (ringed sideroblasts and siderocytes), massive erythroid hyperplasia, and inadequate reticulocyte response; a defect in heme synthesis is probable in these cases, but has never been demonstrated. Ineffective erythropoiesis is highly probable and the survival of circulating red cells is minimally reduced, if at all (KUSHNER et al., 1971). Excess (non-hemoglobin) globin has been demonstrated in the red cells of these patients, probably in the form of heme-free alpha-beta dimers (WHITE et al., 1971).

Thus, hypochromia per se, does not, in any known disease, appear to be associated with a marked decrease in the lifespan of circulating red cells. Yet, in both conditions in which hypochromia is significant and erythroid differentiation does not appear inhibited (severe beta thalassemia and the sideroblastic anemias) massive destruction of bone marrow erythroid precursors seems to take place. These two conditions are also associated with the production of excess globin chain in red cells, though whether or not heme-free globin is present in iron deficient hypochromic red cells is not known.

The production and destruction of excess alpha chain undoubtedly plays an important role in the premature destruction of thalassemic erythroid precursors. But the little we know of red cell physiology in the other hypochromic anemias described above does not exclude the possibilities that reduced cytoplasmic hemoglobin concentration, iron and/or heme accumulation, or the production of globin chains in balanced quantities which are not bound to heme may also shorten the survival of thalassemic bone marrow erythroid cells.

Premature destruction of red cells or their precursors in and out of the bone marrow need not, of course, be the only basis for thalassemic anemia. We have already cited evidence for a late inhibition of the differentiation of thalassemic erythroid precursors (WICKRAMASINGHE et al., 1970), and it may be that abnormalities of globin chain synthesis can affect red cell production as well as red cell survival.

III. The Molecular Basis of the Suppression of Beta Chain Synthesis in Severe Beta Thalassemia

Because the absolute rate of alpha chain synthesis is known to be normal in severe beta thalassemia, and since other red cell proteins appear to be present in normal quantities, it is reasonable to assume that the non-specific components of the machinery of protein synthesis are normal in severe beta thalassemia. That is, no defects in erythroid call ribosomes, tRNA's or other supernatant factors would be anticipated, and the defect must reside in the beta chain mRNA.

The most direct approach to the thalassemia problem would, therefore, be the isolation of normal and severe beta thalassemia beta chain mRNA, and their characterization and comparison. Until quite recently, however, no eukaryotic messenger had been isolated. The criterion for isolation is the stimulation by the putative mRNA fraction of the synthesis of its homologous protein in the presence of heterologous ribosomes and supernatants. This was first achieved by SCHAPIRA and co-workers

(1968) with a 9S RNA fraction and non-specific factors of reticulocytes of two species of rabbit. In 1969, LOCKARD and LINGREL used a mouse 9S RNA to stimulate mouse globin chain synthesis in a cell-free system derived from rabbit reticulocytes. Subsequently, a rabbit messenger — human cell free system has produced globin chain (NIENHUIS et al., 1971) and a considerable amount of information concerning the isolation and characterization of rabbit reticulocyte messenger is accumulating [BLOBEL, 1971 (1, 2); GASKILL and KABAT, 1971]. At the time of this writing, human thalassemic and non-thalassemic mRNA fractions have been used to stimulate predominantly human globin chain synthesis in a rabbit cell-free system (ANDERSON and NIENHUIS, 1971), so that direct information concerning the thalassemic beta chain messenger may be available in the not too distant future.

During the past decade, however, approaches to elucidating the molecular basis for the suppression of beta chain synthesis in severe beta thalassemia have necessarily been indirect; results obtained fall into two general categories:

1. Data which confirm the assumption that the non-specific (i.e. non-mRNA) elements of the translation mechanism in severe beta thalassemia are normal.

2. Data concerning the nature of the defect (i.e. qualitative vs. quantitative) in severe beta thalassemia beta chain messenger.

1. Non-Specific Elements of the Translation Mechanism in Seven Beta Thalassemia

Not long after the demonstration of the suppression of beta chain synthesis in whole severe beta thalassemia reticulocytes, BANK and MARKS [1966 (2)] showed that, although the capacity of a polyribosome fraction from thalassemic reticulocytes to incorporate radioactive amino acids into TCA precipitable material was less than 50% of normal (BURKA and MARKS, 1963), thalassemic ribosomes had a quantitatively normal response to synthetic messenger (poly U). More recently, ANDERSON and co-workers measured the rate of globin chain synthesis by a thalassemic cell-free system, and found a suppression of beta chain synthesis identical to that noted in the whole reticulocytes from which the system was derived (GILBERT et al., 1970). Thay also showed that the thalassemic system produced as much rabbit globin chain as a non-thalassemic human cell-free system when stimulated with rabbit messenger RNA fraction (NIENHUIS et al., 1971).

The initiation of bacterial protein synthesis is known to be mediated by a form of methionyl-tRNA (met-$tRNA_A$) which places an N-formylated methionine in the N-terminal position of proteins, from which it is cleared before completion of the chain. Internal methionines, which are part of the completed protein, are specified by another tRNA (met-$tRNA_F$) with a different base triplet specificity (MARCKER and SANGER, 1964; 1969; CLARK and MARCKER, 1966). A similar mechanism for chain initiation has been demonstrated in reticulocyte and other eukaryote systems [SHAFRITZ and ANDERSON, 1970 (1); JACKSON and HUNTER, 1970; HOUSMAN et al., 1970] except that the methionine of met-$tRNA_F$ is not formylated.

Three soluble initiation factors (M_1, M_2 and M_3), defined by their ability specifically to bind me-$tRNA_M$, to initiate synthetic messenger protein synthesis at low (Mg^H), and to support synthesis of globin on endogenous messenger have been isolated from reticulocyte ribosome extracts (PRICHARD et al., 1970).

Gilbert and co-workers (1970) have shown that the addition of these factors, derived from human non-thalassemic, thalassemic and rabbit reticulocytes to a thalassemic cell free system do not correct the suppression of beta chain synthesis. However, ribosomal initiation factors derived from thalassemic cells, support globin chain synthesis by non-thalassemic and rabbit system normally.

These results suggest that the non-specific components of the initiation mechanism are normal in severe beta thalassemia; they do not, however, exclude an abnormality of the initiation region of the thalassemic beta chain mRNA itself.

2. Data and Hypotheses Concerning the Beta Chain Messenger RNA Abnormality in Severe Beta Thalassemia

The original experiments with whole thalassemic reticulocytes in which suppression of beta chain synthesis was first demonstrated suggested that the number of beta chains synthesized per unit time in the thalassemic cells is diminished. This, in turn, may be the result of the following beta chain mRNA abnormalities:

1. A reduced number of normal beta chain messengers.
2. Beta chain messengers whose initiation region is genetically altered in a manner such that the number of new chains whose synthesis is initiated is reduced, but which translate normally once initiation has occurred.
3. An abnormal beta chain messenger which takes more time to translate given beta chain than a non-thalassemic beta chain messenger.

Because of the obvious importance of translation time in distinguishing between these possibilities, Clegg and co-workers performed the following experiment (1968):

Dintzis (1961) measured translation time of globin chains in rabbit reticulocytes in the following manner:

1. Rabbit reticulocytes were incubated with a very short (30") radioactive leucine pulse.
2. Cells were lysed and ribosome complexes, including all nascent polypeptide chains, were removed from the supernatant by centrifugation.
3. Supernatant globin was isolated, digested with trypsin, and the digest subjected to two-dimensional chromatography.
4. Each peptide was then eluted and counted for incorporated radioactivity.
5. The same experiment was repeated with progressively longer leucine pulses.

Rabbit globin chain sequences were known, and leucine is distributed homogeneously across the chains.

Globin derived from reticulocytes which had received the shortest pulse was found to have incorporated radioactivity, only into its C-terminal peptide; that is, only those globin chains which were so close to completion that they had been released from the ribosomes before cell lysis and centrifugation had incorporated the label.

With progressively longer leucine pulses more and more peptides, progressing sequentially from the C-terminal to the N-terminal region of the chain, were found to be labelled. At three minutes all peptides, including the N-terminal peptide itself, had been labelled, and homogeneous labelling had been achieved. Three minutes, under the conditions of this system, may, therefore, be said to be the translation time for a single globin chain.

CLEGG and his group (1968) performed the analogous experiment with thalassemic and non-thalassemic human reticulocytes; they found that the translation time for a single beta chain in both types of reticulocytes was *the same*. RIEDER (1971) confirmed these experiments for severe beta thalassemic reticulocytes, and also showed that beta chain translation time was normal in beta thalassemia trait and sickle-thalassemia reticulocytes.

Thus, possibility No. 3 above (a beta mRNA whose translation mechanism is defective) appears to be excluded, and the following hypothetical abnormalities of thalassemic beta chain mRNA must be considered:

1. There are a reduced number of functionally normal beta chain messengers either because of:

a) A mutation in a regulatory gene in a hypothetical beta chain operon resulting in a reduced transcription rate of normal beta chain mRNA's, or

b) Some undefined instability of the beta chain messenger which does not interfere with normal initiation and translation but results in a reduction in the number of functional cytoplasmic messengers during erythroid cell maturation.

2. The initiation region of the beta chain mRNA may be abnormal. For example, the first base of the initiation triplet might be altered, allowing for continued binding of met-tRNA$_F$ at a reduced rate. Such a phenomenon is consistent with the wobble hypothesis of base triplet specificity, whereby alteration of the third base does not necessarily alter specificity but only binding kinetics (CRICK, 1966).

No data presently available and known to this author distinguish between these hypotheses. The fact that the number of beta chains per alpha chain synthesized in thalassemic nucleated erythroid precursors is greater than that in reticulocytes (BRAVERMAN and BANK, 1969; SCHWARTZ, 1970; BRAVERMAN et al., 1971) would seem consistent with the unstable messenger hypothesis (1:b above). However, a feedback mechanism whereby the nucleus responds to a reduced cytoplasmic beta chain concentration by increasing (within genetically determined limits) its production of beta chain messengers might account for the relatively greater fall-off in beta chain synthesis in the enucleated reticulocytes (BRAVERMAN, R., 1971).

FUHR and co-workers (1969) have shown that purified ribosomal subunits from non-thalassemic and thalassemic reticulocytes stimulate the synthesis of TCA precipitable material in cell-free systems derived from non-thalassemic reticulocytes; however, cell-free systems derived from thalassemic reticulocytes were not stimulated. These data may mean that the amount of free beta chain mRNA (mRNA not saturated with ribosomes) is reduced in thalassemic reticulocytes, or that binding of ribosomal subunits to beta chain mRNA is less likely in thalassemic cells because of some abnormality of the binding site which may be related to the initiation site. The data, however, do not distinguish between these possibilities.

NATHAN et al. (1971) infer that, if translation time of thalassemic beta chain mRNA is normal (assuming it is normal in all cases), thalassemic beta chain should be manufactured on polyribosomes of the same size as non-thalassemic beta chain, *unless* the rate of beta chain initiation is diminished. They present data which suggest that thalassemic beta chain is manufactured on polyribosomes of normal size; by their hypothesis, these data constitute space circumstantial evidence for a normal rate of initiation in the patients they studied.

At the present time, two general experimental approaches to the problem of thalassemia seem most likely to be fruitful:

1. Completion of the task of isolation and characterization of normal and thalassemic beta chain messenger RNA. This would involve separation of alpha, beta and gamma chain mRNA's and also, if possible, separation of mRNA's from ribosomes.

2. Study of the functional characteristics of the specific initiation region of the thalassemic beta chain messenger.

The elucidation of the defect in beta chain synthesis in thalassemia is likely to increase our understanding of the nature of the quantitative regulation of protein synthesis in eukaryotic cells.

Appendix I: The Alpha Thalassemias

Hemoglobin H disease is a familial hemolytic anemia of Oriental and Mediterranean populations characterized by inclusion bodies in circulating red cells, decreased survival time of circulating red cells, and the presence of an electrophoretically abnormal hemoglobin in red cell lysates (hemoglobin H) which, as noted above, proves to be a tetramer of beta chains. The precipitation of a denatured form of this unstable hemoglobin (hemichromes) is responsible for the red cell inclusion bodies (Rachmilewitz et al., 1969). One parent of such patients ordinarily has the mild anemia and microcytosis characteristic of beta thalassemia trait, but neither in the parents nor the patients themselves are hemoglobins A_2 or F elevated. The other parent's red cell morphology is usually entirely normal. A second electrophoretic hemoglobin variant is often found in patients with hemoglobin H disease, and is also present at higher levels in newborn relatives of these patients: this is hemoglobin Bart's, which proves to be a tetramer of γ chains (Weatherall, 1965, p. 154).

Because of the presence of excess non-alpha chains in these syndromes, the possibility of alpha chain suppression was considered (Ingram and Stretton, 1959). Interactions between the alpha thalassemia trait and structural alpha chain mutants analogous to sickle thalassemia were known: That is, patients were described who had more than 50% hemoglobin Q and hemoglobin T, in whom one parent lacked the mutant hemoglobin but had findings consistent with alpha thalassemia trait. The presence of hemoglobins H and Bart's in propositii and relatives supported the assumption of interaction between a gene which suppressed alpha chain synthesis and an alpha chain mutant. (Weatherall, 1965, p. 154 et seq.).

In 1867, Clegg and Weatherall separated globin chains from hemoglobin H disease reticulocytes incubated with radioactive amino acids, and demonstrated a 50% reduction in alpha chain synthesis. In these cases, the peaks of optical density and radioactivity were similar, indicating that the excess beta chain manufactured is not rapidly destroyed in these cells.

When the reticulocytes of individuals with alpha thalassemia trait are similarly studied, alpha chain synthesis is suppressed on the average by 25%; the hematologically normal parents of patients with hemoglobin H disease shows minimal alpha chain suppression (on the order of 10%) which may indicate a minimally active alpha thalassemia gene (Schwartz et al., 1969).

The interaction of genes which involve moderate and minimal suppression of alpha chain synthesis may not be the only genotype responsible for hemoglobin H disease. EFREMOV and co-workers (1971) have demonstrated interaction between alpha thalassemia trait, and what appear to be alpha chain mutants present at very low levels and perhaps synthesized at similarly low rates, in two siblings with hemoglobin H disease. A situation analagous to the interaction of the Lepore hemoglobin trait (see Appendix II) and beta thalassemia trait may exist in these patients.

There is evidence for the presence of two more non-allelic genetic loci for alpha chain (three types of alpha chain in the same individual — HOLLAN et al., 1970), which provides another possible mechanism for the alpha thalassemia syndromes.

Double heterozygosity for the more active genes may be lethal; in Oriental populations, severely anemia stillborn infants occur who have almost 100% hemoglobin Bart's; their parents both have hematologically detectable alpha thalassemia trait. These fetuses, unlike those with the severe beta thalassemia genotype, are as incapacitated for hemoglobin F as for hemoglobin A production.

The author is not aware of any attempts to sequence the alpha chain produced in these syndromes, so that an electrophoretically silent structural mutation is not completely excluded. We are also not aware of any attempts to elucidate the molecular basis of the alpha chain suppression.

Appendix II: The Lepore-Pylos Hemoglobins

Four abnormal hemoglobins have been isolated from lysates of children of Mediterranean and Papuan origin with the typical severe beta thalassemia syndrome (WEATHERALL, 1965, p. 116). These hemoglobins ($Lepore_{Boston}$; $Lepore_{Hollandia}$; $Lepore_{Cyprus}$; Pylos) have similar electrophoretic mobilities and constitute 10% or less of the total hemoglobin. Both parents of these patients have red cell morphology consistent with beta thalassemia trait. One parent has the abnormal hemoglobin, normal hemoglobin A_2 and a mild (10%, as opposed to 50% in the severe beta thalassemia offspring) elevation of hemoglobin F. The other parent has typical high A_2 beta thalassemia trait with no abnormal hemoglobin. The exception to this genetic arrangement is a Papuan family, in which the affected offspring has severe beta thalassemia with more than 25% $Lepore_{Hollandia}$ hemoglobin, and both parents have less than 10% of this hemoglobin as well. Thus, the offspring with severe beta thalassemia syndrome appear to be homozygotes for the abnormal hemoglobin in this case, and double heterozygotes for beta thalassemia trait and an abnormal hemoglobin which produces thalassemic red cell changes in heterozygotes.

These syndromes appear to be, as it were, the converse of sickle-thalassemia. That is, in the latter the interaction of a thalassemia gene and a gene for a hemoglobin structural mutation interact to produce the homozygous mutant hemoglobin syndrome; the Lepore-Pylos hemoglobins interact with thalassemic genes or with themselves to produce the syndrome of severe homozygous thalassemia (ibid.) Studies of the structure of several Lepore hemoglobins sheds some light on this phenomenon:

Fingerprint chromatography of $Lepore_{Boston}$ revealed a pattern almost indistinguishable from that of hemoglobin A_2; that is, the alpha chains were normal and the non-alpha chain appeared almost identical to delta chain (GERALD et al., 1961). However, BAGLIONI showed (1962) that the C-terminal peptide was identical to that of beta rather than delta chain (WEATHERALL, 1965, p. 116 et seq.).

Hemoglobin Pylos also proved to have a delta chain whose C-terminal region was beta-like while hemoglobin $Lepore_{Hollandia}$ has a non-beta chain the N-terminal half of which is delta-like and the C-terminal half beta-like.

Thus, the mutant chains of the Lepore-Pylos hemoglobins appear to be fusion products of the beta and delta loci. This implies close linkage of these loci and possible partial crossing-over.

RIEDER and WEATHERALL (1965) have shown that the synthetic rate of delta chain in human reticulocytes corresponds to the concentration of hemoglobin A_2; that is, it is manufactured at 1/40th the rate of alpha or beta chain. Synthetic rate studies of the Lepore-Pylos hemoglobins have not yet been performed, but it seems probable that their synthesis is suppressed, in comparison to that of alpha or beta chain, to a degree unique amongst the hemoglobin mutants. This extreme suppression may be due to the incorporation into the Lepore-Pylos non-beta chain gene of a rate determinant corresponding to its delta, rather than its beta region.

Acknowledgements

I am obliged to Drs. W. FRENCH ANDERSON, ARTHUR BANK and JOEL SCHWARTZ, and to Mr. RICHARD BRAVERMAN for their advice and comments on the preparation of this review.

The illustrations are reprinted from the Journal of Molecular Biology, **42**, 57 (1969), by permission of Academic Press, Ltd. London.

References

ANDERSON, W. F , NIENHUIS, A. W.: Pers. communication 1971.

BAGLIONI, C.: Correlations between genetics and chemistry of human hemoglobins. Molecular genetics, Part I, p. 405 (TAYLOR, J. H., Ed.). New York and London: Academic Press 1963.

BANK, A., MARKS, P. A.: (1) Alpha chain synthesis relative to beta chain synthesis in thalassemia major and minor. Nature (Lond.) **212**, 1198 (1966).

BANK, A., MARKS, P. A.: (2) Protein synthesis in a cell free human reticulocyte system: Ribosome function in thalassemia. J. clin. Invest. **45**, 330 (1966).

BANK, A.: Hemoglobin synthesis in beta thalassemia: The properties of the free alpha chain. J. clin. Invest. **47**, 860 (1968).

BANK, A., BRAVERMAN, A. S., O'DONNELL, J. V., MARKS, P. A.: Absolute rates of globin chain synthesis in thalassemia. Blood **31**, 226 (1968).

BANK, A., O'DONNELL, J. V.: Intracellular loss of free alpha chains in beta thalassemia. Nature (Lond.) **222**, 295 (1969).

BANK, A., O'DONNELL, J. V., BRAVERMAN, A. S.: Globin chain synthesis in heterozygotes for beta chain mutations. J. Lab. clin. Med. **76**, 616 (1970).

BARGELLESI, A., PONTREMOLI, S., CONCONI, F.: Absence of beta globin chain synthesis and excess alpha globin chain synthesis in homozygous beta thalassemia. Europ. J. Biochem. **1**, 73 (1967).

BEUTLER, E.: Drug induced hemolytic anemia. Pharmac. Rev. **21**, 73 (1969).

BISHOP, J. O.: Initiation of hemoglobin polypeptide chains. Biochim. biophys. Acta (Amst.) **119**, 130 (1966).

BLOBEL, G.: Release, identification and isolation of messenger RNA from mammalian ribosomes. Proc. nat. Acad. Sci. (Wash.) **68**, 832 (1971).

BOYER, S. H., HATHAWAY, P., GARRICK, M. D.: Modulation of protein synthesis in man: an in vitro study of hemoglobin synthesis by heterozygotes. Cold Spr. Harb. Symp. quant. Biol. **29**, 333 (1964).

Braverman, A. S., Bank, A.: Changing rates of globin chain synthesis during erythroid cell maturation in thalassemia. J. molec. Biol. **42**, 57 (1969).

Braverman, A. S., McCurdy, P. R., Manos, O.: Unpub. data 1971.

Braverman, R. H.: Pers. Communication 1971.

Burka, E. R., Marks, P. A.: Ribosomes active in protein synthesis in human reticulocytes: a defect in thalassemia major. Nature (Lond.) **199**, 706 (1963).

Clark, B. F. G., Marcker, D.: The role of N-formyl methionine RNA in protein synthesis. J. molec. Biol. **17**, 394 (1966).

Clegg, J. B., Weatherall, D. J.: Hemoglobin synthesis in alpha thalassemia (hemoglobin H disease). Nature (Lond.) **215**, 1241 (1967).

Clegg, J. B., Weatherall, D. J., Na-Nakorn, S., Wasi, P.: Hemoglobin synthesis in beta thalassemia. Nature (Lond.) **220**, 664 (1968).

Crick, F. H. C.: Codon-Anticodon pairing: the wobble hypothesis. J. molec. Biol. **19**, 548 (1966).

Dintzis, H. M.: Assembly of the polypeptide chains of hemoglobin. Proc. nat. Acad. Sci. (Wash.) **47**, 247 (1961).

Doyle, D., Schimke, R. T.: The genetic and developmental regulation of hepatic delta aminolevulinate dehydratase in mice. J. biol. Chem. **244**, 5449 (1969).

Efremov, G. D., Wrightstone, R. N., Huisman, T. H. J., Schroeder, W. A., Hyman, C., Ortega, J., Williams, K.: An unusual hemoglobin anomaly and its relation to alpha thalassemia and hemoglobin H disease. J. clin. Invest. **50**, 1628 (1971).

Fessas, P., Loukopoulos, D., Kaltsoya, A.: Peptide analysis of the inclusions of erythroid cells in beta tha assemia. Biochim. biophys. Acta (Amst.) **124**, 430 (1966).

Fuhr, J., Natta, C., Marks, P. A., Bank, A.: Protein synthesis in cell-free systems from reticulocytes of thalassemia patients. Nature (Lond.) **224**, 1305 (1969).

Gabuzda, T. G., Nathan, D. G., Gardner, F. H.: The turnover of hemoglobins A, F and A_2 in the peripheral blood of three patients with thalassemia. J. clin. Invest. **42**, 1678 (1963).

Garrod, A. E.: Inborn errors of metabolism. London: Henry Frowde 1908.

Gerald, P. S.: A human mutation (the Lepore hemoglobinopathy) possibly involving two cistrons. Amer. J. Dis. Chi d. **102**, 514 (1961).

Gaskill, P., Kabat, D.: Unexpectedly arge size of globin messenger ribonuc eic acid. Proc. nat. Acad. Sci. (Wash.) **68**, 72 (1971).

Gilbert, J. M., Thornton, A. G., Nienhuis, A. W., Anderson, W. F.: Cell-free hemoglobin synthesis in beta thalassemia. Proc. nat. Acad. Sci. (Wash.) **67**, 1854 (1970).

Grinstein, M., Robin, M., Bannerman, B. M., Vavra, J. D., Moore, C. V.: Hemoglobin metabolism in thalassemia: in vivo studies. Amer. J. Med. **39**, 18 (1960).

Guidotti, G.: Thalassemia. In: Conf. on Hemoglobin, Arden House. New York: Columbia Univ. 1962.

Heywood, J. D., Karon, M., Weissman, S.: Studies of in vitro synthesis of heterogenic hemoglobins. J. clin. Invest. **43**, 2368 (1964).

Heywood, J. D., Karon, M., Weissman, S.: Asymetrica incorporation of amino acids into the alpha and beta chains of hemoglobin synthesized in tha assemia reticulocytes. J. Lab. clin. Med. **66**, 476 (1965).

Hershko, Ch., Karsai, A., Eylon, L., Izak, G.: The effect of chronic iron deficiency on some biochemical functions of the human hemopoietic tissue. Blood **36**, 321 (1970).

Hillman, R. S., Finch, C. A.: Erythropoiesis. New Engl. J. Med. **285**, 99 (1971).

Hollan, S. R., Brimhall, B., Jones, R. T., Koler, R. D., Stocklen, Z., Szelenyi, J. G.: Multiple alpha chain loci for human hemoglobin. XIIIth Intern. Congress of Hemat. Munich. Abstract, p. 8 (1970).

Housman, D., Jacobs-Lorena, M., Rajbhandary, U. L., Lodish, H. F.: Initiation of hemoglobin synthesis by methionyl-tRNA. Nature (Lond.) **227**, 913 (1970).

Huisman, T. H. J., Schroeder, W. A., Charache, S., Bethlenfalvay, N. C., Bouver, N., Shelton, J. R., Shelton, J. B., Apell, G.: Hereditary persistance of feta hemoglobin. New Engl. J. Med. **285**, 711 (1971).

Ingram, V. M.: Gene mutation in human hemoglobin: the chemical difference between normal and sickle hemoglobin. Nature (Lond.) **180**, 326 (1957).

Ingram, V. M., Stretton, A. O. W.: The genetic basis of the thalassemia diseases. Nature (Lond.) **184**, 1903 (1959).

Jackson, R., Hunter, T.: Role of methionine in the initiation of hemoglobin synthesis. Nature (Lond.) **227**, 672 (1970).

Jacob, F., Monod, J.: Genetic regulatory mechanisms in the synthesis of proteins. J. molec. Biol. **3**, 318 (1961).

Kan, Y. W., Nathan, D. G.: Mild thalassemia: the result of interactions of alpha and beta thalassemia genes. J. clin. Invest. **49**, 635 (1970).

Kushner, J. P., Lee, G. R., Wintrobe, M. M., Cartwright, G. E.: Idiopathic refractory sideroblastic anemia: clinical and laboratory investigation of 17 patients and review of the literature. Medicine (Baltimore) **50**, 139 (1971).

Lockard, R. E., Lingrel, J. B.: The synthesis of mouse hemoglobin beta chains in rabbit reticulocyte cell free system programmed with mouse 9S RNA. Biochem. biophys. Res. Commun. **37**, 204 (1969).

London, L. M., Shemin, D., West, R., Rittenberg, D.: Heme synthesis and red blood cell dynamics in normal humans and in subjects with polycytjemia vera, sickle cell anemia and pernicious anemia. J. biol. Chem. **179**, 463 (1949).

Loria, A., Sanchez-Medal, L., Lisker, R., de Rodriguez, E., Labardini, J.: Red cell life span in iron deficiency anemia. Brit. J. Haemat. **13**, 294 (1967).

Marcker, K., Sanger, F.: N-Formyl sRNA. J. molec. Biol. **8**, 835 (1964).

Modell, C. B., Latter, A., Steadman, J. H., Huehns, E. R.: Hemoglobin synthesis in beta thalassemia. Brit. J. Haemat. **17**, 485 (1969).

Nathan, D. G., Lodish, H. F., Kan, Y. W., Housman, D.: Beta thalassemia and translation of globin messenger RNA. Proc. nat. Acad. Sci. (Wash.) **68**, 2514 (1971).

Nienhuis, A. W., Anderson, W. F.: Isolation and translation of hemoglobin messenger RNA from thalassemia, sickle cell anemia and normal human reticulocytes. J. clin. Invest. **50**, 2458 (1971).

Nienhuis, A. W., Laycook, D. G., Anderson, W. F.: Translation of rabbit hemoglobin messenger RNA by thalassemic and non-thalassemic ribosomes. Nature (Lond.) New Biol. **231**, 205 (1971).

Parker, W. C., Bearn, A. G.: Application of genetic regulatory mechanisms to human genetics. Amer. J. Med. **34**, 680 (1963).

Pauling, L., Itano, H. A., Singer, S. J., Wells, I. C.: Sickle cell anemia, a molecular disease. Science **110**, 543 (1949).

Prichard, P. M., Gilbert, J. M., Shafritz, D. A., Anderson, W. F.: Factors for the initiation of hemoglobin synthesis by rabbit reticulocyte ribosomes. Nature (Lond.) **226**, 511 (1970).

Rachmilewitz, E. A., Peisach, J., Bradly, T. B., Blumberg, W. E.: Role of hemichromes in the formation of inclusion bodies in hemoglobin H disease. Nature (Lond.) **222**, 248 (1969).

Ranney, H. M.: Clinically important variants of human hemoglobin. New Engl. J. Med. **282**, 144 (1970).

Rieder, R. F., Weatherall, D. J.: Studies on hemoglobin biosynthesis: asynchronous synthesis of hemoglobin A and hemoglobin A_2 by erythroid precursors. J. clin. Invest. **44**, 42 (1965).

Rieder, R. F.: Aspects of the structure, synthesis and clinical effects of unstable hemoglobins. Red cell structure and metabolism (Ramot, B., Ed.). New York: Academic Press 1971.

Rieder, R. F.: Pers. communication 1971.

Schapira, G., Dreyfus, J. C., Maleknia, N.: The ambiguities in the rabbit hemoglobin: evidence for a messenger RNA translated specifically into hemoglobin. Biochem. biophys. Res. Commun. **32**, 558 (1968).

Schwartz, E., Kan, Y. W., Nathan, D. G.: Unbalanced globin chain synthesis in alpha thalassemia heterozygotes. Ann. N. Y. Acad. Sci. **165**, Second conf. on the problems of Cooley's anemia. Art. L, p. 288 (1969).

Schwartz, E.: Heterozygous beta thalassemia: balanced globin synthesis in bone marrow cells. Science **167**, 1513 (1970).

SHAFRITZ, D. A., ANDERSON, W. F.: (1) Factor dependent binding of methionyl-tRNAs to reticulocyte ribosomes. Nature (Lond.) **227**, 918 (1970).
SHAFRITZ, D. A., ANDERSON, W. F.: (2) Isolation and partial characterization of reticulocyte factors M_1 and M_2. J. biol. Chem. **245**, 5553 (1970).
SHAHID, M. J., SAHLI, I. T.: Erythrokinetic studies in thalassemia. Brit. J. Haemat. **20**, 75 (1971).
SHEPARD, M. K., WEATHERALL, D. J., CONLEY, C. L.: Semiquantitative estimation of distribution of fetal hemoglobin in red cell populations. Bull. Johns Hopk. Hosp. **110**, 293 (1962).
SMITH, E. W., TORBERT, J. V.: Two abnormal hemoglobins with evidence for a new genetic locus for hemoglobin formation. Bull. Johns Hopk. Hosp. **102**, 38 (1958).
SMITH, E. W., KREVANS, J. R.: Clinical manifestations of hemoglobin C disorders. Bull. Johns Hopk. Hosp. **104**, 17 (1959).
STANBURY, J. B., WYNGAARDEN, J. B., FREDRICKSON, D. S.: The metabolic basis of inherited disease. New York: McGraw-Hill 1966.
TOMKINS, G. M., GELEHRTER, T. D., GRANNER, D., MARTIN, D., Jr., SAMUELS, H. H., THOMPSON, E. B.: Control of specific gene expression in higher organisms. Science **166**, 1474 (1969).
VIGI, V., VOLPATO, S., GABURRO, D., CONCONI, F., BARGELLESI, A., PONTREMOLI, S.: The correlation between red cell survival and excess of alpha globin synthesis in beta thalassemia. Brit. J. Haemat. **16**, 25 (1969).
WEATHERALL, D. J.: The thalassemia syndromes. Philadelphia: F. A. Davis Comp. 1965. (A good deal has been learned about the biochemical basis of the thalassemias since the publication of Dr. WEATHERALL's monograph, he and his co-workers having been responsible for a great deal of the new information. Yet *The Thalassemia Syndromes* remains the standard account of the clinical and genetic aspects of these diseases, and much of this review could hardly have been written without it.)
WEATHERALL, D. J., CLEGG, J. B., NAUGHTON, M. A.: Globin synthesis in thalassemia: an in vitro study. Nature (Lond.) **208**, 1061 (1965).
WHITE, J. M., BRAIN, M. C., ALI, M. A. M.: Globin synthesis in sideroblastic anemia. Brit. J. Haemat. **20**, 263 (1971).
WICKRAMASINGHE, S. N., MCELWAIN, T. J., COOPER, E. H., HARDISTY, R. M.: Proliferation of erythroblasts in beta thalassemia. Brit. J. Haemat. **19**, 719 (1970).

Addendum

Since the completion of this Review, a novel methodology has been used to quantitate thalassemia mRNA more directly. KACIAN and coworkers (1972), using a viral reverse transcriptase, employed globin mRNA as an in vitro template for the synthesis of DNA, and demonstrated the complimentarity of the DNA product to the DNA product to the template material by hydridization studies with the same mRNA fractions. Using a modification of their original technique, GAMBINO and coworkers (1973) then demonstrated that, in beta thalassemia reticulocyte extracts, the amount of RNA available for hybridization with the DNA product of globin mRNA template was significantly less than that in non-thalassemuc cell extracts. These results have been confirmed by FORGET and co-workers (1973); they constitute the first direct evidence for a diminished *quantity* of beta globin mRNA in beta thalassemia.

Both groups also demonstrated a diminished quantity of mRNA in alpha thalassemia (hemoglobin H disease) reticulocytes, using the same techniques.

Forget, B. G., Housman, D., Skoultchi, A., Benz, Jr., E. J.: Quantitative deficiency of chain specific globin messenger ribonucleic acid in the thalassemia syndromes. J. Clin. Invest. **52** (Abstracts) (1973) (in press).

Gambino, R., Kacian, D., Ramirez, F., Dow, L. W., Grossbard, E., Natta, C., Spiegelman, S., Marks, P. A., Bank, A.: Decreased globin messenger RNA in thalassemia by hybridization and biologic activity assays. Ann. New York Acad. Sci. 3rd Conference on Cooley's Anemia (1973) (in press).

Kacian, D. L., Spiegelman, S., Bank, A., Terada, M., Metafora, S., Dow, L., Marks, P. A.: In vitro synthesis of DNA components of human genes for globins. Nature New Biol. **235**, 167 (1972).

The Mitochondrial DNA of Malignant Cells

CLAUDE A. PAOLETTI and GUY RIOU

I. Introduction

Mitochondria have multiple functions: they control the energetic balance of animal cells and are thus essential for their survival; they are involved in the metabolism of sugars and lipids and are responsible for the synthesis of some polynucleotides and proteins. The mitochondria are the targets for the thyroid hormones.

Several lines of indirect argument suggest that mitochondria could be involved in the differentiation process of complex multicellular organisms (SLONIMSKI, 1956). They carry a specific macromolecular equipment made up of 60-55 S ribosomes consisting of subunits of 45-40 S and 35-30 S, DNA, several tRNAs and ribosomal RNAs (4-12-16 S but probably no 5 S RNA), tRNA synthetases, and DNA and RNA polymerases.

This biochemical semi-autonomy determines the genetic individuality responsible for non-Mendelian cytoplasmic heredity.

Two independent sets of arguments strongly suggest that mitochondrial DNA (mtDNA) has some genetic functions:

a) in animal cells, it is transcribed into both heterogeneous messenger-like RNA, and stable and discrete ribosomal 4-12-16 S RNAs (20 to 25% of the total genome);

b) in yeast, some cytoplasmic "petite" mutations contain mtDNA which is modified in its average base composition (MOUNOULOU, JACOB and SLONIMSKI, 1966) and in the base pair sequences (BERNARDI et al., 1970).

Mutations in mtDNA are probably inherited maternally in animal cells since DAWID and BLACKLER (1972) have demonstrated that the hybrid progeny of a mating of *Xenopus laevis* with *X. Mülleri* contains only the mtDNA of the maternal parent.

However, the architecture and the functional role of mitochondria are certainly under dual nuclear and mitochondrial control.

This duality is suggested by several lines of evidence. A direct one indicates that some of the mitochondrial enzymes obey a Mendelian inheritance pattern; in animals, this has been demonstrated for malate dehydrogenase and for the malic enzyme of the mouse (SHOWS, CHAPMAN and RUDDLE, 1970). An indirect one is the size of mtDNA, which can code for only about 30 proteins of molecular weight 20,000 daltons, according to the rules of the genetic code, whereas there are more than 70 enzymes, 23 protein species in the inner membranes, and 12 protein species in the outer membranes of the mitochondria of rat liver.

It is not known in detail which mitochondrial functions are under the control of mtDNA. However, the assembly of a functional cytochrome oxidase in yeast and in animal cells and of a functional rutamycin (oligomycin)-sensitive ATPase requires polypeptides, some of which are synthesized on mitochondrial ribosomes.

mtDNA has also been assumed to be responsible for the synthesis of (a) some structural parts of the mitochondrial membranes which could be common to other membranes in animal cells (Attardi, 1969), (b) some ribosomal proteins in *Neurospora* and in *Tetrahymena*, and (c) a replication factor, a product of mitochondrial protein synthesis which is required for the correct replication of mtDNA (Williamson et al., 1971 and Weislogel and Butow, 1970, 1971).

It has repeatedly been claimed that specific alterations of some mitochondrial functions characterize malignant cells or can be involved in oncogenesis. The theory of Warburg (1956) which implies that the oxidative metabolism of malignant cells is fundamentally altered and that a fermentation process takes over the generation of energy is no longer accepted. Nevertheless, alterations of the mitochondrial respiratory functions are observed in most tumor cells, although no conclusive general explanation for this impairment has thus far been provided (Wenner, 1967).

On the other hand, a well-developed hypothesis concerning the origin of malignancy has been based on alterations of cellular membranes, including those of mitochondria (Wallach, 1968). This hypothesis requires more direct evidence although there is some experimental support for it, such as the abnormalities observed in the mitochondrial membranes of rat minimal hepatoma (Chang, Schnaitman and Morris, 1971).

The involvement of mitochondria in the cancer process was again suggested by the striking observations of Clayton and Vinograd (1967) who established the existence of topological anomalies in the mtDNA of human leukemic leukocytes. These observations have now been extended to several other human and experimental tumors.

This review attempts to describe and discuss the latter phenomenon with respect to its origin and specificity in neoplasia in the light of the few guidelines so far established relative to the biochemical genetics and functions of mitochondria in normal and tumor cells.

II. Morphology of Mitochondria of Malignant Cells

The shape and structure of the chondriomal apparatus in tumor cells can only be clearly visualized by electron microscopy; it has been described by Bernhard (general reviews 1963 and 1969).

The structure and intracellular arrangement of mitochondria are variable and depend on the growth stages and environmental conditions of the cells (Frederic, 1958). It is therefore, difficult to interpret the many morphological changes of the chondriomal apparatus in tumor cells. Such changes are by no means specific to the cancer process and are probably secondary phenomena, triggered by vascular or nutritive disturbances associated with neoplastic growth. Moreover, they are not stable and disappear when the cells are grown *in vitro*.

Nevertheless, an interesting exception to this rule was reported by Bernhard and Tournier (1966): hamster cells, transformed by an oncogenic strain of adenovirus 12, reveal atypical mitochondria, characterized by exceptional size or by abnormal aspects of their internal structure. These alterations persist after many serial passages *in vitro*. However, this observation is limited to that particular strain of adenovirus.

There is good evidence for the multiplication of some plant RNA viruses within mitochondria, probably inside the inter-membrane space [Tobacco mosaic virus: RALPH and CLARK (1966); bean yellow mosaic virus: WEINTRAUB, RAGETLI and JOHN (1966); tobacco rattle virus: HARRISON and ROBERTS (1968) and HARRISON, STEFANAC and ROBERTS (1970); cucumber green mottle mosaic virus: HATTA et al. (1971)].

Recently, two groups attempted to demonstrate intramitochondrial replication of Rous sarcoma virus. GAZZOLO et al. (1969) published electron micrographs showing intramitochondrial bodies in a hamster cell line transformed by this virus (SCHMIDT-RUPPIN strain). They were interpreted as viral nucleocapsids. MACH and KARA (1971) and KARA et al. (1971) prepared from Rous sarcoma extracts mitochondrial preparations which display a high titer of infectious virus and liberated material identified as subviral protein particles. They revealed, under the electron microscope, organized forms which they described as virions. The criteria for purification of the mitochondria were not rigorously met in these experiments which therefore require confirmation.

Abnormal bodies, associated with the mitochondria of tumor cells have been reported in several instances: they are located near the mtDNA in Ehrlich ascites cells (NASS and NASS, 1964). They appear as paraviral formations in avian erythroblastosis (BENEDETTI and BERNHARD, 1958), as glycogen inclusions in Warthin's parotid tumors [TANDLER and SHIPKEY, 1964 (2)], as paracrystalline proteins (HRUBAN, SWIFT and RECHCIGL, 1965), and as irregular formations in rat hepatoma (SVOBODA, 1964).

The number of mitochondria per tumor cell is variable; it may be higher or lower than in normal cells, as far as can be evaluated by present methods. Generally, as the tumor develops, fewer mitochondria are found, although some secreting tumors, such as thyroid, parathyroid, parotid and kidney tumors, display eosinophilic and granular cells containing very large numbers of mitochondria, mostly giant ones. In some instances, this number is so great as to give the cell observed under the electron microscope the appearance of a bag filled with mitochondria. These cells are called, depending on their origin, oxyphil oncocytes, Hurthle or Askenazy cells [OBERLING, RIVIERE and HAGUENAU, 1959; ROTH, OLEN and HANSEN, 1962; TANDLER and SHIPKEY, 1964 (1); BECHER, 1964; ERICSON, SELJELID and ORRENIUS, 1966; SCHIEFER, HUBNER and KLEINSASSER, 1968 and ASKEW et al., 1971). Oncocytoma contain a modified mtDNA (see below III—1).

III. Size and Structure of Mitochondrial DNA in Malignant Cells

Exhaustive general reviews have been devoted to mtDNA [NASS, 1969 (4); BORST, 1969; KROON, 1969; BORST and KROON, 1969; SWIFT and WOLSTENHOLME, 1969; ASHWELL and WORK, 1970; RABINOWITZ and SWIFT, 1970; SCHATZ, 1970; BORST, 1972).

The mtDNA of non-malignant animal cells consists of double-stranded covalently closed circles (Fig. 1); in contrast, the mtDNA of some other eukaryotes (plants) is linear. In most cases, these circles are found to be highly twisted ,unless nicked during extraction. The length of these circular molecules is remarkably constant among animal species: it varies between 4.55 and 5.55 μ. These molecules are circular monomers and make up usually more than 90%[1] of the total mtDNA. In addition, circles overlapping at two cross-over zones are regularly seen (1 to 10% of total mtDNA) (Fig. 1, items 2 and 3; Table 1). They have been interpreted as catenated interlocked molecules, made up of imbricated monomers (Clayton et al., 1968; Hudson et al., 1968). In the opinion of the authors no compelling argument has yet

Table 1. Frequency of monomers and catenated oligomers in mtDNA of non-malignant cells (expressed as % by weight of extracted mtDNA). Circular dimers were not observed except in rat liver (0.1 %) by Kirschner et al. quoted by Wolstenholme, Koike and Renger (1970)

Species	Organ	Number of molecules scored	Catenated oligomers (%)	References
		A. Grown in vivo		
Man	Leukocytes	1926	1.4	Clayton and Vinograd, 1967
		2707	1.7	Clayton et al., 1968
Mouse	Liver	200—500	4.3	Nass, 1970
	Leukocytes	200—500	4.2	Nass, 1970
	Leukocyte mononucleosis	200—500	7.3	Nass, 1970
	Embryo	2300	7.4	Clayton et al., 1968
Rabbit	Brain	1762	10.3	Clayton et al., 1968
	Bone marrow	1845	9.6	Clayton et al., 1968
	Kidney	2817	6.5	Clayton et al., 1968
	Liver	2855	5.2	Clayton et al., 1968
Guinea pig	Brain	3287	8.4	Clayton et al., 1968
	Liver	2210	8.9	Clayton et al., 1968
Chick	Bone marrow	150	5.1	Riou and Lacour, 1971
	Liver	761	3.8	Riou and Lacour, 1971
		B. Grown in vitro		
Hamster	Kidney (BHK_{21}/C_{13})			
	log phase	200—500	10.7	Nass, 1970
	confluent	200—500	17.0	Nass, 1970
	Embryo	1511	6.0	Riou and Delain, 1971
Chick	Embryo fibroblasts			
	log phase	200—500	10.7	Nass, 1970
	confluent	200—500	17.0	Nass, 1970

been offered to support this interpretation. The spatial arrangement of the two duplexes at each cross-over zone cannot be ascertained by presently available techniques of electron microscopy. One cross-over zone could be a four-stranded structure joining the circular duplexes through covalent bonds. The other cross-over zone could occur through mechanical and random folding of the molecule on itself. This arrangement would be identical to the figure-of-eight structure generated *in vitro*

[1] Percentages are expressed in weight throughout this review.

by CLAYTON et al. (1970) after denaturation and renaturation of a mixture of unicircular monomers and dimers. A similar structure has recently been described by BOURGAUX et al. (1971). According to these authors, the replicating DNA of polyoma virus contains an unexpectedly high proportion of molecules made up of two joined circles and considered as almost 100% replicated but not separated. Whatever the

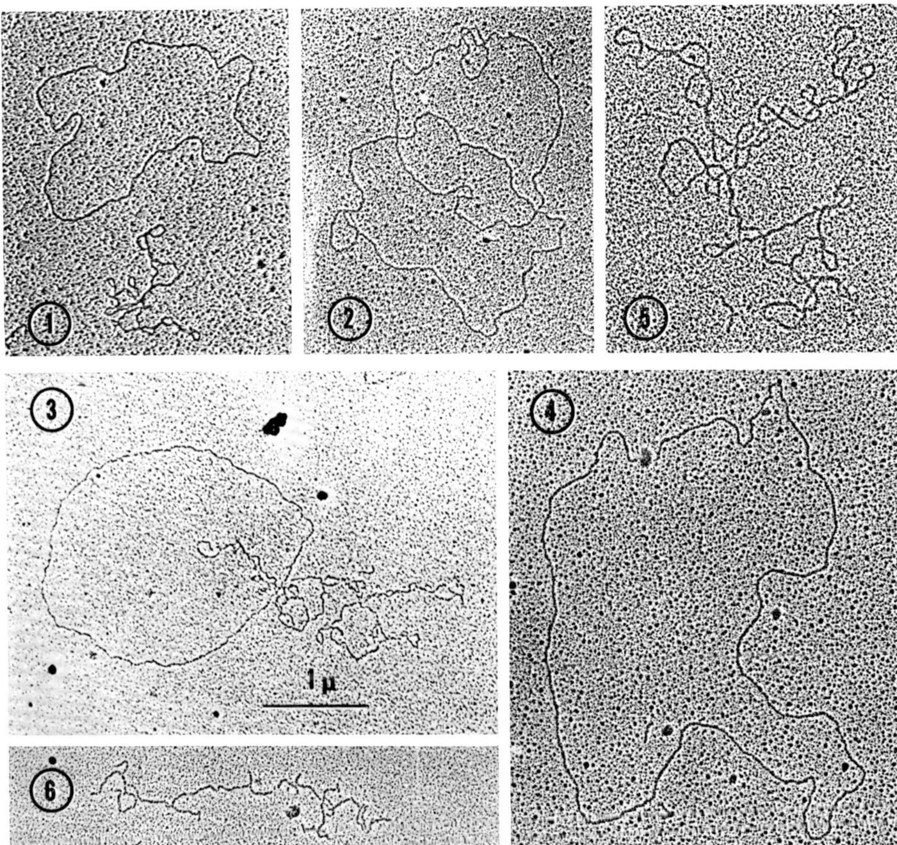

Fig. 1. Electron micrographs of mtDNA from leukocytes of human myeloid chronic leukemia. 1: Relaxed monomer; twisted monomer. 2: Catenated dimer. 3: Catenated trimer. 4: Relaxed circular dimer. 5: Twisted dimer. 6: Highly twisted dimer. × 20.000. Data from this laboratory. (PAOLETTI, RIOU and PAIRAULT, 1972)

actual structure of these apparently catenated oligomers, we have used this term for them throughout this review.

Double-branched, circular molecules, which are usually interpreted as partially replicated forms according to Cairn's model, are very rarely seen (KIRCHNER, WOLSTENHOLME and GROSS, 1966; PAOLETTI, RIOU and PAIRAULT, 1972) and (Fig. 2, item 5; RIOU and PAOLETTI, 1973) and (Fig. 4 item a).

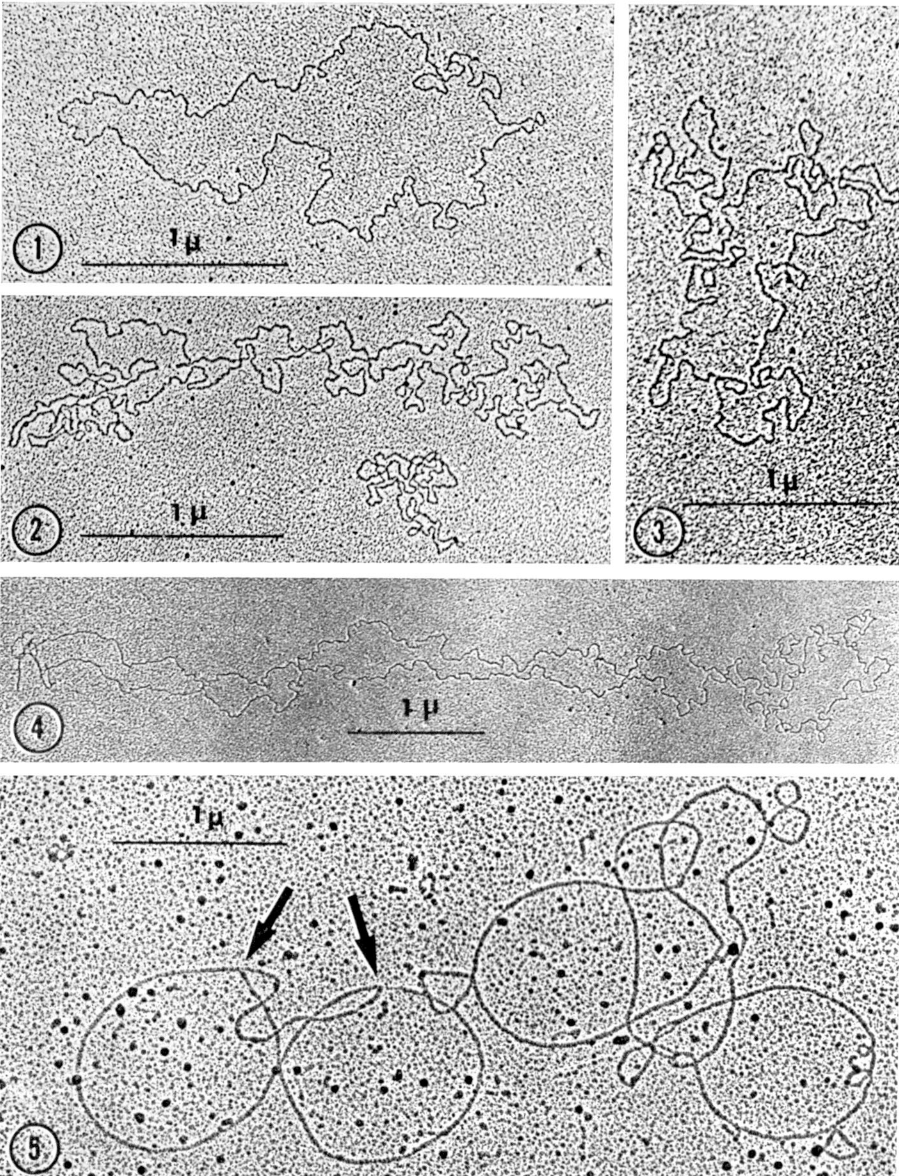

Fig. 2. Electron micrographs of mtDNA from human thyroid. 1: Circular dimer. 2: Twisted monomer and catenated tetramer made up of two circular dimeric subunits. 3: Circular trimer. 4: Catenated hexamer (three interlocked circular dimeric subunits). 5: DNA extruded from osmotically shocked mitochondria. Replicative form of the monomeric part of a catenated tetramer. The arrows indicate two branching points on the same structure. The distances between them are 1.65 μ, 3.35 μ and 3.75 μ

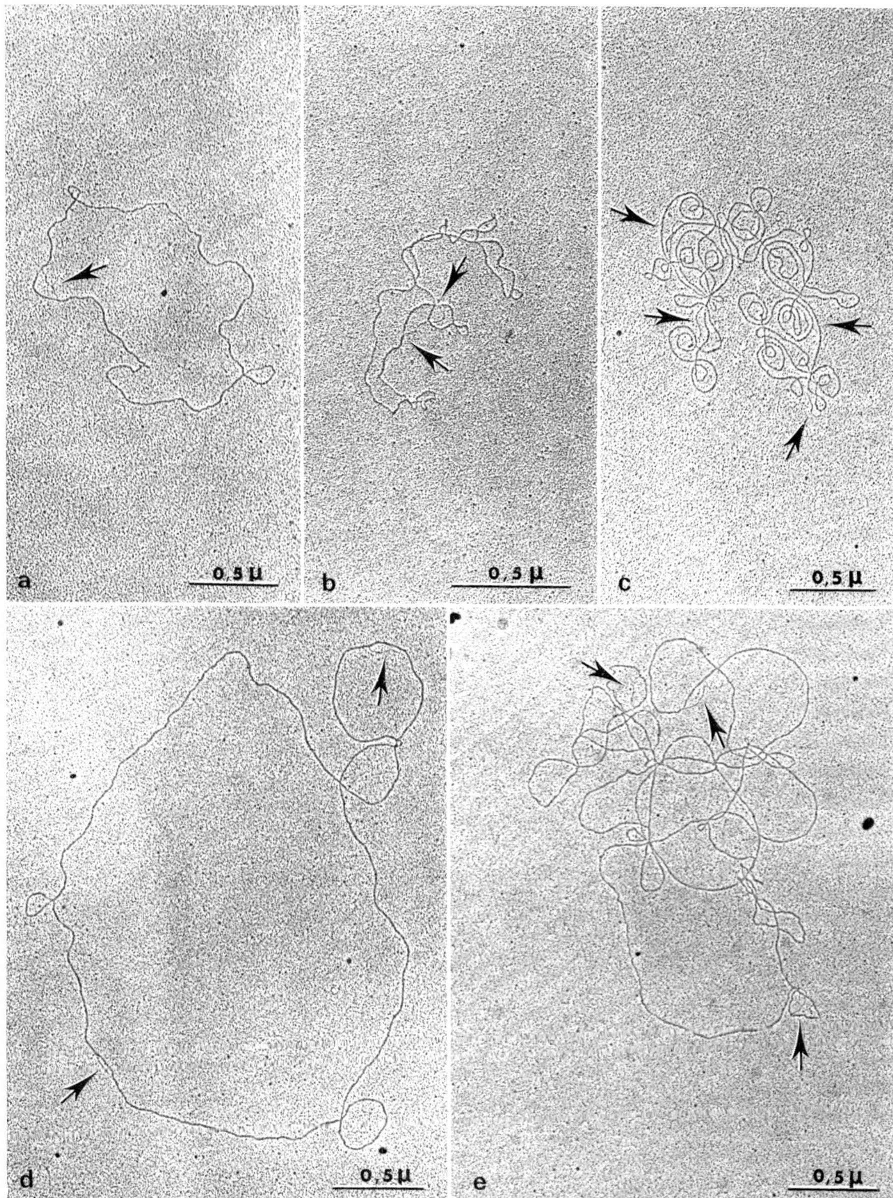

Fig. 3a—e. Electron micrographs of D-loops in mt DNA from thyroid human oncocytoma. a Monomer with one D-loop; b monomer with one expanded D-loop; c two catenated circular dimers with four D-loops; d circular dimer with two D-loops separated from each other by a distance of 5 μm; e complex circular oligomer with three D-loops. Arrows indicate the D-loop structure. (Riou and Paoletti, 1973)

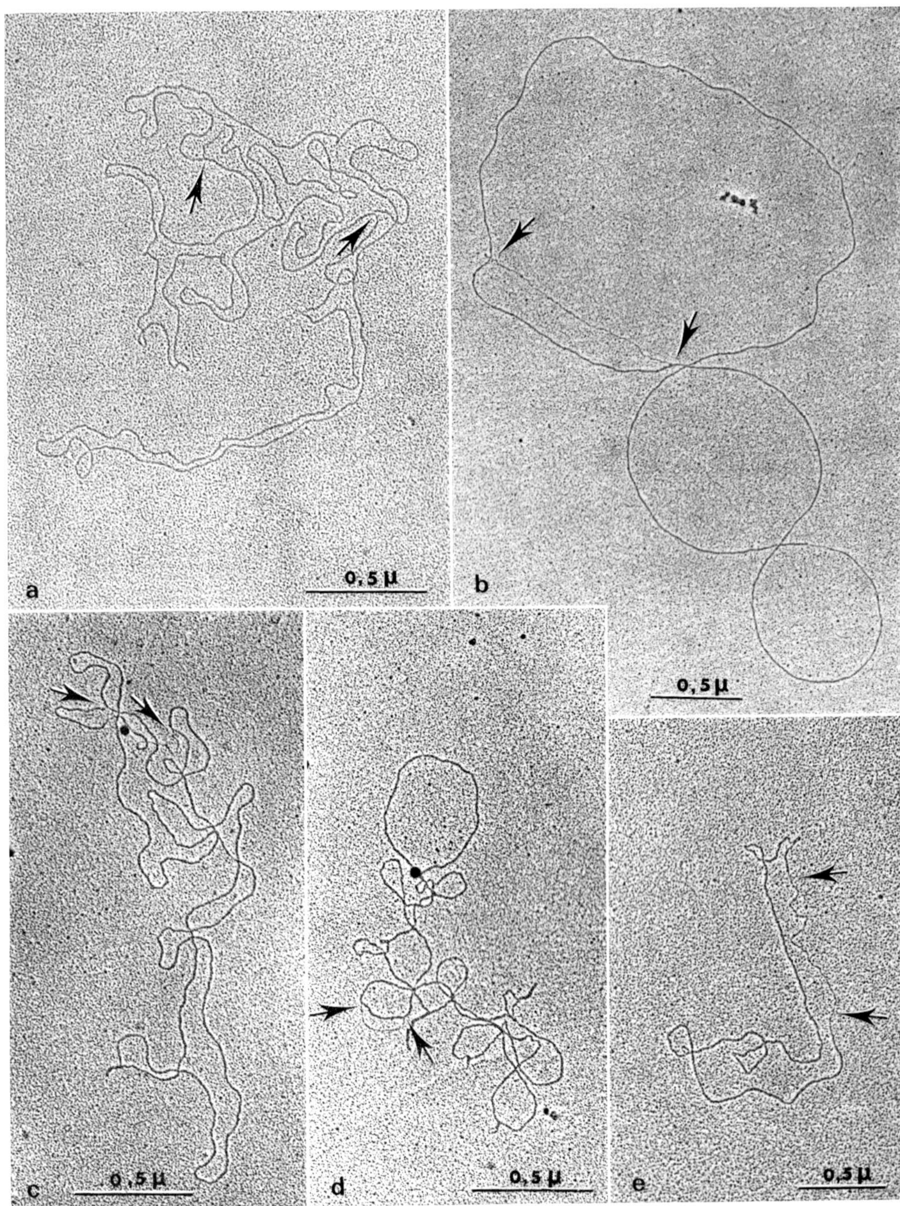

Fig. 4a—e. Electron micrographs of mt DNA molecules from human oncocytoma, assumed to be replication. a Circular dimer in replication. The replicated parts is catenated with a monomer; b and d circular dimer with an expanded D-loop; c circular dimer with an expanded loop. The loop is partly double-stranded; e grapped monomer. (RIOU and PAOLETTI, 1973)

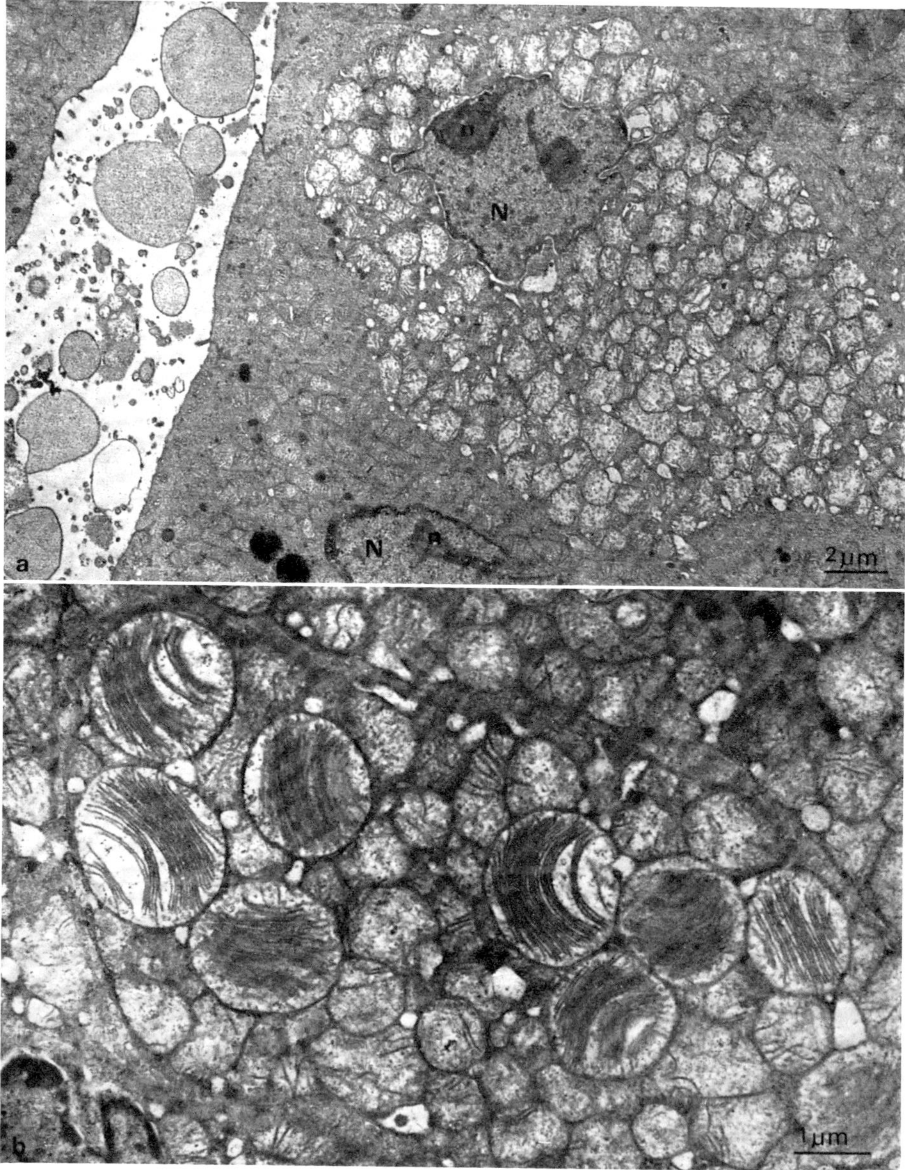

Fig. 5a—b. Electron micrographs of human parotid oncocytoma. Histological sections. a lower magnification; b higher magnification. Giant round mitochondria with lamellar cristae

In some molecules, circles with a length of single-stranded DNA attached at two places on the duplex are seen (Fig. 3). Actually, this structure is a closed circular duplex to which a short single strand (E-strand) is hydrogen-bonded, displacing one of the strands to form a displacement loop (D-loop) [Kasamatsu, Robberson and

Vinograd, 1971; Ter Schergget and Borst, 1971 (1 and 2)]. D-loops have also been seen in human mtDNA (Paoletti, Riou and Pairault, 1972). The D-loops have a mean length of about 3% of the genome size. Their frequency can be high (20 to 60% of the monomers). They are assumed to be part of a replicative structure of mtDNA. Besides D-loops, expanded D-loops, gapped molecules have been described and related to the replication of mtDNA extracted from human oncocytoma (Figs. 3—6) (see complementary references).

The most interesting feature of mtDNA of malignant cells was first observed in 1967 by Clayton and Vinograd in human leukemic leukocytes whose mtDNA

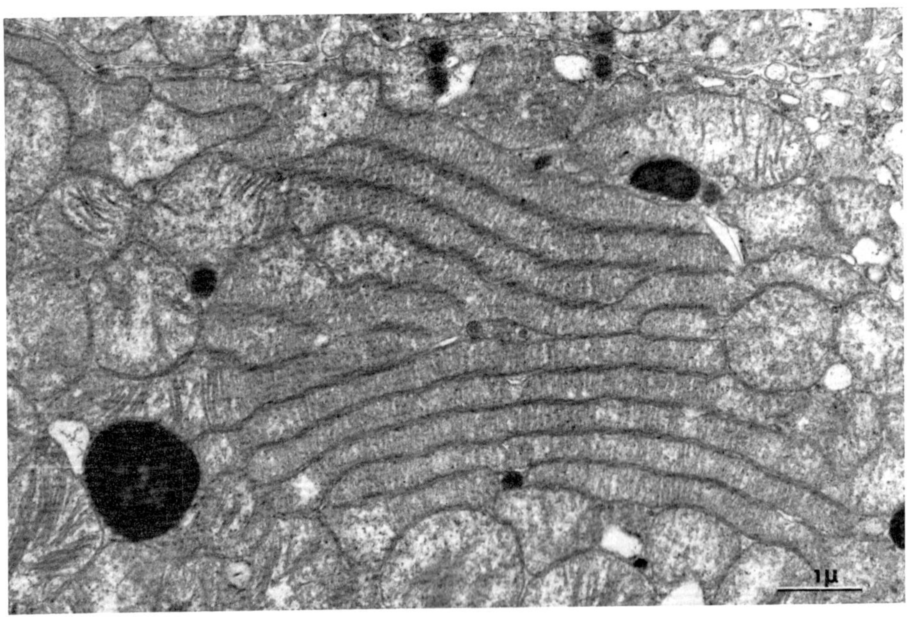

Fig. 6. As Fig. 5, with giant tubular mitochondria

contains a relatively high proportion of multiple-length uni-circular oligomers (dimers, trimers ···) which are not found in non-malignant controls (Fig. 1 items 4, 5 and 6). Similar observations were subsequently made in other human and experimental tumors. This review will be mainly devoted to these unicircular oligomers.

The apparent number of twists per unit length of mtDNA of different origins is not constant, as was previously thought (reviewed in Borst, 1972). The twists are considered as negative (i.e. one twist would be unwound by the unwinding of one turn of the Watson-Crick helix) the biological significance of the twisting is still obscure. Hence a systematic study of this physical property of mtDNA from malignant cells is not warranted.

1. Mitochondrial DNA of Human Tumors

The data, obtained from the leukocytes of patients with acute and chronic granulocytic leukemia, are summarized in Tables 2 and 3. The uni-circular dimers and circular oligomers, which are not found in controls (Table 1), can represent in some cases about 50% of total weight. The proportion of catenated oligomers, which is already high in non-malignant cells, is frequently increased, although this trend cannot be taken as a general rule.

Table 2. Frequency of monomers and oligomers in mtDNA of human leukemic leukocytes (expressed as % by weight of extracted mtDNA)

		Chronic		Acute
		VINOGRAD et al. (3[a])	RIOU et al. (3[a])	VINOGRAD et al. (2[a])
Monomers		54—78	65—73	43—64
Dimers	Circular	14—31	12—20	29—48
	Catenated	6— 7	6—10	4— 6
Higher oligomers		<1— 8	5— 8	<1— 5

[a] Number of patients examined. Data from CLAYTON and VINOGRAD (1967); CLAYTON et al. [1969 (2)] and PAOLETTI and RIOU (1971).

Table 3. Frequency of monomers and oligomers in mtDNA of leukocytes of leukemic patients treated with chemical drugs (expressed as % in weight of extracted mtDNA)

		Chronic		Acute	
		VINOGRAD et al. (6-)	RIOU et al. (6[a])	VINOGRAD et al. (3[a])	RIOU et al. (1[a])
Monomers		79—94	81—90	94—96	83
Dimers	Circular	2— 7	1— 2	2—21	7
	Catenated	4—11	8—15	4— 5	7
Higher oligomers		<3	1— 3	<1	3

[a] Number of patients examined. Data from CLAYTON and VINOGRAD (1967), CLAYTON et al. [1969 (2)] and PAOLETTI and RIOU (1971).

One patient with chronic lymphocytic leukemia had 3.5% of uni-circular dimers in the mtDNA from lymphocytes. A similar value was found twice in this laboratory in another case of lymphatic leukemia (unpublished results).

Uni-circular dimers have also been observed in mtDNA from human solid tumors (CLAYTON, SMITH and VINOGRAD, 1969) but details are not yet available. In two nephroblastomas and two neuroblastomas examined in this laboratory no uni-circular dimers were found; however, the patients had been irradiated and treated with drugs before the removal of the tumors (unpublished data). One epithelioma of the thyroid was found to contain uni-circular dimers (Table 6); however, non-malignant thyroids also contain these forms (PAOLETTI et al., 1972), as discussed in Section VIII-1.

In a non-malignant oncocytoma of the parotid gland we found a very high concentration of circular dimers and circular oligomers (84% in mtDNA). In another non-malignant tumor of the salivary glands, a papillary cysto-adeno-lymphoma, we found 67% of circular dimers and oligomers (Riou et Paoletti, 1973). Oncocytoma (oxyphil cell adenoma) of the major salivary glands is a rare, usually benign lesion. Its morphology has been described above (Section II and Figs. 5 and 6).

2. Mitochondrial DNA of Experimental Tumors

a) Spontaneous and Chemically Induced Tumors

Content of Uni-Circular Dimers. The mtDNA extracted and purified from L cells (mouse fibroblasts) at the exponential stage as well as directly observed in situ after osmotic shock of the mitochondria contains some 5 to 8 %, of uni-circular dimers of $9.77 \pm 0.12\ \mu$ contour length. There are also 2 to 10%, of catenated oligomers, made up of 1 to 4 monomers [Nass, 1969 (3)]. The proportion of uni-circular dimers can be drastically changed by modifying the growth conditions of the cells or by adding drugs (see Section VIII-2). For instance, when harvested in the stationary, rather than in the exponential phase, L cells yield up to 60% circular dimers. The superhelix density of this DNA, 7.2 twists per μ, is the same as that of normal mtDNA.

Heat or alkaline denaturation, as well as sedimentation behavior do not reveal any abnormality of this mtDNA.

After being grown *in vitro* for many generations, hamster embryo cells contain few circular dimers and an increased proportion of catenated dimers and oligomers. These cells are malignant, since they are able to induce tumors on transplantation. In contrast, cells which are not yet malignant do not contain circular dimers after the first passage (Table 4, columns EHA and EHB).

Inaba (1967) described in mtDNA of rat ascitic hepatoma AH 130 circular molecules heterogeneous in size (1 μ to 10 μ) with a mean length of 4.3 μ, that is, shorter than normal. One should be cautious in interpreting this result, since the purification procedure may not have been satisfactory (see Appendix I-5). Most of the molecules were open circles and could therefore have occurred through a folding process similar to the one described by Thomas et al. (1970).

Content of Catenated Oligomers. There have been reports that the mtDNA of some experimental tumors contains a high proportion of catenated molecules and that the degree of catenation (number of interlocked closed circles per catenated entity) is sometimes increased from the most frequent value of 2 to higher values. This is true of hamster embryo cells after many passages *in vitro* (16%) (Table 4), mouse Ehrlich ascites tumor (38%), Chang solid hepatoma (up to 21%) (Wolstenholme, Koike and Renger, 1970), and L cells harvested at the stationary stage of growth (33%) [Nass, 1969 (3)]. However, the mtDNA of experimental tumors does not usually contain such a high proportion of catenated oligomers. Other examples are 10% of such forms in HeLa cells (Hudson and Vinograd, 1967), 8.2 to 10.1% in Novikoff ascites (Wolstenholme, Koike and Renger, 1970), and only 3% in L cells harvested during the exponential phase of growth [Nass, 1969 (3)]. For controls, see Table 1.

Content of Replicating Structures. Wolstenholme, Koike and Renger (1970), reported that 0.6 to 2.6% of double-forked replicative structures occurred in Chang solid hepatoma and in Novikoff ascites, that is, 6 to 28 times more than in mtDNA of normal rat liver (21 such structures were found in about 10,000 rat-liver mtDNA molecules examined). These replicative forms are said to be found on the denser side of the lighter band after cesium chloride-ethidium bromide (CsCl-EB) gradient centrifugation of mtDNA. The presence or absence of EB does not make any difference. One interpretation of this phenomenon is that the first portion of the mtDNA molecule to be replicated is a region rich in guanylic and cytidylic acids. No systematic report has yet been published on the presence and frequency of D-loops in mtDNA of cancer cells.

Table 4. Frequency of monomers and oligomers in mtDNA of cells transformed by adenoviruses and SV 40, and of cells spontaneously transformed (expressed as % by weight of extracted mtDNA)

		T. Adv 7 (1116)[a]		T. Adv 12 (1562)[a]		EHSVi (1694)[a]		EHB (1769)[a]		EHA (1511)[a]	
		no.	DNA %	no.	DNA %	no.	DNA %	no.	DNA %	no.	DNA %
Monomers		923	63.9	1270	64.3	1529	80.9	1620	83.4	1467	94.0
Dimers	Catenated	68	9.4	94	9.5	97	10.3	127	13.1	40	5.1
	Circular	40	5.5	50	5.1	5	0.5	2	0.2	0	—
	Circular or catenated	18	2.5	83	8.4	35	3.7	0	—	0	—
Trimers	Catenated	30	6.9	27	4.1	25	4.0	15	2.3	3	0.6
	Circular or catenated	24	5.0	13	2.0	0	—	0	—	0	—
Oligomers (circular or catenated)		13	6.8	25	6.6	3	0.6	5	1.0	1	0.3

[a] T. Adv 7 and T. Adv 12 cell lines were obtained from subcutaneous tumors induced by adenovirus 7 and adenovirus 12 in newborn Syrian hamster. The number of subcultures was, respectively, 181—184 and 78—87. EHSVi is a cell line originating from Syrian hamster embryo cells transformed *in vitro* by induced SV 40. The number of subcultures was 78—98. EHB consists of hamster embryo fibroblasts spontaneously transformed after serial passages *in vitro*. The number of subcultures was 90—100. EHA consists of hamster embryo fibroblasts cultured *in vitro* and collected after the first generation.
Total molecules scored: number plus.
Riou and Delain (1971).

Partially Single-Stranded, Circular Molecules. Wolstenholme, Koike and Renger (1970) also reported a class of circular molecules distinguished by the presence of a single-stranded region, preferentially found in mtDNA of tumor cells. The frequency of these molecules in mtDNA of normal rat liver is 0.09%, whereas it increases up to 2.8% in Novikoff ascites, and up to 15% in Chang solid hepatoma. These molecules are said to have a higher average density in CsCl gradient than replicating molecules, which supports the view that they may be partially replicated molecules in which one of the strands of the parent double helix has, in whole or in part, failed to replicate. (See complementary references and Fig. 4.)

b) Virus-Induced Tumors

Adenoviruses. The results obtained in this laboratory by Riou and Delain (1971) are summarized in Table 4. The proportion of circular oligomers is relatively high (around 10%) in hamster embryo cells after transformation by adenoviruses 7 and 12. This phenomenon is observed after 80 to 100 serial passages *in vitro*. Freshly transplanted, non-transformed cells yield mtDNA with no uni-circular oligomers, although the same cells can develop a very small proportion of circular oligomers in their mtDNA after serial transfers (more than 90 transfers are required). There is also a marked increase in the frequency of the catenated oligomers in transformed cells.

Table 5. Frequency of monomers and oligomers in mtDNA of cells transformed by AMV (expressed as % by weight of extracted mtDNA)

	Monomers		Catenated oligomers							
			Dimers		Trimers		Tetramers		Pentamers	
	no.	DNA %	no.	DNA %	no.	DNA %	no.	DNA %	no.	DNA %
AMV	1676	74.4	246	21.9	17	2.3	7	1.2	1	0.2
Liver[a]	510	95.1	13	4.9						
Bone marrow[a]	150	94.9	4	5.1						
Liver[b]	761	96.2	15	3.8						

Transformed cells are myeloblasts. Control cells are from liver and bone marrow of normal chicken.
no. = Number of molecules scored.
- Average values of three experiments.
[a] Liver and bone marrow of 3—10 days old chickens.
[b] Liver of 2 months old chickens.
Three circular dimers only were observed in AMV transformed cells.
Riou and Lacour (1971).

Polyoma Virus and Simian Virus 40. Results obtained in this laboratory (Table 4) show only a slight difference, if any, between the content of circular dimers of SV 40-transformed hamster cells, and that of untransformed control cells after many serial passages *in vitro*. The proportion of catenated forms is not significantly different from that of the control cells. We found 32 molecules between 2 and 3.5 μ long among 1726 circular molecules scored in SV 40 transformed cells. These molecules might be of non-mitochondrial origin.

The mtDNA of polyoma-transformed BHK cells at confluence are substantially enriched in interlocked dimers and oligomers (33.6%) (Nass, 1970). In control cells, these values are 10% in the log phase, and 17.4% at confluence.

Avian Myeloblastosis Virus (AMV). The results obtained in our laboratory are summarized in Table 5 (Riou and Lacour, 1971). Only three circular dimers were found out of 1947 mtDNA molecules scored, but 25.6% of the DNA was made up of catenated oligomers, a value well above the control which is less than 5% in this case.

Rous Sarcoma Virus. This RNA virus does not influence the content of catenated dimers and oligomers in mtDNA of BHK cells (NASS, 1970). This virus has been said to replicate inside the mitochondria (see above II).

IV. Synthesis of Mitochondrial DNA in Malignant Cells: Content, Rate and Mechanism

1. Content of mtDNA in Malignant Cells

mtDNA represents only 1 to 5‰ of the total DNA in most animal cells. The DNA content of mitochondria is expressed as μg DNA per μg of mitochondrial protein; the range is 0.24 to 0.8 for the heart, liver and kidney of several mammals, 0.9 to 1.8 for rapidly growing cells (L cells, placenta and mice, rat or hamster embryos) and 2.5 to 8.2 for malignant cells (various experimental tumors, either spontaneous or induced by viruses or chemicals). The data can be found in M. M. K. NASS' general review [1969 (4)]. Other references are: WUNDERLICH, SCHUTT and CORAFFI, 1966), INABA (1967) and LEFFLER et al.(1970).

The amount of DNA per mitochondrion is not precisely known; it could vary according to the origin of the tissues, the stage of the cell cycle, and the cellular environment. In most cases, there appear to be between 2 and 6 molecules of DNA per organelle.

These observations suggest that the amount of DNA per mitochondrion is greatly increased in actively dividing cells, and especially in malignant cells. Some cytochemical work led to a similar conclusion (NASS, NASS and AFZELIUS, 1965; SWIFT et al., 1968).

However, caution is needed in this matter: most analytical methods for measuring DNA and protein in mitochondria yield results which are not wholly reliable. There are a number of reasons for this: poor accuracy, lack of specificity, non-selective extraction of mtDNA, contamination by nuclear DNA, interference by membranes or subcellular contaminants, and heterogeneity of cell populations. Moreover, the proportion of mtDNA in cells varies widely. These variations are usually thought to reflect changes in the number of mitochondria per cell, but other explanations have not been ruled out.

2. Rate of Synthesis of mtDNA in Malignant Cells

Very few data are available as to the rate of synthesis of mtDNA in malignant cells. VESCO and BASILICO (1971) showed that polyoma virus can stimulate the incorporation of ^{3}H-thymidine into mtDNA of confluent cultures of 3T3 mouse cells when nuclear DNA synthesis is maximally derepressed, that is, 24 to 34 h after infection. This phenomenon does not require viral DNA replication and must therefore be caused by an early viral function. The same conclusion was reached for SV 40 by LEVINE (1971).

3. Mechanism of Synthesis of mtDNA in Malignant Cells

The mode of replication of mtDNA is still not certain. Replicative or repair mechanisms, or both, are equally plausible. In yeast, the synthesis of some mtDNA accompanies the induction of mitochondrial respiratory functions (Mounoulou, Perrodin and Slonimski, 1968; Rabinowitz et al., 1969; Guerineau, Buffenoir and Paoletti, 1973).

Moreover, there is turnover of mtDNA in mammalian organs [Gross et al., 1969 (1); Gross, 1971], in monkey-kidney cells or mouse 3T3 cells at confluence *in vitro* (Levine, 1971; Vesco and Basilico, 1971) and probably in yeast (Sarkar and Poddar, 1965).

Many questions can be raised as to the molecular mechanisms which underly the formation of uni-circular oligomers[2]. They are often thought to originate either from disturbances in the replication of the monomeric DNA (Kiger and Sinsheimer, 1969; Goebel and Helinski, 1969) or from impairment of recombination between circular monomers (Hudson and Vinograd, 1967). The data obtained with non-animal organisms, such as bacteria and viruses (see Appendix II), suggest that either hypothesis is equally plausible to explain the occurrence of oligomers in mtDNA, and more generally in any circular DNA. Micrographs of circular dimers of mtDNA, and of phage λ DNA in the process of replication, as well as labeling experiments of colicigenic-factor DNA, favor the replication hypothesis. On the other hand, the fact that circular dimers form from monomers when the replicative process of phage S 13 is blocked by mutation, and the way in which the bacterial recombination system influences the appearance of these dimers (phage S 13 and λdV plasmid DNA) strongly support the recombination hypothesis.

For animal cells, two sets of data pertaining to the formation of the uni-circular dimers are available.

a) D-loops (see III) have been seen on uni-circular dimers of mtDNA from cells liver (Robberson and Clayton, 1972), from bovine and human thyroid (Paoletti, Riou and Pairault, 1972), and from human parotid oncocytoma (Riou et Paoletti, see III-1). These observations might imply that the circular dimers are able to replicate according to a semi-conservative process. Some dimers display two D-loops per molecule.

b) Labeling experiments with L cells have been carried out by M. M. K. Nass (1969): The rational and main results are schematically presented in Fig. 7. mtDNA was ^{3}H-labelled during the exponential phase of growth when circular monomers are most abundant. After reaching the stationary phase, the cells were exposed to a ^{14}C:labelled nucleotide precursor; the circular dimers, which are formed in high numbers in this phase, were found to contain both ^{3}H and ^{14}C. Similar results are obtained when the cells are shifted back to the exponential stage in the presence of a third label (^{32}P). Unfortunately, these results do not permit clearcut conclusions since the dimers (or monomers) are formed from preexisting monomers (or dimers) as well as from *de novo* precursors. They could thus equally well be due to turnover of mtDNA (see above), i.e. an exchange of labeled material between DNA molecules, either as lengths of polynucleotides or as nucleotides, after breakdown of parent molecules.

Although a DNA-dependent DNA polymerase has been isolated from rat-liver mitochondria (Meyer and Simpson, 1970), this enzymatic activity has not yet been

[2] Several models have been proposed to explain the replication and recombination processes at the molecular level (see Richardson, 1969; Becker and Hurwitz, 1971; Davern, 1971). A detailed discussion of these models would go beyond the purpose of this review.

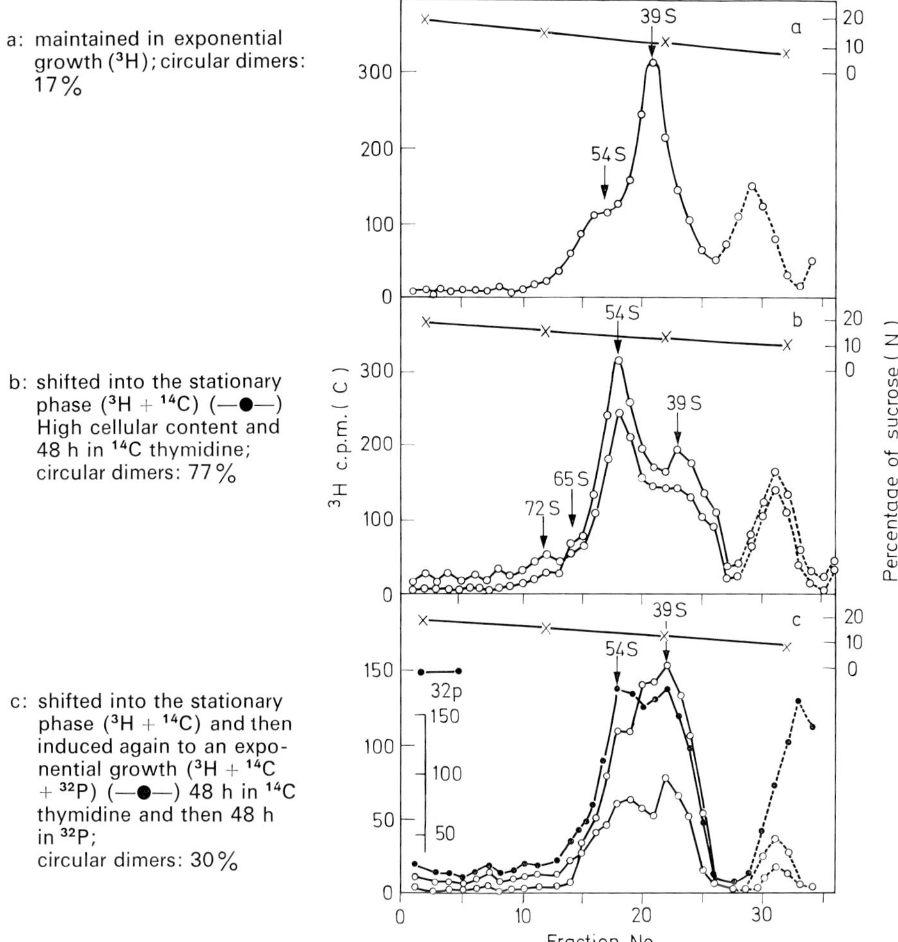

Fig. 7. Metabolic fate of circular monomers and dimers of mtDNA. Mouse fibroblasts (L-cells) in exponential growth (majority of circular monomers) or at the stationary phase (majority of circular dimers). Date from Nass [1969 (3)]. The cells are pre-labeled by ^{3}H-thymidine (-o-) for 48 h (exponential growth). Then they are: A: maintained in exponential growth (^{3}H); circular dimers: 17%. B: shifted into the stationary phase (^{3}H + ^{14}C) (-o-). High cellular content and 48 h in ^{14}C thymidine; circular dimers: 77%. C: shifted into the stationary phase (^{3}H + ^{14}C) and then induced again to an exponential growth (^{3}H + ^{14}C + ^{32}P) (-o-) 48 h in ^{14}C thymidine and then 48 h in ^{32}P; circular dimers: 30%. Closed circular molecules of mtDNA are extracted from the lower band of a CsCl-EB gradient and centrifuged in a sucrose gradient. 39 S and 54 S peaks are respectively made of circular monomers and of a large majority of circular dimers with few catenated dimers, as shown by EM

studied in tumor cells. The existence in yeast of a "replication factor" — a product of mitochondrial protein synthesis required for the correct replication of mtDNA — has been postulated from indirect arguments by Williamson et al. (1971) and by Weislogel and Butow (1970, 1971). It is not known whether this factor exists in animal-cell mitochondria. It might be rewarding to evaluate the data and hypothesis pertaining to the regulation of mtDNA synthesis in yeast in an attempt to comprehend how this regulation works in normal and malignant cells. In this context, two attractive hypothesis have been put forward to explain the changes in mtDNA which accompany the induction of the cytoplasmic "petite" mutation in yeast[3].

Hypothesis a). A structural change in mtDNA following mutation could result in the selection of a particular region as the preferred template. After several rounds of replication, the base composition of this region would be favored (Carnevali, Morpurgo and Tecce, 1969).

This hypothesis has received some support from recent work (Arca, Caneva and Frontali, 1970). Acridine is an efficient inducer of "petite" mutations in yeast. DNA, synthesized *in vitro* by *E. coli* polymerase in the presence of acridine with yeast mtDNA as template, displays a significant increase in the ratio AT/GC from 80% to 92.2% AT. However, this result was obtained, not with a *mitochondrial* polymerase, but with a *bacterial* enzyme, which does not necessarily have the same mechanism of action.

Indirect support was also obtained by Spiegelman and his collaborators (Mills Peterson and Spiegelman, 1967). They performed an extracellular Darwinian selection *in vitro* by selecting the first molecules of $Q\beta$ phage RNA synthesized *in vitro* by the $Q\beta$ replicase at each round of replication and using these molecules for the next round. This stepwise process resulted in an increase in the overall rate of synthesis of RNA and a shortening of the molecules owing to the elimination of sequences not required for replication, that is, those coding for the virion envelopes and for the synthetase. After the 74th round, 83% of the RNA genome had been eliminated (that is, 550 nucleotides were left of the, 3330 present at the beginning of the experiment), and the *rate* of synthesis of the remaining segment of RNA was 15 times faster than the initial rate. The simplest explanation of these results is that a mutation had been selected related to a change in the structure of the polymer. This structural change would favor the interaction of the nucleic acid with the replicase.

Hypothesis b). The "petite" induction is accompanied by several anomalies in the structure of the mitochondrial membranes. Therefore, one can postulate some modifications of the environmental conditions of the DNA-polymerizing enzymes inside the mitochondria which could change the nature of the product of these enzymes. This hypothesis is indirectly supported by the results of Burd and Wells (1970) on unprimed *(de novo)* incorporation of complementary deoxyribonucleoside triphosphates into double-stranded polynucleotides of defined repeating sequences by DNA polymerases of either *E. coli* or *Micrococcus luteus*. Small changes in the reaction conditions (such as pH and metal ions) can modify the enzyme and markedly influence the nucleotide sequence of the polymeric product.

V. Is the Informational Content of Mitochondrial DNA Modified in Malignant Cells?

The sequence homology of rat mtDNA from liver and kidney is complete, as estimated by DNA-DNA hybridization studies (quoted in Borst, 1972). Although no systematic study has yet been undertaken to extend this observation to mtDNA

[3] Mechanisms other than modifications of the DNA-polymerizing enzymes could explain these changes, such as gene amplification, or mitochondrial gene translocation.

from various organs in a single species, this conclusion can probably be generalized since other physical characteristics of mtDNA (size and mean base composition) are constant whatever the organ from which it is extracted.

No difference has been found in the mean base composition (buoyant density) of mtDNA from tumor cells as compared to that of normal mtDNA. Differences such as that between mtDNA of mouse fibroblasts (L cells) (buoyant density = 1.698) and mtDNA of mouse-liver cells (1.701) [NASS, 1969 (2)] are probably within the limits of error of the method.

However, since buoyant density is only a crude function of base *composition* and not of base *sequence*, one cannot rule out the existence of differences between mtDNA from normal and cancer cells in such features as point mutations, short deletions or insertions, or base methylation. The informational content of mtDNA of normal and malignant cells has not yet been investigated using more refined methods, with the exception of the work of SAMBROOK et al. (1968). These authors found no detectable change in the base sequences of cytoplasmic DNA extracted from SV 40-transformed and normal hamster cells.

ODA and his coworkers (ODA, 1968; TAKE, 1969; ODA et al., 1970; YAMAMOTO and ODA, 1970) claimed that the average contour length of mtDNA from a number of tumors is somewhat lower than in mtDNA from control tissues. These results could be due to artifacts of extraction (see Appendix I) since the authors also found in their mtDNA preparations various small circles whose average size was between 0.5 and 1.5 μ. These circles are reminiscent of those described in HeLa cells by VINOGRAD's group (HUDSON and VINOGRAD, 1967), hence not derived from mitochondria.

The denaturation mapping technique of INMAN (1966) could be fruitfully used in this field. It does not involve any preliminary separation of the monomers and dimers or any reannealing after denaturation. Preliminary results by WOLSTENHOLME, KOIKE and RENGER (1970), indicate that three regions of preferential strand separation are located in one half of each molecule of rat-liver mtDNA. However, this technique could be difficult to use because of sequence homogeneity of mtDNA, as reflected by the sharpness of its melting curve.

The 5 and 10 microns circular duplexes of mtDNA from human leukemic leukocytes were separated and characterized by preparative centrifugation in CsCl-EB (CLAYTON, DAVIS and VINOGRAD, 1970). The buoyant densities of each of these species in neutral CsCl, as well as those of the corresponding heavy and light strands in alkaline CsCl, were indistinguishable. These results indicate that the overall base composition does not differ in the monomers and dimers. The heteroduplexes, obtained after reannealing the separated light and heavy strands of the monomers and dimers, were examined in a buoyant density gradient and under the electron microscope. It was concluded that heterogeneous regions, insertions, or deletions exceeding more than 50 to 100 nucleotides in length, did not occur in these heteroduplexes. Thus, the dimer appears to be a circular concatanamer of the two monomer molecules, which are connected in a head to tail structure.

These data suggest that the dimers do not carry information different from that carried by the monomers, and must therefore code for the same proteins, if any. However, a similar study on the relationship between mtDNA from leukemic patients and that from normal subjects remains to be undertaken.

A provocative review by GRAUSE (1968) expresses some unorthodox views on the variations in the informational content of DNA in bacteria.

VI. Are the Changes in Mitochondrial DNA of Malignant Cells Under Genetic Control?

All mtDNA from human myeloid leukemia cases so far examined contains circular dimers. This molecular abnormality provides a new marker for myeloid leukemia, which is at least as reliable as the two previously known markers: partial deletion or complete absence of chromosome 21[4], called Philadelphia (Ph^1), and a fall in the level of alkaline phosphates in the leukocytes. The correlation between the Ph^1 chromosome anomaly and the disease is not absolute since, according to different authors, between 66% and 100% of patients are Ph^1-positive. One cannot assume that the presence of circular oligomers is a genetic marker nearly because it characterizes mtDNA, which is a genetic material. No analyses of human pedigrees, such as those under way for Ph^1, have yet been undertaken for the topological disturbances of mtDNA described here, and there are no data to support the idea that they could be the result of a somatic mutation. Nevertheless, one is tempted to interpret these observations in the light of the new perspectives offered by recent developments in cytoplasmic genetics, which rely on extrapolation to animal cells of the important concepts recently deduced from the yeast *Saccharomyces cerevisiae* (COEN et al., 1970). These concepts include the probable coexistence of several populations of mitochondria, the identification of the chromosomal functions which control the maintenance and perpetuation of the mitochondrial population at the right level, the demonstration of recombination in mtDNA, the discovery of cytoplasmic mutants (some of them characterized by changes in the physico-chemical properties of the mtDNA), and the recognition of the polar effects of certain mitochondrial mutations. Experiments in mitochondrial genetics are very difficult to perform with animal cells. Such cells are necessarily aerobic and their chondriome cannot therefore be manipulated as easily as the yeast chondriome, the latter organism being only facultatively aerobic. Moreover, the presence of many copies of the same DNA in animal cells must ensure great stability of genetic markers. A mutation occurring in one of these copies would not be phenotypically expressed unless a process of preferential replication or asymetric recombination favored the mutated copy.

The absence of sexual exchanges between somatic cells precludes direct analysis of the determinants of cytoplasmic heredity. Nevertheless, the technique of cell hybridization provides an opportunity to overcome this difficulty, as long as mitochondrial genetic markers can be either individualized in a normal situation, or generated by spontaneous or induced mutations. In this respect, mtDNA itself could be a candidate as a biochemical marker of the mitochondria of a given animal species, since its length and buoyant density are slightly but significantly in several species. For instance, the mean length of monomers is $4.77 \pm 0.02\ \mu$ in L cells, and $5.06 \pm 0.11\ \mu$ in chicken liver. When mixed and scored, these two types of molecules appear as two different populations on a histogram [NASS, 1969 (2)]. Similar mixing experiments have also demonstrated differences in the contour length of mtDNA

[4] Actually, this chromosome is probably number 22 (O'RIORDAN et al., 1971) and different from the one involved in Down's syndrome.

from urodele and anuran amphibia (Wolstenholme and Dawid, 1968). The existence of circular oligomers in some cells (L cells in stationary phase) cannot be conveniently used, since their number varies with the environment. Similarly, the buoyant density of mtDNA varies according to species: for instance, $\varrho = 1.705$ to 1.706 for man (Clayton and Vinograd, 1967; Paoletti, Riou and Pairault, 1972) and $\varrho = 1.698$ for mice [Nass, 1969 (2)][5]. These types of mtDNA can readily separated after extraction from hybrid human 1/N mouse cells (Attardi and Attardi, 1972; Clayton et al., 1971).

Obtaining animal-cell mutants by chemical treatments and identifying cytoplasmic mutations will be very difficult. Most such mutations should be lethal, since they interfere with mitochondrial functions which are essential for cell survival. In spite of these obstacles, it might be possible to obtain mutants resistant either to some specific inhibitors of mitochondrial protein synthesis, such as chloramphenicol or erythromycin, or to inhibitors of oxidative phosphorylation, or to amino-acid analogues.

Wagner (1969) published a general review on the genetics and cytogenetics of mitochondria. Borst (1972) recently made an excellent summary of the contribution of mitochondrial genetics to the identification of genes on mtDNA.

VII. Are the Changes in Mitochondrial DNA Related to Some Energy Imbalance in Mitochondria of Malignant Cells?

Warburg's original proposal (1956), that the neoplastic process can be attributed solely to disturbances of the energy metablolism in mitochondria, is not tenable as such. In minimal-deviation tumors, the energy derived from glycolysis is often small compared with that derived from respiration. However, it is commonly believed that a high rate of glycolysis (both aerobic and anaerobic) is one of the striking biochemical characteristics of the majority of cancer cells, whatever their origin, localization, and type (for discussion, see Wenner, 1967).

More specifically, it has been found that, in giant mitochondria isolated from the oncocytes of human adenolymphomas, the P:O ratio is lowered (Schiefer et al., 1968). These mitochondria display a loose coupling of oxidative phosphorylation and a low sensitivity to oligomycin. Along the same line, a marked reduction of mitochondrial ATPase (activated by uncouplers of oxidative phosphorylation) has been demonstrated in rapidly growing hepatoma cells (Emmelot et al., 1959; Devlin and Pruss, 1962) and, more interestingly, in a slowly growing minimal hepatoma (Pedersen et al., 1971). It follows that the reversibility of the energy transfer system, and the unknown mechanism by which this energy is utilized to produce ATP, are impaired in mitochondria of some malignant cells.

A puzzling and still unexplained observation was reported by Lorans and Tournier (1964). These authors obtained a strain of hamster cells transformed by adenovirus 12, which showed such energy impairment that the respiratory pigment, hemin had to be added to the nutrient for these cells.

---

[5] The standard DNA for measuring buoyant density is usually provided by *E. coli* (1.710 g/ml); if the crab poly d (AT) is chosen as the standard, a value of 1.700 g/ml would be found instead of 1.705 g/ml.

Altogether, one is led to the conclusion that the integration of energy functions inside the mitochondria of transformed cells is in most cases seriously impaired, even though each function studied separately *in vitro* is not disturbed.

Can this impairment affect the fate of mtDNA? A positive answer to this question might be expected if there were any requirements concerning the proper type and correct channeling of energy in the metabolism and biosynthesis of mtDNA. Actually, several independent lines of evidence impressively converge to suggest that there are such requirements and that they are stringent. Consequently, if the requirements were not fulfilled in neoplastic cells, this could be the cause of some of the changes in mtDNA. This evidence is as follows:

a) Oxidative metabolism occurs inside the mitochondrial matrix on the inner membrane, and mtDNA is probably located on the same substructure [Kislev, Swift and Bogorad, 1965; Nass, 1969 (1)] together with chloroplast DNA, probably forming a membrane complex (Woodcock and Fernandez-Moran, 1969; Bisalputra and Burton, 1970). Besides this postulated spatial relationship between mtDNA and the mitochondrial membranes, it has been suggested that there is a metabolic link between them. Kalf and Faust (1969) found that the inner membranes of rat-liver mitochondria, prepared by the digitonin method, are able to incorporate ^{3}H-dATP into mtDNA. It has, moreover, been suggested that a major unit of the inner mitochondrial membrane, including mtDNA, turns over as an entity (Gross et al., 1969). Finally, it has been reported that the changes in the turnover rate of both mtDNA and mitochondrial proteins are of the same order of magnitude in liver mitochondria after administration of thyroid hormone (Gross, 1971). Similarly, in other systems, there is a close spatial association between DNA and membranes; thus, the sites of chromosomal DNA replication in bacteria are associated with the cell membrane where the enzymes of the respiratory chain are also located (Ryter, Hirota and Jacob, 1968). In yeast, mtDNA was shown to be associated with a membrane mesosome-like structure (Yotsuyanagi, 1966).

b) Respiration is coupled, in some mandatory fashion, with replication of double-stranded bacterial and mtDNA.

In bacteria, carbon monoxide and cyanide inhibit replication (Cairns, Denhardt and Burgess, 1968). It has also been shown that the addition of azide prevents DNA synthesis in *B. subtilis* (Ganesan and Lederberg, 1965). The first steps of the synthesis of the ϕX 174 DNA are not blocked by cyanide or CO. Therefore, this inhibition is not related to the two reductive steps involved in the metabolic pathways of the nucleoside phosphate precursors, that is, reduction of the ribonucleotides and methylation of deoxyuridylic acid. Neither is it associated with ATP synthesis, energy-linked translocation, mere stimulation of the respiration, or activation of the oxidation-reduction state of an electron carrier, since uncouplers of oxidative phosphorylation fail to prevent DNA replication or to influence the respiration rate (Howland and Hughes, 1969). Howland and Hughes assumed that a flow of charges, carried by protons from the sites of DNA replication toward the cell exterior, is associated with strand separation prior to copying.

When anaerobically grown cells of *Saccharomyces cerevisiae*, are exposed to oxygen, an immediate burst of mtDNA synthesis, which represents less than 35% of mtDNA. is triggered (Mounoulou, Perrodin and Slonimski, 1968; Rabinowitz et al., 1969), There is also a lengthening of the mtDNA molecules (Guerineau, Buffenoir and

PAOLETTI, 1973). This oxygen-induced DNA synthesis is inhibited by high concentrations of glucose. Investigations along this line, with normal and cancer cells, are suggested by the well-known CRABTREE effect (1929), which consists of a transitory decrease in the cellular oxygen uptake caused by the addition of glucose to respiring malignant tissue.

c) Several important enzymes of DNA metabolism, most of them presumably involved in replication or recombination, require ATP for activity. They are: T4-induced ligase (WEISS and RICHARDSON, 1967) and rabbit-spleen and bone-marrow ligases (LINDAHL and EDELMAN, 1968); T4-induced kinase (RICHARDSON, 1965); the nucleases: *E. coli* restriction endonuclease (MESELSON and YUAN, 1968), *Micrococcus lysodeikticus* exonuclease (HOUT et al., 1970) and endonuclease (ANAI, HIRAHASHI and TAKAGI, 1970; ANAI et al., 1970), *E. coli* exonuclease (BUTTIN and WRIGHT, 1968; OISHI, 1969; BARBOUR and CLARK, 1970; WRIGHT, BUTTIN and HURWITZ, 1971) and the endoexonuclease involved in genetic recombination (GOLDMARK and LINN, 1970). These enzymes interact not only with DNA but with ATP as well, their activity resulting in pyrophosphate exchange and/or ATP splitting in ADP + Pi. A T4-induced ATPase (> ADP + Pi) requires DNA as cofactor, but does no attack it (DEBRECENI, BEHME and EBISUZAKI, 1970).

No ATP-dependent polymerase has yet been isolated. However, toluene-treated cells of *E. coli* (MOSES and RICHARDSON, 1970; MORDOH, HIROTA and JACOB, 1970) and azide-poisoned, pyrophosphate-permeable cells of *B. subtilis* (GANESAN, 1971) are capable of continued DNA synthesis in the presence of the four deoxyribonucleotide triphosphates. This synthesis is strongly activated by ATP and therefore requires energy.

d) There is a close correlation between the turnover rate of mtDNA and mitochondrial activity. In rats, this turnover is most rapid in the heart, where mitochondrial energy consumption is very high, and slowest in the brain, where mitochondrial activity is low (GROSS et al., 1969).

Even though their molecular mode of action remains unknown, the thyroid hormones are known to be involved in regulating the basal rate of oxygen consumption and the respiration-dependent swelling of mitochondria. Again, there is correlation between these phenomena and the fate of mtDNA since (a) the injection of thyroxine in rats induces a rapid increase in mtDNA turnover in liver; (b), in contrast, mtDNA turnover in the liver of thyroidectomized animals is about twice as slow as in normal rats (GROSS, 1971). In connection with the preceding observations, let us recall that the only non-malignant animal tissue in which the occurrence of circular dimers is well documented is the thyroid (PAOLETTI, RIOU and PAIRAULT, 1972), although no other abnormality has so far been ascribed to thyroid mitochondria (TYLER and GONZE, 1967).

VIII. Are the Changes Observed in Mitochondrial DNA of Malignant Cells Specific to Malignancy?

1. Uni-Circular Dimers and Oligomers are Found in mtDNA of Non-Malignant Cells

With the exception of an unpublished result by KIRCHNER et al., quoted in WOLSTENHOLME, KOIKE and RENGER (1970) (one uni-circular dimer out of one thousand molecules of rat-liver mtDNA), circular oligomers have always been thought to be absent in mtDNA extracted from non-malignant cells (published data are summarized in Table 1). This is no longer a general rule, since normal beef thyroid and non-malignant human thyroid have these forms in mtDNA (PAOLETTI, RIOU and PAIRAULT, 1972) (Table 6). The proportion of uni-circular dimers is between 10 and 37%. Uni-circular trimers as well as higher oligomers are also present. The largest circular molecule was 34 μ long. A majority of uni-circular oligomers was found in two human oncocytomas (see above III-1). However, these tumors cannot be considered truly malignant and the oncocytes are found in normal tissues.

Up to now, it has proved impossible to generate circular oligomers in normal eukaryotic cells merely by changing their environmental or metabolic usual conditions. Transformation is one way for generating these circular oligomers; no systematic search for circular oligomers has been carried out on cells infected by non-oncogenic viruses. NASS (1970) was able to double the proportions of circular and catenated dimers and higher oligomers which normally exist in L cells by infecting them with the non-oncogenic Mengo virus. Another way for obtaining the circular oligomers is to use some toxics; GUERINEAU, GUERINEAU and GROSSE (personal communication, 1973), working in our laboratory, have been able to induce the formation of oligomeric classes of uni-circular mtDNA in the livers of rats treated per os by cuprizone; very interestingly, this DNA is also modified in size, being either shorter or longer than mtDNA extracted from untreated rats.

Table 6. Frequency of monomers and oligomers in mtDNA of human thyroid (malignant and non-malignant) and of beef thyroid (expressed as % by weight of extracted mtDNA)

		No. of molecules scored	mtDNA %			
			Monomers	Dimers		Higher oligomers
				Catenated	Circular	
Human thyroid	"Normal"[a]	481	26.2	15.6	48.4	9.8
	"Normal"[b]	820	55.8	12.5	25.2	6.5
	Follicular adenoma	687	47.5	13.0	25.0	14.5
	Follicular poly-adenoma	308	32.3	18.6	37.1	12.0
	Thyroiditis	451	73.6	6.4	11.7	8.3
	Hyperthyroidism	173	40.4	9.9	32.0	17.7
	Epithelioma	540	52.0	12.0	22.1	13.9
Beef thyroid	Normal	220	70.0	7.7	17.8	4.5
	Normal	293	84.1	5.5	10.1	0.3

[a] Histologically normal glandular cells surrounding adenoma.
[b] Excision after autopsy (acute lymphoblastic leukemia with aplasia).
Data from PAOLETTI, RIOU and PAIRAULT (1972).

Great caution must be exercised in defining normal versus malignancy. For instance, the spontaneous transformation of cells cultivated *in vitro* is well documented (see general review by SANFORD, 1967) and should be taken into account when interpreting results obtained with mtDNA extracted from cells grown *in vitro*. This is demonstrated by the data obtained in this laboratory with adenovirus-transformed cells (Table 4). The mtDNA of "control" hamster embryo cells after many passages *in vitro* contain a few uni-circular dimers, and these cells are capable of eliciting tumor growth after grafting.

The occurrence of circular oligomers is wide-spread: they are found in eukaryotic organisms other than animals, in bacteria (episomes), in viruses including bacteriophages. Although not directly related to mtDNA, this question is discussed in Appendix II.

2. There is No Correlation between Malignancy and the Occurrence or Content of Abnormal Molecular Forms in mtDNA

Several tumors do not yield uni-circular dimers. This negative observation was made in (a) human tumors: 2 cases of nephroblastoma and 2 cases of neuroblastoma (unpublished data from this laboratory); (b) tumor cells of human origin grown *in vitro*: HeLa cells (Radloff, Bauer and Vinograd, 1967; Hudson and Vinograd, 1967), Burkitt lymphoma cells (Nass, 1970); (c) rodent experimental tumors: Ehrlich ascites of mice, grown either intraperitoneally or under the skin (Nass, 1970; Wolstenholme, Koike and Renger, 1970); Chang solid hepatoma and Novikoff ascitic tumors of rats (Wolstenholme, Koike and Renger, 1970) and of mice (CL and GH-1) (Nass, 1970).

We cannot attribute more than indicative significance to such a negative correlation, since there is always a possibility of some loss of mtDNA during extraction. This question will be discussed at length in Appendix I (preparation of mtDNA).

However, Clayton and Vinograd (1969) have shown for three cases of human chronic myeloid leukemia that the content of uni-circular dimers in mtDNA falls very sharply after drug therapy (myleran and 6-mercaptopurine). In one case of acute myeloid leukemia, the blood concentration of myeloblasts did not change, while the content of abnormal mtDNA fell to about 2% of the initial value. Such a decrease could be due either to the replacement of a population of myeloblasts rich in abnormal mtDNA by a population containing fewer abnormal forms, or to active elimination of abnormal mtDNA through the action of the drug. The important point is that the morphologic criteria of malignancy were preserved while the biochemical alteration disappeared.

Moreover, many unrelated modifications in the environment of mouse fibroblasts (L cells) grown *in vitro*, sharply modify their content of mtDNA circular dimers. This content is usually around 5% when the cells are in the exponential phase of growth, but it rises to 53%, 52% and 68 to 73% when the cells are deprived of methionine or phenylalanine or are treated with cycloheximide, respectively. When the cells are shifted to the stationary phase, this figure goes as high as 82% without any other treatment. These changes are reversible. The content in uni-circular dimers is not modified by chloramphenicol, actinomycin, or colchicine [Nass, 1969 (3) and 1970]. Nass (1970) also reported an increase in the proportion of uni-circular dimers and oligomers after various treatments of L cells: this increase was small with vinblastine but greater with puromycin, rifampicin, and hydroxyurea. However, attempts to build up the content of uni-circular dimers and oligomers in chick embryo fibroblasts and baby hamster kidney (BHK) cells were unsuccessful, except with cycloheximide or through the effect of confluency.

Altogether, these data suggest that the occurrence of mitochondrial uni-circular dimers could be under the control of a cytoplasmic protein.

In any case, this phenomenon depends on the metabolic state and the environmental conditions of the cells and involves mechanisms which are neither understood nor fully controllable. In this connection, one should mention the observation of LEDUC, BERNHARD and TOURNIER (1966). According to these authors, the timing of the cyclic appearance and disappearance of atypical mitochondria, rich in DNA fibers in adenovirus 12-induced hamster tumor cells varies with the environmental conditions. They are abundant in young cultures, then vanish, and subsequently reappear.

IX. Concluding Remarks

The history of cancer biochemistry is replete with phenomena which it was hoped would prove specific for malignancy until a careful check of controls was carried out. Since no unambiguous definition of cancer can be offered, that is, no clear-cut and general border line can be drawn between malignancy and normalcy no control is really satisfactory. At best, good probabilities can be obtained.

The demonstration of the occurrence of uni-circular dimers in the mtDNA of malignant cells, although well and cautiously documented (CLAYTON and VINOGRAD, 1967), once more obeyed this rule and proved a delusive research path, since the whole set of data points to the non-specificity of the changes in size (occurrence of circular dimers and oligomers) or in the topology (increase in content of catenated circles) of mtDNA of malignant cells. Uni-circular oligomers are found in non-malignant cells (human and beef thyroid) or in tumors not considered as truly malignant (oncocytomas), whereas they are absent in several human and experimental tumors.

Such oligomers are also found in several eukaryotic organisms other than animal cells, in bacteria (episomes), and in animal and bacterial viruses (see Appendix II).

Moreover, their percentage in malignant cells is highly variable and can be drastically modified, depending on cell type, by changing the environment (*in-vitro* culture favors their appearance) and the growth conditions (confluency results in an increase) or by interfering with fundamental biological functions. These functions can differ as widely as mitotic spindle formation, cytoplasmic protein synthesis, and transcription and synthesis of DNA.

Furthermore, no indication has been obtained of a change in the informational content of mtDNA from tumor cells. The uni-circular dimers present in the mtDNA of human myeloid leukemia are probably made up of two normal monomers linked head to tail. Therefore, the malignant transformation would affect the replication or recombination of mtDNA, which would retain its base sequence and be normal in other respects. It is not unlikely that such disturbances could be a secondary consequence of some imbalance of the mitochondrial energy functions which characterizes most cancer cells. All in all, one cannot escape the conclusion that research on mtDNA will do more to further the understanding of some basic cellular functions than to interfere with the cancer process.

Appendix I: Isolation and Examination of Mitochondrial DNA

a) Preparation of Subcellular Fractions Enriched in Mitochondria

Since mtDNA represents only a minor part of the total cell DNA (about 1 to 5‰ in most tissues), any procedure aiming at its isolation involves first of all the preparation of subcellular fractions enriched in mitochondria. The techniques used for the preparation and examination of mitochondria of different origin, from both normal and neoplastic tissues, have been described in detail in Methods in Enzymology, volume X.

b) Isolation of Total mtDNA

Several methods are available among which the ethidium-bromide (EB) CsCl gradient is the most frequently used.

Equilibrium Centrifugation in Ethidium-Bromide CsCl Gradient. This technique is so far the most reliable and efficient procedure for the isolation of covalently closed circles, that is, almost entirely mtDNA.

It is now agreed that certain molecules of flat steric structure can be intercalated between the base pairs of DNA (LERMAN, 1961).

Among these molecules, the phenanthridines, whose prototype is EB, have been extensively studied since the initial report on the parameters of their interaction with DNA (LE PECQ, YOT and PAOLETTI, 1964; see also general review by LE PECQ, 1971). VINOGRAD and his collaborators (RADLOFF, BAUER and VINOGRAD, 1967) have described a method for the isolation of covalently closed circular DNA, which takes advantage of the differential affinity of EB for different forms of DNA. At saturation, closed circular DNA binds less EB than nicked or linear DNA. Since the binding of EB to DNA results in a decrease in buoyant density of the dye-complexed polymer, closed circles remain denser than nicked circles or open molecules, thus forming heavier bands in CsCl density gradients (Fig. 8).

The difference in buoyant density between linear and nicked circular DNA on the one hand and closed circular DNA on the other is linearly related to the initial superhelix density of the closed molecules. The buoyant density also depends on the base composition and the amount of EB bound to DNA, which itself varies with ionic strength, temperature, and dye concentration (BAUER and VINOGRAD, 1968 and 1970). This method is highly selective and sensitive because the quantum yield of EB fluorescence is greatly increased after binding to DNA (LE PECQ, YOT and PAOLETTI, 1964; LE PECQ and PAOLETTI, 1967) thus allowing the detection of amounts of DNA as small as 1 μg per band in a preparative gradient. Propidium iodide, an analogue of ethidium bromide, permits even better separation (by about 80%) of closed circles from the remaining DNA (HUDSON et al., 1969).

It must be pointed out that any increase of the superhelix density will allow mtDNA to bind more EB and consequently to band closer to that of open DNA and eventually coincide with it. Since some drugs are able to modify mtDNA superhelix density (SMITH, JORDAN and VINOGRAD, 1971), care must be taken in estimating the amount of mtDNA from a CsCl-EB gradient. This could account for the discrepancies between the results obtained by NASS (1970) and SMITH, JORDAN and VINOGRAD (1971) from mammalian cells treated *in vitro* by EB.

Equilibrium Centrifugation in CsCl Gradient. This method utilizes differences in base composition which exist in some species (for example, man) between nuclear and mtDNA. Usually this method is not sensitive enough, but the technique can be improved by the use of fixed-angle rotors which have a higher resolution power, and/or by the addition of some heavy metals which bind preferentially to base pairs, such as Hg to AT, or Ag to GC (NANDI, WANG and DAVIDSON, 1965; DAVIDSON et al., 1965). The influence of base composition on buoyant density is thus amplified upon metal binding. This method has not yet been used for the fractionation of mtDNA. It has, however, been useful in such instances, as the isolation of nuclear DNA from kinetoplast DNA in Trypanosoma (RIOU and PAOLETTI, 1967) or the

Fig. 8. mtDNA from human leukemic cells after centrifugation in a CsCl-EB density gradient. Lower band: closed circular mtDNA molecules; Upper band: relaxed circular mtDNA molecules and linear molecules of nuclear origin. Photographed under UV light. Unpublished data from this laboratory. Method of RADLOFF et al. (1967)

fractionation of human DNA (CORNEO, GINELLI and POLLI, 1970). Its potential importance would be to distinguish mtDNAs not only according to structure, but also base composition and sequence.

Chromatography on Hydroxyapatite. This method allows fractionation of the DNA according to its secondary structure (BERNARDI et al., 1968 and BERNARDI, 1969). It is currently being used for the isolation of native yeast mtDNA (BERNARDI et al., 1970) though no major difference between the secondary structures of this DNA and other DNAs in yeast is known. Chromatography on hydroxyapatite does not work with mtDNAs from animal cells. However, BOURGAUX-RAMOISY, VAN TIEGHEM and BOURGAUX (1967) have reported that the chromatographic behavior of the twisted forms of a closed circular DNA (polyoma virus DNA) is different from the behavior of nicked and linear forms of the same DNA, although this difference is not large enough to permit complete purification of the closed circles.

The reason why hydroxyapatite chromatography allows a satisfactory resolution between yeast mtDNA and nuclear DNA and not between DNA from animal cells remains obscure, but is probably related to differences in base sequences.

LEFFLER et al. (1970) have taken advantage of the preferential affinity of hydroxyapatite for double-stranded DNA to separate the more easily renatured mtDNA from denatured nuclear DNA at a carefully selected temperature.

Isolation of Open Circular DNA. The most relevant physical feature of open circular DNA molecules is their sedimentation rate, which is 13 to 14% above that of linear molecules (HERSHEY, BURGI and INGRAHAM, 1963; WANG and DAVIDSON, 1966). A method of isolation proposed by FUKE and THOMAS (1970) takes advantage of the circular character of open DNA molecules rather than of their hydrodynamic properties. After trapping in gelled agar, circular DNA molecules at least several microns long, are not topologically free to diffuse while linear DNA molecules can do so.

c) Isolation of Circular Oligomers

The circular dimers of mtDNA from human leukemic patients have been separated by sedimentation in EB-CsCl (CLAYTON, DAVIS and VINOGRAD, 1970); they were, however, still contaminated with 12% monomers, and 3% ambiguous and catenated dimers.

d) Yields of Extraction of mtDNA

Several artifacts may arise during the extraction and examination of mtDNA. One could imagine that the abnormal circular dimers and oligomers found in mtDNA of malignant cells might also exist in normal cells from which they would be lost during extraction. For instance, M. M. K. NASS [1969 (3)] observed that the frequency of circular dimers in mtDNA of L cells fell by half if the cells were immediately extracted rather than washed, centrifuged and resuspended. The exact yield of mtDNA extraction cannot be ascertained because there is no reliable method for measuring with precision the concentration of mtDNA in a biological preparation.

However there is no conclusive evidence that the yield of extraction of mtDNA varies according to the type of tissue or to the nature of mtDNA. NASS [1969 (3)] observed that this yield was only slightly increased (15 to 20%) when the same extraction procedure was applied to L cells whose content of circular dimers varied from 4% to 80%. Moreover, leukocytes of human leukemia have been found to contain up to 48% of circular dimers in their mtDNA; it seems improbable that, if these molecules had been present in normal tissues in such a high proportion, they would have remained undetected.

There are several endonucleases in animal cells [LINDAHL, GALLY and EDELMAN, 1969 (1 and 2)]. One of them is found in the mitochondria (CURTIS, BURDON and SMELLIE, 1966) probably located on the outer membranes (MORAIS, 1969; MORAIS and DE LAMIRANDE, 1970). It is therefore surprising that one can recover most or all mtDNA in the form of closed circular molecules. This is probably due to the fact that the mitochondrial nucleases are bound to membranes: their enzymatic activities may be latent and revealed only after sundry treatments, mainly with detergents (C. PAOLETTI, unpublished data). Nevertheless, the risk of loss of closed circular molecules is high, since a single endonucleolytic break per molecule is sufficient to eliminate the molecules from the lower band of a CsCl-EB gradient. The resulting circles are found in the higher band together with some contaminating nuclear DNA. The proportion of closed circles is thus always underestimated. For example it was found that a given mtDNA contained 23% by weight of circular dimers when directly extruded from osmotically shocked mitochondria, but only 16% when purified by a CsCl-EB gradient centrifugation (Paoletti, Riou and Pairault, 1972). The longer the molecule, the higher the probability of at least one break for any given endonucleolytic activity and consequently the greater the chance that these molecules will be lost.

The possibility of an artefactual transformation of linear DNA molecules into circular ones during extraction should also be kept in mind although it is unlikely in the case of mtDNA of animal cells. Thomas et al. (1970) have shown that such a transformation is possible for the nuclear DNA of certain eukaryotes after partial hydrolysis with *E. coli* exonuclease III or λ exonuclease. This could occur during the extraction of mtDNA, since similar exonucleases have been found in mammalian tissues (Georgatsos and Symeonidis, 1965; Lindahl, Gally an Edelman, 1969 (1 and 2)]. DNA polymerases and ligases capable of repairing and closing open circles have also been found. Nevertheless, such events would lead to populations of molecules heterogeneous in size. This has not been reported in the literature devoted to mtDNA of tumor cells, with the exception of one report by Inaba (1967).

e) The Problem of Contamination of mtDNA by Other DNAs

Nuclear DNA. Many authors use pancreatic DNase to eliminate nDNA from mitochondrial suspensions, assuming that the organelles are impermeable to this enzyme (Rabinowitz et al., 1965) and that the mtDNA is therefore not affected. We avoid this treatment in our laboratory because the mitochondria of a number of tissues, such as salamander liver (Wolstenholme and Dawid, 1966), sheep heart (Kroon et al., 1966), and human leukemic leukocytes (personal observation) seem to be permeable to pancreatic DNase, probably because they have been partially damaged during their purification.

In most cases, the subcellular fraction still contains some nuclear DNA as a contaminant, though in a much reduced proportion (usually about 1:1).

Nuclear Satellite DNA. Kalf and Grece (1966) as well as Corneo, Ginelli and Polli (1967) have described in sheep heart a heavy type of mtDNA (buoyant density higher than 1.714: normal density 1.701) assumed to be of mitochondrial origin. Kroon et al. (1966) could not confirm these observations which they attributed to contamination of the mitochondria by satellite nuclear DNA.

A similar explanation can be proposed for the data reported by Koch and Stokstad (1967), who measured a buoyant density of 1.688 g/ml for a DNA extracted from the mitochondrial fraction of human liver cells grown *in vitro*. On the other hand, Corneo (1967) isolated from the total DNA of human bone marrow and lymph nodes a "satellite" DNA with a buoyant density 1.687 g/ml, that is, identical with the previous value.

Closed Circular DNA of Unknown Origin. With the aid of CsCl-EB gradient centrifugation one can select DNA molecules on the basis of closed, circular, and double-stranded structure. DNAs of different origin could, of course, fulfil these conditions, although no DNA has yet been described in animal cells which is equivalent to that of bacterial episomes or viral replicative forms. Nevertheless, in the first report of their method, VINOGRAD and his collaborators described the isolation from HeLa cells of circular DNA molecules which were not mitochondrial, varying in length from 0.2 to 3.5 μ (RADLOFF, BAUER and VINOGRAD, 1967). A similar type of DNA has also been described in boar-sperm cells (0.5 to 9.7 μ[6], in African monkey kidney cells (BSC-1) (0.1 to 1.5 μ) (RUSH, EASON and VINOGRAD, 1971) and in hamster embryo fibroblasts transformed by SV 40 (2.0 to 3.5 μ) (RIOU and DELAIN, 1971) (see III-B 2 above). Another eukaryotic organism, yeast, yields a multimeric population of closed circles of non-mitochondrial origin (HOLLENBERG, BORST and VAN BRUGGEN, 1970; GUERINEAU et al., 1971; STEVENS and MOUSTACHI, 1971).

mtDNA is probably not the only cytoplasmic DNA of animal cells. For example, a membrane-associated DNA was found in the cytoplasm of human lymphocytes (LERNER, MEINKE and GOODSTEIN, 1971; HALL et al., 1971). It has also been shown that the mere successive transfer of mouse-embryo liver cells is accompanied by the appearance of a cytoplasmic DNA, amounting to 20% of total DNA and displaying the same buoyant density as nuclear DNA (WILLIAMSON, 1970).

f) Examination of mtDNA

Electron microscopy is a powerful method for studying the size and topology of DNA (for general reviews, see KLEINSCHMIDT, 1967; LANG et al., 1967; BUJARD, 1970). A quantitative estimate of the frequency of the different forms of mtDNA circles (monomers, dimers, oligomers, either circular or catenated) is obtained by measuring the contour length of molecules on photographs and plotting the values on a histogram. A less time-consuming procedure is to classify the molecules as monomers, dimers, and oligomers as they appear on the fluorescent screen under the electron microscope.

As previously mentioned, another way of making such an estimate [NASS, 1969 (3); GOEBEL and HELINSKI, 1968) is band sedimentation in sucrose gradients, since the sedimentation coefficients of the linear and nicked, circular, or closed circular DNA monomers and multimers are different (HUDSON, CLAYTON and VINOGRAD, 1968). A systematic comparison of the two techniques has established that they yield identical results, at least for yeast mtDNA (GUERINEAU, BUFFENOIR and PAOLETTI, 1973). Such a comparison seemed necessary after the report that the S values obtained for mtDNA ribosomal RNAs could depend on the method used (EDELMAN et al., 1971).

The problems encountered in scoring DNA molecules on the screen of an electron microscope and classifying them into different types have been extensively discussed by HUDSON and VINOGRAD (1967), CLAYTON et al. (1968) and GORDON and WARNER (1970). For instance, two overlapping monomers may be scored as one dimer. A dimer can be unambiguously classified as uni-circular when no cross-over point divides the molecules into two circular structures. Moreover, the frequency of each type of molecule should not change when the densities of the molecules of the grid are modified. Finally, HUDSON and VINOGRAD (1967) reported that DNA molecules on a surface layer appear to repel each other, thus avoiding overlap on specimen grids,

[6] Unfortunately the number of recorded molecules was small.

unless very high DNA concentrations are employed. This observation was confirmed in other laboratories. The catenated circular oligomers are usually interpreted, after VINOGRAD, as consisting of free interlocked circles. Such an interpretation may not always be correct, as previously discussed (see above III).

The extensive twisting of mtDNA, which makes it difficult to distinguish between uni-circular and catenated circles can be eliminated by nicking the twisted circles with a haplotomic (active on one strand only) endonuclease, such as pancreatic DNAse, or with X-rays. A single break on one strand provides a point around which the two strands can rotate, thus removing the topological constraint responsable for the twisting. The molecules can then be observed on the screen as untwisted structures (Fig. 1). The methods used for opening closed circles with X-rays or pancreatic DNase have been described by FREIFELDER (1965), and PAOLETTI, LE PECQ and LEHMAN (1971), respectively. The amount of enzyme or the dose of irradiation to be given is a critical factor, particularly when the circles are small: if too low, not all molecules will be untwisted; if too high, some molecules will be linear and shortened.

Appendix II: DNA Circular Oligomers Other Than Mitochondrial

It is worth while briefly to review what is presently known about circular dimers and oligomers in DNA of different origins, from bacteriophages to eukaryotic monocellular organisms as this will help us to understand the conditions and mechanismf which control the origin and maintenance of these forms in the mitochondria os animals cells.

Caution is necessary for any generalization in this field. For instance, inhibitors of specific metabolic pathways yield completely different results depending on the systems they act upon. Chloramphenicol has been shown to increase the proportion of dimers and oligomers in the DNA of the colicigenic factor EI present in *Proteus mirabilis* (GOEBEL and HELINSKI, 1968), but not in L cells [NASS, 1969(3)]. Again cycloheximide cannot elicit an accumulation of intracellular circular oligomers of SV 40 (KIT and NAKAJIMA, 1971) whereas it does with mtDNA of L cells [NASS, 1969 (3)]. A general review of circular DNAs has recently been published (HELINSKI, 1971).

a) DNA Circular Oligomers of Eukaryotic Organisms Other Than Animal Cells

A population of closed circular DNA molecules, made up of oligomers of different sizes (from monomer of 2.3 μ to hexamers), has been extracted from *Saccharomyces cerevisiae*. The buoyant density of this DNA is identical to that of nuclear DNA; its intracellular localization remains unknown, but it is probably non-mitochondrial (HOLLENBERG, BORST and VAN BRUGGEN, 1970; GUERINEAU et al., 1971; STEVENS and MOUSTACCHI, 1971). Since such a DNA is extracted from an eukaryotic organism living in an environment considered normal, one cannot assume in this case that the presence of circular oligomers is elicited by disturbances in the cell metabolism.

Uni-circular oligomers were described by RIOU and DELAIN (1969) in another eukaryote. *Trypanosoma* possesses a giant organelle, the kinetoplast, which belongs to the mitochondrial apparatus and contains a large amount of DNA (up to 20% of total cell DNA). This DNA is composed of small closed circles between 0.30 μ and 0.70 μ long, depending on the strain. Some catenated molecules are found. Uni-circular dimers are normally rare. After treatment of *Trypanosoma cruzi* with low concentrations of ethidium bromide (EB), the kinetoplasts contain up to 30% of circular oligomers, ranging from dimers to pentamers. However, EB, whose action is not yet understood at the molecular level, does not seem to be able to increase the content of uni-circular oligomers in mouse L cells, in spite of damage to mitochondrial morphology (NASS, 1970). This intercalating dye specifically blocks mtDNA synthesis and induces its degradation in yeast (GOLDRING et al., 1970; PERLMAN and MAHLER, 1971), and presumably in L cells (NASS, 1970), although this later results has been challenged (SMITH, JORDAN and VINOGRAD, 1971). Uni-circular dimers extracted from the kinetoplast of EB-treated *Trypanosoma* yield Cairns replicative structures (unpublished results from this laboratory).

b) DNA Circular Oligomers of Bacterial Episomes

Colicigenic Factor E1. When the colicigenic factor E*1* is transferred to *Proteus mirabilis* by conjugation, its DNA is recovered as a mixture of uni-circular monomers and oligomers (dimers and trimers), whereas it is composed almost exclusively of monomers in its usual host *Escherichia coli* (GOEBEL and HELINSKI, 1968; BAZARAL and HELINSKI, 1968). However, a few rare uni-circular dimers can be extracted from E1 in *E. coli* (INSELBURG and FUKE, 1970) or in a DNA-free form of *E. coli*, called minicells (INSELBURG and FUKE, 1971). In this case, the dimers could be recovered from a sucrose gradient but were not found in the lower band of a CsCl-EB gradient. GOEBEL and HELINSKI (1968) observed a striking increase in multiple circular DNA forms of the bacterial plasmid in *P. mirabilis* after chloramphenicol treatment or amino-acid starvation. This is in contrast to observations of NASS [1969 (3)], who found that chloramphenicol was unable to modify the frequency of circular oligomers in mtDNA of mouse fibroblasts (see Section VIII-2).

The colicigenic factor E1 also forms circular oligomers in a mutant of *E. coli* thermosensitive for DNA synthesis, grown at a non-permissive temperature (GOEBEL, 1970).

Using a radioactive labeling procedure similar to that reported by M. M. K. NASS [1969 (3)] for L cells (Fig. 7), GOEBEL and HELINSKI (1968) followed the formation of monomers and oligomers after conjugation transfer. They concluded that both pathways, recombination of monomers and oligomer replication, can explain the formation of the oligomers. However, this conclusion must be taken as cautiously as the one offered by NASS for data on L cells (see IV-3).

The interpretation of such data is further obscured by the high background of oligomeric forms of Col E1 DNA in *P. mirabilis*. In more recent work, GOEBEL (1971) has offered a more convincing argument in favor of errors in replication as the true explanation of the origin of the multiple-length circles. When Col E1 is grown at 43 °C in *E. coli* ts/Cr 34/43 (whose DNA synthesis is thermosensitive), the plasmid DNA contains a high proportion of circular oligomers. Monomers of Col E1 DNA, labeled by a one-minute pulse prior to shifting to the non-permissive temperature, are more often converted to oligomeric forms than monomers which were exposed to label for a much longer time. Thus, the newly-made monomers, presumably still in the process of replication, were preferentially converted to oligomers. Thus the occurrence of the multiple-length circles cannot be accounted for solely by random recombination between prexisting smaller circular DNA molecules, although these results do not exclude such a mechanism.

In addition, GOEBEL established that the formation of oligomers from monomers occurs in wild type as well as in two different recombination deficient strains.

Finally, it has been shown (GOEBEL and HELINSKI, 1968) that the proportion of circular oligomers in the DNA of the colicigenic factor transferred into *Proteus mirabilis* is greatly but reversibly increased after addition of chloramphenicol or deprivation of amino-acids, and noticeably decreased after shifting from the exponential to the stationary phase.

Drug Resistance Factors. The DNA of the plasmid responsible for penicillin resistance in *Staphylococcus aureus* contains a low proportion (0.8%) of circular dimers (RUSH et al., 1969).

The DNA of the R-factor does not contain multiple-length circular molecules, but it provides an interesting system to study the interactions of a population of closed circles of different sizes.

It has been established that in *P. mirabilis* this DNA is composed of three independently replicating species of closed circular DNA which can be distinguished through their buoyant density, whereas a single molecular class is obtained from *E. coli*. The predominant molecular species present in *E. coli* corresponds to the largest of the three forms found in *P. mirabilis* (size 33 μ) and may represent a composite of the two smallest forms (26 to 31 μ and 4 to 6 μ), assembled through a recombinational event. It is thought that the smallest unit, which cannot be transferred as such, is the drug-resistance unit, while the 26 to 31 μ molecule, which can be transferred to recombination-deficient bacteria without carrying the drug-resistance marker, is the transfer unit (COHEN and MILLER, 1970).

Other Bacterial Plasmids. COZZARELLI, KELLY and KORNBERG (1968) described in *E. coli 15* a plasmid of unknown origin. LEE and DAVIDSON (1970) have found 5% by number of circular oligomers in the DNA of this plasmid.

c) DNA Circular Oligomers of Animal Viruses

No circular dimers or oligomers have been found in DNA encapsulated in virions. However, they have been described in free intra-cellular viral DNA.

Polyoma Virus. Mouse-embryo 3T3 cells, grown *in vitro*, can be transformed by a thermosensitive mutant of polyoma virus (Ts-a) when infected at a non-permissive temperature (37 °C). At this temperature, the transformed cells, called Ts-a-3T3, do not yield virions; on shifting to the permissive temperature (31 °), virions are formed. They contain the usual viral circular closed DNA (MW $= 3 \times 10^6$). In addition, the cells also harbor non-encapsulated viral DNA containing about 40% twisted dimers and trimers. When infected by the wild-type virus or the Ts-a mutant at permissive temperature, the cells contain less than 1% of dimers and no trimers. The circular DNA dimers are infectious (CUZIN et al., 1970). MEINKE and GOLDSTEIN (1971) have demonstrated that multimeric forms of polyoma DNA are also produced during the replication of wild-type polyoma. They are mostly catenated dimers but some uni-circular dimers are also found. Labelling experiments indicate that these forms are probably on the normal pathway of DNA replication. Such a finding implies that Ts-a mutant has a lesion which bears primarily on the normal replication of the viral DNA.

Simian Virus 40. The intracellular SV 40 DNA, isolated between 48 and 64 h after infection of African green monkey kidney cells (BSC-1) grown *in vitro*, contain about 1% of uni-circular dimers (Rush, Eason and Vinograd, 1971). A similar frequency of uni-circular trimers was also found, but some of the latter forms could have been contaminated with mtDNA. After 90 h of infection, the level of dimers and trimers decreased to 0.2%. Like Rush and his collegues, Jaenisch and Levine 1971 (1) reported the existence of 1 to 2% of circular and interlinked oligomeric SV 40 DNAs in infected cells. Some of the rings are up to six times the size of the monomer. Each class of oligomers in infectious SV 40 dimers have a relative specific infectivity of about 0.5 when compared, on a weight basis, to SV 40 monomers. Since only monomeric DNA is packaged into virions, conversion of the infecting dimers to monomers during the dimer infection process must be inferred. It might occur either as a result of the process of dimer replication or from a rarer recombination event arising in a large dimer pool. However, this conversion of dimers to monomers has a low efficiency since a high level of SV 40 DNA dimers observed soon after infection with the dimeric DNA [Jaenisch and Levine, 1971 (2)]. This indicates that dimeric molecules of SV 40 can replicate; this observation does not favor a circular dimeric intermediate in the normal replication process of SV 40 DNA.

The frequency of circular dimers and trimers is unaffected by the presence of ethidium bromide (EB) during the infectious cycle (Rush et al., 1971). However, when African green monkey cells are grown *in vitro* and infected by SV 40 virions in the presence of EB, the superhelix density of the intracellular closed DNA is homogeneous and identical to that of the viral DNA, which is unaffected by the presence of the dye. In the absence of EB, the mean superhelix density of intra-cellular closed DNA is heterogeneous and approximately three-fourths as large as the superhelix density of the viral DNA. These results could be explained by assuming that EB interferes with a nicking-closing cycle which leads to the formation of the viral DNA from the intracellular forms (Eason and Vinograd, 1971). A similar explanation has been offered by Smith, Jordan and Vinograd (1971) for explaining the changes in the superhelix density of mtDNA from HeLa and SV 3T3 cells treated with EB.

Kit and Nakajima (1971) could not induce an accumulation of multiple uni-circular forms of SV 40 in monkey (CV-I) cultures treated with cycloheximide, a well known inhibitor of cell-free protein synthesis, which secondarily decreases the rate of DNA synthesis (Kit et al., 1969). In contrast, cycloheximide increases the proportion of uni-circular oligomers in the mtDNA of mouse fibroblasts [Nass, 1969 (3)] (see above III-2).

d) DNA Circular Oligomers of Bacteriophages

Phage λ: Wild-Type. Intra-cellular circular dimers of phage λ DNA have been reported (Weissbach, Bortl and Salzman, 1968; Kiger and Sinsheimer, 1971). Moreover, two double-length rings with tail shorter than the ring have also been observed (Kiger and Sinsheimer, 1971). The finding suggests that, in this system, the circular dimers are capable of replication, as also indicated by data on the circular dimers of the kinetoplasts and of mitochondria (see above IV-3).

λ dv Mutant. Matsubara and Kaiser (1968) have characterized a plasmid λ dv (defective virulent) derived from phage λ, which persists in bacteria non-permissive for λ (see also Lieb, 1971). This plasmid contains only that part of the genome responsible for the replication of the phage DNA and its regulation (Genes CI, V1, V2, V3, O and P) and consists of closed circular DNA with a molecular weight of 8.6×10^6 daltons. It also contains up to 20% of molecules twice as long (tetramers) After transfer to rec^- bacteria, the plasmid DNA exists only as monomers (4.3×10^6 daltons). In rec^+ bacteria, the monomers disappear, and dimers as well as higher oligomers are found. This explains why in rec^+ wild-type bacteria, the λ dv DNA is too large for the number of genes it contains.

Phage S 13. Bacteriophage S 13 has a replicative form (RF) DNA, as does ϕX 174 bacteriophage. Its genetic map is circular, and made up of at least seven complementation groups. It is known that the primary mechanism for recombination of phage S 13 is under the control of the host rec A gene (Baker, Doniger and Tessman, 1971). Very few circular oligomers have been found in RF DNA (Gordon, Rush and Warner, 1970).

Two experiments favor the recombination model to explain the formation of circular dimers.

Mutations in gene IV of S 13 (which corresponds to gene VI of ϕX 174) prevent replication of viral RF DNA. Nevertheless, uni-circular dimers appear in the parental RF DNA extracted from bacteria infected at an high multiplicity (Gordon, Rush and Warner, 1970) with the phages mutated in the replication functions. This observation indicates that the formation of dimers does not require any previous replication. Other experiments indicate that the occurrence of dimers can result from recombinational events: after simultaneous infection with two thermosensitive mutants, Rush and Warner [1968 (1 and 2)] isolated dimers whose phenotype was wild, as shown by infection of spheroplasts. The simplest explanation of this result is that monomers of the two mutants were occasionally joined into dimers, resulting in a wild phenotype.

Other Phages. Some circular oligomers (up to tetramers and even higher oligomer have been recorded in the RF DNA of ϕX 174 (Rush et al., 1967) of P 22 (Rhoades and Thomas, 1968) and of M 13 (Jaenisch, Hofschneider and Preuss, 1969). Since the dimers are able to infect spheroplasts (Rush and Warner, 1967) they certainly contain at least the information coded for by the monomers.

Acknowledgement

We thank Drs. J. B. Le Pecq, G. Orth and M. Guerineau for providing useful information and comments. We are indebted to Dr. Y. Lanni for critical reading of the manuscript.

The experimental work performed in our laboratory was supported by the CNRS, INSERM, the Ligue Nationale Française contre le Cancer and the Fondation pour l'Aide a la Recherche Médicale.

The expert technical assistance of Miss M. Gabillot is gratefully acknowledged.

References

Anai, M., Hirahashi, T., Tagaki, Y.: A desoxyribonuclease which requires nucleoside triphosphate from *Micrococcus lysodeikticus*. I. Purification and characterization of the desoxyribonuclease activity. J. biol. Chem. **245**, 767 (1970).

Anai, M., Hirahashi, T., Yamanaka, M., Takagi, Y.: A desoxyribonuclease which requires nucleoside triphosphate from *Micrococcus lysodeikticus*. II. Studies on the role of nucleoside triphosphate. J. biol. Chem. **245**, 775 (1970).

Arca, M., Caneva, R., Frontali, L.: Effect of acridines on DNA synthesis in vitro: base composition of product and inhibition in the presence of different primers. Biochim. biophys. Acta (Amst.) **217**, 548 (1970).

Ashwell, M., Work, T. S.: The biogenesis of mitochondria. Ann. Rev. Biochem. **39**, 251 (1970).

Askew, J. B., Fechner, R. E., Bentick, D. C., Jenson, A. B.: Epithelial and myoepithelial oncocytes. Arch. Otolaryng. **93**, 46 (1971).

Attardi, G., Attardi, B.: The informational role of mitochondrial DNA. Park City International Symposia on Problems in Biology. I — RNA in development (1969).

Attardi, B., Attardi, G.: Fate of mitochondrial DNA in human mouse somatic cell hybrids. Proc. Nat. Acad. Sci. (Wash.) **69**, 129 (1972).

Baker, R., Doniger, J., Tessman, I.: Roles of parental and progeny DNA in two mechanisms of phage S 13 recombination. Nature (Lond.) New Biol. **230**, 23 (1971).

Barbour, S. D., Clark, A. J.: Biochemical and genetic studies of recombination proficiency in *Escherichia coli*. I. Enzymatic activity associated with rec B^+ and rec C^+ genes. Proc. nat. Acad. Sci. (Wash.) **65**, 955 (1970).

Bauer, W., Vinograd, J.: The interaction of closed circular DNA with intercalative dyes. I. The superhelix density of SV 40 DNA in the presence and absence of dye. J. molec. Biol. **33**, 141 (1968).

Bauer, W., W., Vinograd, J.: The interaction of closed circular DNA with intercalative dyes. III. Dependence of the buoyant density upon superhelix density and base composition. J. molec. Biol. **54**, 281 (1970).

Bazeral, M., Helinski, D. R.: Circular DNA forms of colicinogenic factors E_1, E_2 and E_3 from *Escherichia coli*. J. molec. Biol. **36**, 185 (1968).

Becher, M.: Elektronenmikroskopische Untersuchungen an Onkozyten eines Adenolymphoms. Acta biol. med. germ. **13**, 615 (1964).

Becker, A., Hurwirtz, J.: Current thoughts on the replication of DNA. Progress in Nucleic Acid Research and Molecular Biology **11**, 423 (1971).

Benedetti, E. L., Bernhard, W.: Recherches ultrastructurales sur le virus de la leucemie erythroblastique du poulet. J. Ultrastruct. Res. **1**, 309 (1958).

Bernardi, G.: Chromatography of nucleic acids on hydroxyapatite. I. Chromatography of native DNA. Biochim. biophys. Acta (Amst.) **174**, 423 (1969).

Bernardi, G., Carnevali, F., Nicolaieff, A., Piperno, G., Tecce, G.: Separation and characterization of a satellite DNA from a yeast cytoplasmic "petite" mutant. J. molec. Biol. **37**, 493 (1968).

Bernardi, G., Fuares, M., Piperno, G., Slonimski, P. P.: Mitochondrial DNA's from respiratory sufficient and cytoplasmic respiratory-deficient mutant yeast. J. molec. Biol. **48**, 23 (1970).

Bernhard, W.: Some problems of fine structure in tumor cells. Progr. exp. Tumor Res. (Basel) **3**, 1 (1963).

Bernhard, W.: Ultrastructure of the cancer cell. In: Handbook of molecular cytology, p. 687 (Lima de Faria, A., Ed.). Amsterdam, London: Publishing Company 1969.

Bernhard, W., Tournier, P.: Modification persistante des mitochondries dans des cellules tumorales de Hamster transformées par l'adenovirus 12. Int. J. Cancer **1**, 61 (1966).

Bisalputra, T., Burton, H.: On the chloroplast DNA membrane complex in Sphacelaria sp. J. Microscopie **9**, 661 (1970).

Borst, P.: Biochemistry and function of mitochondria. In: Handbook of molecular cytology, p. 914 (Lima de Faria, A., Ed.). Amsterdam, London: Publishing Company 1969.

Borst, P.: Mitochondrial nucleic acids. Ann. Rev. Biochem. **41**, 333 (1972).

Borst, P., Kroon, A. M.: Mitochondrial DNA: physico-chemical properties, replication and genetic function. Int. Rev. Cytol. **26**, 107 (1969).

Bourgaux, P., Bourgaux-Ramoisy, D., Seilers, P.: The replication of the ring-shaped DNA of polyoma virus. II. Identification of molecules at various stages of replication. J. molec. Biol. **59**, 195 (1971).

Bourgaux-Ramoisy, D., Van Tieghem, N., Bourgaux, P.: Fractionation of polyoma virus DNA on hydroxypatite: dependence on tertiary structure. J. gen. Virol. **1**, 589 (1967).

Bujard, H.: Electron microscopy of single-stranded DNA. J. molec. Biol. **49**, 125 (1970).

Burd, J. F., Wells, R. D.: Effects of incubation conditions of the nucleotide sequences of DNA products of unprimed DNA polymerase reactions. J. molec. Biol. **53**, 435 (1970).

Buttin, G., Wright, M. R.: Enzymatic DNA degradation in *E. coli.* — Its relationship to synthesis processes at the chromosome level. Cold Spr. Harb. Symp. quant. Biol. **33**, 259 (1968).

Cairns, J., Denhardt, T. D.: Effect of cyanide and carbon monoxide on the replication of bacterial DNA in vivo. J. molec. Biol. **36**, 335 (1968).

Carnevali, F., Morpurgo, G., Tecce, G.: Cytoplasmic DNA from "petite" colonies of *Saccharomyces cerevisiae:* a hypothesis on the nature of the mutation. Science **163**, 1331 (1969).

Chang, L. O., Schnaitman, C. A., Morris, H. P.: Comparison of the mitochondrial membrane proteins in rat liver and hepatoma. Cancer Res. **31**, 108 (1971).

Clayton, D. A., Vinograd, J.: Circular dimer and catenated forms of mitochondrial DNA in human leukaemic leukocytes. Nature (Lond.) **216**, 652 (1967).

Clayton, D. A., Smith, C. A., Jordan, J. M., Teplitz, M., Vinograd, J.: Occurrence of complex mitochondrial DNA in normal tissues. Nature (Lond.) **220**, 976 (1968)

Clayton, D A , Smith, C A., Vinograd, J.: Complex mitochondrial DNA in normal and neoplastic tissue. Fed. Proc. **48**, 532 (1969).

Clayton, D. A., Vinograd, J.: Complex mitochondrial DNA in leukemic and normal human myeloid cells. Proc. nat. Acad. Sci. (Wash.) **62**, 1077 (1969).

Clayton, D. A., Davis, R., Vinograd, J.: Homology and structural relationships between the dimeric and monomeric circular forms of mitochondrial DNA from human leukemic leukocytes. J. molec. Biol. **47**, 137 (1970).

Clayton, D. A., Teplitz, R. L., Nabholz, M., Dovey, H., Bodmer, W.: Mitochondrial DNA of human mouse cell hybrids. Nature (Lond.) **234**, 560 (1971).

Coen, D., Deutsch, J., Netter, P., Petrochilo, E., Slonimski, P. P.: Mitochondrial genetics: methodology and phenomenology. In: Control of organelle development. Symp. Soc. exp. Biol. XXIV. Cambridge: Univ. Press 1970.

Cohen, S. N., Miller, C. A.: Non-chromosomal antibiotic resistance in bacteria. II. Molecular nature of R factors isolated from *Proteus mirabilis* and *Escherichia coli.* J. molec. Biol. **50**, 671 (1970).

Corneo, G., Ginelli, E., Polli, E.: A satellite DNA isolated from human tissues. J. molec. Biol. **23**, 619 (1967).

Corneo, G., Ginelli, E., Polli, E.: Repeated sequences in human DNA. J. molec. Biol. **48**, 319 (1970).

Cozzarelli, N. R., Kelly, R. G., Kornberg, A.: A minute circular DNA from *Escherichia coli* 15. Proc. nat. Acad. Scie. (Wash.) **60**, 992 (1968).

Crabtree, H. G.: Observations on the carbohydrate metabolism of tumours. Biochem. J. **23**, 536 (1929).

Curtis, P. J., Burdon, M. G., Smellie, R. M. S.: The purification from rat liver of a nuclease hydrolysing ribonucleic acid and deoxyribonucleic acid. Biochem. J. **98**, 813 (1966).

Cuzin, F., Vogt, M., Dickman, M., Berg, P.: Induction of virus multiplication in 3 T 3 cells transformed by a thermosensitive mutant of polyoma virus: II. Formation of oligomeric polyoma DNA molecules. J. molec. Biol. **47**, 317 (1970).

Davern, C.: Molecular aspects of genetic recombination. Progress in Nucleic Acid Research and Molecular Biology **11**, 229 (1971).

DAVIDSON, N., WIDHOLM, J., NANDI, U. S., JENSEN, R., OLIVERA, B. M., WANG, J. C.: Preparation and properties of native Crab dAT. Proc. nat. Acad. Sci. (Wash.) **53, 111** (1965).

DAWID, I., BLACKLER, A. W.: Maternal and Cytoplasmic inheritance of mitochondrial DNA in Xenopus. Develop. Biol. **29**, 152 (1972).

DEBRECENI, N., BEHME, J., EBISUZAKI, K.: A DNA dependent ATPase from *E. coli* infected with bacteriophage T 4. Biochem. biophys. Res. Commun. **41**, 115 (1970).

DENHARDT, D. T., BURGESS, A. B.: DNA replication in vitro. I. Synthesis of single-stranded ϕX 174 DNA in crude lysates of Ø × 174 infected *E. coli*. Cold Spr. Harb. Symp. quant. Biol. **33**, 449 (1968).

DELVIN, T. M., PRUSS, M. P.: Proc. Amer. Ass. Cancer **3**, 315 (1962).

EASON, R., VINOGRAD, J.: Superhelix density heterogeneity of intracellular Simian virus 40 deoxyribonucleic acid. J. Virol. **7**, 1 (1971).

EDELMAN, M., VERMA, I. M., HERZOG, R., GALUN, E., LITTAUER, U. Z.: Physicochemical properties of mitochondrial ribosomal RNA from fungi. Europ. J. Biochem. **19**, 372 (1971).

EMMELOT, P., BOS, J. C., BROMBACHER, P. S., HAMPE, J. F.: Studies on isolated tumour mitochondria: biochemical properties of mitochondria from hepatomas with special reference to a transplanted rat hepatoma of the solid type. Brit. J. Cancer **13**, 348 (1959).

ERICSSON, J. L. E., SELJELID, R., ORRENIUS, S.: Comparative light and electron microscopic observations of the cytoplasmic matrix in renal carcinomas. Virchows Arch. path. Anat. **341**, 204 (1966).

FREDERIC, J.: Recherches cytologiques sur le chondriome normal ou soumis à l'experimentation dans des cellules vivantes cultivées in vitro. Thèse Université de Liège, Faculté de Médecine 1958.

FREIFELDER, D.: Mechanism of inactivation of coliphage T 7 by X-rays. Proc. nat. Acad. Sci. (Wash.) **54**, 128 (1965).

FUKE, M., THOMAS, C. A., Jr.: Isolation of open circular DNA molecules by retention in agar gels. J. molec. Biol. **52**, 395 (1970).

GANESAN, A. T.: Adenosine-triphosphate dependent synthesis of biologically active DNA by azide poisoned bacteria. Proc. nat. Acad. Sci. (Wash.) **68**, 1296 (1971).

GANESAN, A. T., LEDERBERG, J.: A cell membrane bound fraction of bacteria DNA. Biochem. biophys. Res. Commun. **18**, 824 (1965).

GAZZOLO, C., DE THE, G., VIGIER, P., SARNA, P. S.: Présence de particules à l'aspect de nucléocapsides associées aux mitochondries dans des cellules de Hamster transformées par le virus de Rous. C. R. Acad. Sci. (Paris) **268**, 1668 (1969).

GEORGATSOS, J. G., SYMEONIDIS, A.: A deoxyribonuclease from mammary tumours of C3H mice preferentially hydrolysing heat denatured DNA. Nature (Lond.) **206**, 1362 (1965).

GOEBEL, W.: Replication of the colicinogenic factor col E 1 in two temperature sensitive mutants of *E. coli* defective in DNA replication. Europ. J. Biochem. **5**, 311 (1970).

GOEBEL, W.: Formation of complex col E 1 DNA by replication. Biochim. biophys. Acta (Amst.) **232**, 32 (1971).

GOEBEL, W., HELINSKI, D. R.: Generation of higher multiple circular DNA forms in bacteria. Proc. nat. Acad. Sci. (Wash.) **61**, 1406 (1968).

GOLDMARK, P. J., LINN, S.: An endonuclease activity from *Escherichia coli* absent from certain rec-strains. Proc. nat. Acad. Sci. (Wash.) **67**, 434 (1970).

GOLDRING, E., GROSSMAN, L., KRUPNICK, D., CRYER, D. R., MARMUR, J.: The "petite" mutation in yeast. I. Loss of mitochondrial DNA during petite induction with ethidium bromide. J. molec. Biol. **52**, 323 (1970).

GORDON, C. N., RUSH, M., WARNER, R. C.: Complex replicative form molecules of bacteriophages ϕX 174 and S 13 su 105. J. molec. Biol. **47**, 495 (1970).

GRAUSE, G. F.: Alterations of DNA base composition in bacteria. Progress in Nucleic Acid Research and Molecular Biology (DAVIDSON, J. N., COHN, W. E., Eds.). **8**, 49 (1968).

GROSS, N. J., RABINOWITZ, M.: Synthesis of new strands of mitochondrial and nuclear deoxyribonucleic acid by semiconservative replication. J. biol. Chem. **244**, 1563 (1969).

GROSS, N. J., NICHOLAS, J., GODFREY, S., GETZ, G. S., RABINOWITZ, M.: Apparent turnover of mitochondrial deoxyribonucleic acid and mitochondrial phospholipids in the tissues of the rat. J. biol. Chem. **244**, 1552 (1969).

GROSS, N. J.: Control of mitochondria turnover under the influence of thyroid hormone. J. Cell Biol. **48**, 29 (1971).

GUERINEAU, M., GRANDCHAMP, C., PAOLETTI, C., SLONIMSKI, P.: Characterization of a new class of circular DNA molecules in yeast. Biochem. biophys. Res. Commun. **42**, 550 (1971).

GUERINEAU, M., BUFFENOIR, C., PAOLETTI, C.: Mitochondrial DNA during the differentiation of the respiratory functions in yeast. In preparation.

HALL, M R., MEINKE, W., GOLDSTEIN, D. A., LERNER, R. A.: Synthesis of cytoplasmic membrane-associated DNA in lymphocyte nucleus. Nature (Lond.) New Biology **234**, 227 (1971).

HARRISON, B. D., ROBERTS, I. M.: Association of tobacco rattle virus with mitochondria. J. gen. Virol. **3**, 121 (1968).

HARRISON, B. D., STEFANAC, Z., ROBERT, I. M.: Role of mitochondria in the formation of X-bodies in cells of *Nicotiana clevelandii* infected by tobacco rattle viruses. J. gen. Virol. **6**, 127 (1970).

HATTA, T., NAKAMOTO, T., TAKAGI, Y., USHIYAMA, R.: Cytological abnormalities of mitochondria induced by infection with cucumber green mottle mosaic virus. Virology **45**, 272 (1971).

HELINSKI, D. R., CLEWEL, D. B.: Circular DNA. Ann. Rev. Biochem. **40**, 899 (1971).

HERSHEY, A. D., BURGI, E., INGRAHAM, L.: Cohesion of DNA molecules isolated from phage lambda. Proc. nat. Acad. Sci. (Wash.) **49**, 748 (1963).

HOLLENBERG, C. P., BORST, P., VAN BRUGGEN, E. F. J.: Mitochondrial DNA — A 25 μ closed circular duplex DNA molecules in wild type yeast mitochondria structure and genetic complexity. Biochim. biophys. Acta (Amst.) **209**, 1 (1970).

HOUT, A., OOSTERBAAN, R. A., POUWELS, P. H., DE JONGE, A. J. R.: Purification of an ATP-dependent nuclease from *Micrococcus lysodeikticus*. Biochim. biophys. Acta (Amst.) **204**, 632 (1970).

HOWLAND, J. L., HUGHES, W. T.: Suggested role of respiration in bacterial DNA replication. Biochem. biophys. Res. Commun. **37**, 106 (1969).

HRUBAN, Z., SWIFT, H., RECHCIGL, M.: Fine structure of transplantable hepatomas of the rat. J. nat. Cancer Inst. **35**, 459 (1965).

HUDSON, B., VINOGRAD, J.: Catenated circular DNA molecules in HeLa cell mitochondria. Nature (Lond.) **216**, 647 (1967).

HUDSON, B., CLAYTON, D. A., VINOGRAD, J.: Complex mitochondria DNA. Cold Spr. Harb. Symp. quant. Biol. **33**, 435 (1968).

HUDSON, B., UPHOLT, W. B., DEVINNY, J., VINOGRAD, J.: The use of an ethidium bromide analogue in the dye buoyant density procedure for the isolation of closed circular DNA: the variation of the superhelix density of mitochondrial DNA. Proc. nat. Acad. Sci. (Wash.) **62**, 813 (1969).

INABA, K.: Nucleic acids and protein synthesis in cancer cell mitochondria. I. Nucleic acids in rat hepatoma mitochondria. Acta med. Okayama **21**, 297 (1967).

INSELBURG, J., FUKE, M.: Replicating DNA: structure of colicin factor E. Science **169**, 590 (1970).

INSELBURG, J., FUKE, M.: Isolation of catenated and replicating DNA molecules of colicin factor E 1 from minicells. Proc. nat. Acad. Sci. (Wash.) **68**, 2839 (1971).

JAENISCH, R., HOFSCHNEIDER, P., PREUSS, A.: Isolation of circular DNA by zonal centrifugation — Separation of normal length, double length and catenated M 13 replicative form DNA and of host-specific "episomal" DNA. Biochim. biophys. Acta (Amst.) **190**, 88 (1969).

JAENISCH, R., LEVINE, A. J.: (1) DNA replication in SV 40 infected cells. V. Circular and catenated oligomers of SV 40 DNA. Virology **44**, 480 (1971).

JAENISCH, R., LEVINE, A. J.: (2) Infection of primary African green monkey cells with SV 40 monomeric and dimeric DNA. J. molec. Biol. **61**, 735 (1971).

KALF, G., GRECE, A.: The isolation of deoxyribonucleic acid from lamb heart mitochondria. J. biol. Chem. **241**, 1019 (1966).

KALF, G. F., FAUST, A. S.: The inner membrane of the rat liver mitochondria as the site of incorporation of radioactively labeled precursor into nucleic acids and proteins in vitro. Arch. Biochem. Biophys. **134**, 103 (1969).

KARA, J., MACH, O., CERVA, H.: Replication of Rous sarcoma and the biosynthesis of the oncogenic subviral ribonucleoprotein particles ("virosomes") in the mitochondria isolated from Rous sarcoma tissue. Biochem. biophys. Res. Commun. **44**, 162 (1971).

KASAMATSU, H., ROBBERSON, D. L., VINOGRAD, J.: A novel closed-circular mitochondrial DNA with properties of a replicate intermediate. Proc. nat. Acad. Sci. (Wash.) **68**, 2252 (1971).

KIGER, J. A., SINSHEIMER, R. L.: DNA of vegetative bacteriophage lambda. VI. Electron microscope studies of replicating lambda DNA. Proc. nat. Acad. Sci. (Wash.) **68**, 112 (1971).

KIRSCHNER, R. M., WOLSTENHOLME, D. R., GROSS, N. J.: Replicating molecules of circular mitochondrial DNA. Proc. nat. Acad. Sci. (Wash.) **60**, 1466 (1968).

KIT, S., KURIMURA, T., DE TORRES, R. A., DUBBS, D. R.: Simian virus 40 deoxyribonucleic acid replication. I. Effect of cycloheximide on the replication of SV 40 deoxyribonucleic acid in monkey kidney cells and in heterokaryons of SV 40 transformed and susceptible cells. J. Virol. **3**, 25 (1969).

KIT, S., NAKAJIMA, K.: Analysis of the molecular forms of simian virus 40 deoxyribonucleic acid synthesized in cycloheximide-treated cell cultures. J. Virol. **7**, 87 (1971).

KLEINSCHMIDT, A. K.: Structure aspects of the genetic apparatus of viruses and cells. In: Molecular Genetics, Part II, 47 (TAYLOR, J. H., Ed.). New York: Academic Press 1967.

KOCH, J., STOKSTAD, E. L. R.: Incorporation of (^{3}H) thymidine into nuclear and mitochondrial DNA in synchronized mammalian cells. Europ. J. Biochem. **3**, 1 (1967).

KROON, A. M.: DNA and RNA from mitochondria and chloroplasts (biochemistry). In: Handbook of Molecular Cytology, p. 943 (LIMA DE FARIA, A., Ed.). Amsterdam, London: Publishing Company 1969.

KROON, A. M., BORST, P., VAN BRUGGEN, E. F. J., RUTTENBERG, J. C. M.: Mitochondrial DNA from sheep heart. Proc. nat. Acad. Sci. (Wash.) **56**, 1836 (1966).

LANG, D., BUJARD, H., WOLFF, B., RUSSEL, D.: Electron microscopy of size and shape of viral DNA in solutions of different ionic strengths. J. molec. Biol. **23**, 163 (1967).

LEDUC, E. H., BERNHARD, W., TOURNIER, P.: Cyclic appearance of atypical mitochondria containing DNA fibers in cultures of an adenovirus 12-induced hamster tumor. Exp. Cell Res. **42**, 597 (1966).

LEE, C. S., DAVIDSON, N.: Physicochemical studies on the minicircular DNA in *E. coli* 15. Biochim. biophys. Acta (Amst.) **204**, 285 (1970).

LEFFLER, A. T., CRESKOFF, E., LUBORSKY, S. W., MCFARLAND, V., MORA, P. T.: Isolation and characterization of rat-liver mitochondria DNA. J. molec. Biol. **48**, 455 (1970).

LE PECQ, J. B.: Use of ethidium bromide for separation and determination of nucleic acids of various conformational forms and measurements of their associated enzymes. Methods of biochemical analysis, p. 20 (GLICK, D., Ed.). New York: John Wiley Sons Publ. 1971.

LE PECQ, J. B., YOT, P., PAOLETTI, C.: Interaction du bromhydrate d'ethidium (BET) avec les acides nucleiques (AN). Etude spectrofluorimetrique. C. R. Acad. Sci. (Paris) **259**, 1786 (1964).

LE PECQ, J. B., PAOLETTI, C.: A fluorescent complex between ethidium bromide and nucleic acids. Physical-chemical characterization. J. molec. Biol. **27**, 87 (1967).

LERMAN, L. S.: Structural considerations in the interaction of DNA and acridines. J. molec. Biol. **3**, 18 (1961).

LERNER, R. A., MEINKE, W., GOODSTEIN, D. A.: Membrane associated DNA in the cytoplasm of diploid human lymphocytes. Proc. nat. Acad. Sci. (Wash.) **68**, 1212 (1971).

LEVINE, A. J.: Induction of mitochondrial DNA synthesis in monkey cells infected by simian virus 40 and (or) treated with calf serum. Proc. nat. Acad. Sci. (Wash.) **68**, 717 (1971).

LIEB, M.: λ mutants which persist as plasmids. J. Virol. **6**, 218 (1970).

LINDAHL, T., EDELMAN, G. M.: Polynucleotide ligase from myeloid and lymphoid tissues. Proc. nat. Acad. Sci. (Wash.) **61**, 680 (1968).

LINDAHL, T., GALLY, J. A., EDELMAN, G. M.: (1) Desoxyribonuclease IV: a new exonuclease from mammalian tissues. Proc. nat. Acad. Sci. (Wash.) **62**, 597 (1969).

LINDAHL, T., GALLY, J. A., EDELMAN, G. M.: (2) Properties of desoxyribonuclease III from mammalian tissues. J. biol. Chem. **244**, 5014 (1969).

Lorans, G., Tournier, P.: Un facteur de croissance nécessaire a la culture in vitro des cellules de tumeurs induites chez les hamsters par l'adenovirus 12. C. R. Acad. Sci. (Paris) **258**, 386 (1964).

Mach, O., Kara, J.: Presence of Rous sarcoma virus inside the mitochondria isolated by zonal and differential centrifugation from Rous sarcoma cells. Folia biol. (Praha) **17**, 65 (1971).

Matsubara, K., Kaiser, A. D.: λ dV: an autonomously replicating DNA fragment. Cold Spr. Harb. Symp. quant. Biol. **33**, 769 (1968).

Meinke, W., Goldstein, D. A.: Studies on the structure and formation of polyoma DNA replicative intermediates. J. molec. Biol. **61**, 543 (1971).

Meselson, M., Yuan, R.: DNA restriction enzyme from *E. coli*. Nature (Lond.) **217**, 1110 (1968).

Meyer, R. R., Simpson, M. V.: Deoxyribonucleic acid biosynthesis in mitochondria. Purification and general properties of rat-liver mitochondrial deoxyribonucleic acid polymerase. J. biol. Chem. **245**, 3426 (1970).

Mills, D. R., Peterson, R. L., Spiegelman, S.: An extracellular darwinian experiment with a self-duplicating nucleic acid molecule. Proc. nat. Acad. Sci. (Wash.) **58**, 217 (1967).

Morais, R.: Studies on the localization of rat liver mitochondrial 5' endonuclease. Biochim. biophys. Acta (Amst.) **189**, 38 (1969).

Morais, R., De Lamirande, G.: Studies on the intra-cellular and intramitochondrial distribution of 5' endonuclease in regenerating rat liver. Biochim. biophys. Acta (Amst.) **209**, 145 (1970).

Mordoh, J., Hirota, Y., Jacob, F.: On the process of cellular division in *Escherichia coli*. V. Incorporation of desoxynucleoside triphosphates by DNA thermosensitive mutants of *Escherichia coli* also lacking DNA polymerase activity. Proc. nat. Acad. Sci. (Wash.) **67**, 773 (1970).

Moses, R. E., Richardson, C. C.: Replication and repair of DNA in cells of *Escherichia coli* treated with toluene. Proc. nat. Acad. Sci. (Wash.) **67**, 674 (1970).

Mounoulou, J. C., Jakob, H., Slonimski, P. P.: Mitochondrial DNA from yeast "petite" mutants: specific change of buoyant density corresponding to different cytoplasmic mutations. Biochem. biophys. Res. Commun. **24**, 218 (1966).

Mounoulou, J. C., Perrodin, G., Slonimski, P. P.: Specific synthesis of a small part of mitochondrial DNA concomitant with the onset of the oxygen-induced development of mitochondria. In: Biochemical aspects of the biogenesis of mitochondria (Slater, E. C., Tager, J. M., Papa, S., Quagliariello, E., Eds.). Adriatica Bari 1968.

Nandi, U. S., Wang, J. C., Davidson, N.: Separation of deoxyribonucleic acids by Hg (II) binding and Cs_2SO_4 density gradient centrifugation. Biochemistry **4**, 1687 (1965).

Nass, M. M. K.: (1) Mitochondrial DNA. I. Intra-mitochondrial distribution and structural relations of single and double length circular DNA. J. molec. Biol. **42**, 521 (1969).

Nass, M. M. K.: (2) Structure and physicochemical properties of isolated DNA. J. molec. Biol. **42**, 529 (1969).

Nass, M. M. K.: (3) Reversible generation of circular dimers and higher multiple forms of mitochondrial DNA. Nature (Lond.) **223**, 1124 (1969).

Nass, M. M. K.: (4) Mitochondrial DNA: Advances, problems and goals. Science **165**, 25 (1969).

Nass, M. M. K.: Abnormal DNA patterns in animal mitochondria: ethidium bromide-induced breakdown of closed circular DNA and conditions leading to oligomers accumulation. Proc. nat. Acad. Sci. (Wash.) **67**, 1926 (1970).

Nass, S., Nass, M. M. K.: Intra-mitochondrial fibers with deoxyribonucleic acid characteristics: observations of Ehrlich ascites tumor cells. J. nat. Cancer Inst. **33**, 777 (1964).

Nass, M. M. K., Nass, S., Afzelius, B. A.: The general occurrence of mitochondrial DNA. Exp. Cell Res. **37**, 516 (1965).

Oberling, Ch., Riviere, M., Haguenau, F. R.: Ultrastructure des epitheliomas à cellules claires du rein (hypernéphromes ou tumeurs de Grawitz) et son implication pour l'histogénèse de ces tumeurs. Bull. Cancer **46**, 356 (1959).

Oda, T.: Circular DNA's from tumor cell mitochondria and nuclei. J. Cell Biol. **39**, 173a (1968).

ODA, T., OMURA, S., YAMAMOTO, S., NISHIDA, S., HIRATA, S.: Circular DNA's from HeLa cell nuclei and mitochondria. Acta med. Okayama **24**, 405 (1970).

OSHI, M.: An ATP-dependent deoxyribonuclease from *E. coli* with a possible role in genetic recombination. Proc. nat. Acad. Sci. (Wash.) **64**, 1292 (1969).

O'RIORDAN, M. L., ROBINSON, J. A., BUCKTON, K. E., EVANS, H. S.: Distinguishing between the chromosomes involved in Down's syndrome (trisomy 21) and chronic myeloid leukaemia (Ph 1) by fluorescence. Nature (Lond.) **230**, 167 (1971).

PAOLETTI, C., RIOU, G.: Le DNA mitochondrial des cellules malignes. Bull. Cancer **57**, 301 (1970).

PAOLETTI, C., LE PECQ, J. B., LEHMAN, I. R.: The use of ethidium bromide circular DNA complexes for the fluorimetric analysis of breakage and joining of DNA. J. molec. Biol. **75**, 55 (1971).

PAOLETTI, C., RIOU, G., PAIRAULT, J.: Circular oligomers in mitochondrial DNA from non malignant thyroid glands. Proc. nat. Acad. Sci. (Wash.) **69** 847 (1972).

PEDERSEN, P. L., ESKA, T., MORRIS, H. P., CATTERALL, W. A.: Deficiency of uncoupler stimulated adenosine triphosphatase activity in tightly coupled hepatoma mitochondria. Proc. nat. Acad. Sci. (Wash.) **68**, 1079 (1971).

PERLMAN, P. S., MAHLER, H.: Molecular consequences of ethidium bromide mutagenesis. Nature (Lond.) New Biol. **231**, 12 (1971).

RABINOWITZ, M., SINCLAIR, J., DE SALLE, L., HASELKORN, R., SWIFT, H. H.: Isolation of desoxyribonucleic acid from mitochondria of chick embryo heart and liver. Proc. nat. Acad. Sci. (Wash.) **53**, 1126 (1965).

RABINOWITZ, M., GETZ, G. S., CASEY, Y., SWIFT, H.: Synthesis of mitochondrial and nuclear DNA in anaerobically grown yeast during the development of mitochondrial function in response to oxygen. J. molec. Biol. **41**, 381 (1969).

RABINOWITZ, M., SWIFT, H.: Mitochondrial nucleic acids and their relation to the biogenesis of mitochondria. Physiol. Rev. **50**, 376 (1970).

RADLOFF, R., BAUER, W, VINOGRAD, J.: A dye-buoyant density method for the detection and isolation of closed circular duplex DNA: the closed circular DNA in HeLa cells. Proc. nat. Acad. Sci. (Wash.) **57**, 1514 (1967).

RALPH, R. K., CLARK, M. F.: Intracellular location of double-stranded plant viral ribonucleic acid. Biochem. biophys. Acta (Amst.) **119**, 29 (1966).

RHOADES, M., THOMAS, C.: The P22 bacteriophage DNA molecule. II. Circular intracellular forms. J. molec. Biol. **37**, 41 (1968).

RICHARDSON, C. C.: Phosphorylation of nucleic acid by an enzyme from T 4 bacteriophage-infected *E. coli*. Proc. nat. Acad. Sci. (Wash.) **54**, 158 (1965).

RICHARDSON, C. C.: Enzymes in DNA metabolism. Ann. Rev. Biochem. **38**, 795 (1969).

RIOU, G., DELAIN, E.: Abnormal circular DNA molecules induced by ethidium bromide in the kinetoplast of trypanosoma cruzi. Proc. nat. Acad. Sci. (Wash.) **64**, 618 (1969).

RIOU, G., DELAIN, E.: Mitochondrial DNA from cells transformed by adenoviruses and SV 40. Biochimie **53**, 831 (1971).

RIOU, G., LACOUR, F.: Mitochondrial DNA from cells transformed by myeloblastosis virus. Biochimie **53**, 47 (1971).

RIOU, G., PAOLETTI, C.: Preparation and properties of nuclear and satellite deoxyribonucleic acid of *Trypanosoma cruzi*. J. molec. Biol. **28**, 377 (1967).

RIOU, G., PAOLETTI, C.: Mitochondrial DNA from benign tumors. In preparation.

ROTH, S. I., OLEN, E. M. D., HANSEN, L. S.: The eosinophilic cells of the parathyroid (oxyphil cells) salivary (oncocytes) and thyroid (Hurthle cells) glands. Lab. Invest. **11**, 933 (1962).

RUSH, M. G., KLEINSCHMIDT, A. K., HELLMANN, W., WARNER, R. C.: Multiple length rings in preparation of ϕX 174 replicative form. Proc. nat. Acad. Sci. (Wash.) **58**, 1676 (1967).

RUSH, M. G., WARNER, R. C.: (1) Multiple length rings of ϕX 174 and S 13 replicative forms. III. A possible intermediate in recombination. J. biol. Chem. **243**, 4821 (1968).

RUSH, M. G., WARNER, R. C.: (2) Molecular recombination in a circular genome ϕX 174 and S 13. Cold Spr. Harb. Symp. quant. Biol. **33**, 459 (1968).

RUSH, M. G., GORDON, C. N., NOVICK, R. P., WARNER, R. C.: Penicillinase plasmid DNA form *Staphylococcus aureus*. Proc. nat. Acad. Sci. (Wash.) **63**, 1304 (1969).

RUSH, M. G., EASON, R., VINOGRAD, J.: Identification and properties of complex forms of SV 40 DNA isolated from SV 40 infected African green monkey (BSC-I) cells. Biochim. biophys. Acta (Amst.) **228**, 585 (1971).

RYTER, A., HIROTA, Y., JACOB, F.: DNA membrane complex and nuclear segragation in Bacteria. Cold Spr. Harb. Symp. quant. Biol. **33**, 669 (1968).

SAMBROOK, J., WESTPHAL, H., SRINIVASAN, P. R., DULBECCO, R.: The integrated state of viral DNA in SV 40-transformed cells. Proc. nat. Acad. Sci. (Wash.) **60**, 1288 (1968).

SANFORD, K. K.: Spontaneous neoplastic transformation of cells in vitro: some facts and theory. Nat. Cancer Inst. Monogr. **26**, 387 (1967).

SARKAR, S. H., PODDAR, R. K.: Non-conservation of H^3-thymidine label in the DNA of growing yeast cells. Nature (Lond.) **207**, 550 (1965).

SCHATZ, G.: Biogenesis of mitochondria. In: Membranes of mitochondria and chloroplasts, p. 251 (RACKER, E., Ed.). Van Nostrand Reinhold 1970.

SCHIEFER, M. G., HUBNER, G., KLEINSASSER, O.: Riesenmitochondrien aus Onkocyten menschlicher Adenolymphome. Isolierung, morphologische und biochemische Untersuchungen. Virchows Arch. path. Anat. **1**, 230 (1968).

SHOWS, T. B., CHAPMAN, V. M., RUDDLE, F. H.: Mitochondrial malate dehydrogenase and malic enzyme: Mendelian inherited electrophoretic variants in the mouse. Biochem. Genet. **4**, 707 (1970).

SLONIMSKI, P.: Adaptation respiratoire: developpement du système hemoprotéique induit par l'oxygene, p. 242. Proc. Third Intern. Congr. Biochem. Brussels. New York: Academic Press 1956.

SMITH, CH. A., JORDAN, J. M., VINOGRAD, J.: In vivo effects of intercalating drugs on the superhelix density of mitochondrial DNA isolated from human and mouse cells in culture. J. molec. Biol. **59**, 255 (1971).

STEVENS, B. J., MOUSTACCHI, E.: ADN satellite et molecules circulaires torsadées de petite taille chez la levure *Saccharomyces cerevisiae*. Exp. Cell Res. **64**, 259 (1971).

SVOBODA, D. J.: Fine structure of hepatomas induced in rats with p-dimethyl-aminoazobenzene. J. nat. Cancer Inst. **33**, 315 (1964).

SWIFT, H., WOLSTENHOLME, D. R.: Mitochondria and chloroplasts: nucleic acid and the problem of biogenesis (genetic and biology). In: Handbook of molecular cytology, p. 972 (LIMA DE FARIA, A., Ed.). Amsterdam, London: Publishing Company 1969.

SWIFT, H., SINCLAIR, H. H., STEVENS, B. J., RABINOWITZ, M., GROSS, N.: Studies on size characteristics of mitochondrial DNA. In: Biochemical aspects of the biogenesis of mitochondria, p. 71 (SLATER, E. C., TAGLER, J. M., PAPA, S., QUAQLIARIELLO, E., Eds.). Adriatica Ed. Bari 1968.

TAKE, S.: DNA's from human hepatoma and gastric cancer mitochondria. Acta med. Okayama **23**, 465 (1969).

TANDLER, B., SHIPKEY, F. H.: (1) Ultrastructure of Warthin's tumor. I. Mitochondria. J. Ultrastruct. Res. **11**, 292 (1964).

TANDLER, B., SHIPKEY, F. H.: (2) Ultrastructure of Warthin's tumor. II. Crystalloids. J. Ultrastruct. Res. **11**, 306 (1964).

TER SCHEGGET, J., BORST, P.: DNA synthesis by isolated mitochondria. I. Effect of inhibitors and characterization of the product. Biochem. biophys. Acta (Amst.) **246**, 239 (1971).

TER SCHEGGET, J., BORST, P.: DNA synthesis by isolated mitochondria. II. Detection of product DNA hydrogen bonded to closed duplex circles. Biochim. biophys. Acta (Amst.) **246**, 249 (1971).

THOMAS, C. A., Jr., HAMKALO, B. A., MISRA, D. N., LEE, C. S.: Cyclization of eucaryotic deoxyribonucleic acid fragments. J. molec. Biol. **51**, 621 (1970).

TYLER, D. D., GONZE, J.: The preparation of thyroid and thymus mitochondria. In: Methods in enzymology, Vol. X, p. 101 (ESTABROOK, R. W., PULMAN, M. E., Eds.). New York: Academic Press 1967.

VESCO, C., BASILICO, C.: Induction of mitochondrial DNA synthesis by polyoma virus. Nature (Lond.) **229**, 336 (1971).

WAGNER, R. P.: Genetics and phenogenetics of mitochondria. Science **163**, 1026 (1969).

WALLACH, D. F. H.: Cellular membranes and tumor behavior: a new hypothesis. Proc. nat. Acad. Sci. (Wash.) **61**, 868 (1968).

WANG, J. C., DAVIDSON, N.: Thermodynamic and kinetic studies on the interconversion between the linear and circular forms of phage lambda DNA. J. molec. Biol. **15**, 111 (1966).

WARBURG, O.: On the origin of cancer cells. Science **123**, 309 (1956).

WEINTRAUB, M., RAGETLI, H. W. J., JOHN, V. T.: Fine structural changes in isolated mitochondria of healthy and virus-infected *Vicia faba L.* Canad. J. Bot. **44**, 1017 (1966).

WEISLOGEL, P. O., BUTOW, R. A.: Low temperature and chloramphenicol induction of respiratory deficiency in a cold-sensitive mutant of *Saccharomyces cerevisiae*. Proc. nat. Acad. Sci. (Wash.) **67**, 52 (1970).

WEISLOGEL, P. O., BUTOW, R. A.: The fate of mitochondrial membrane proteins and mitochondrial deoxyribonucleic acid during "petite" induction. J. biol. Chem. **246**, 5513 (1971).

WEISS, B., RICHARDSON, C. C.: Enzymatic breakage and joining of deoxyribonucleic acid. I. Repair of single strand breaks in DNA by an enzyme system from *E. coli* infected with T 4 bacteriophage. Proc. nat. Acad. Sci. (Wash.) **57**, 1021 (1967).

WEISSBACH, A., BARTL, P., SALZMAN, L. A.: The structure of replicative lambda DNA. Electron microscopic studies. Cold Spr. Harb. Symp. quant. Biol. **33**, 525 (1968).

WENNER, C. E.: Progress in tumor enzymology (NORD, F. F., Ed.). Advanc. Enzymol. **29**, 321 (1967).

WILLIAMSON, R.: Properties of rapidly labelled deoxyribonucleic acid fragment isolated from the cytoplasm of primary cultures of embryonic mouse liver cells. J. molec. Biol. **51**, 157 (1970).

WILLIAMSON, D. H., MAROUDAS, N. G., WILKIE, D.: Induction of the cytoplasmic petite mutation in *Saccharomyces cerevisiae* by the antibacterial antibiotics erythromycin and chloramphenicol. Mol. gen. Genet. **111**, 209 (1971).

WOLSTENHOLME, D. R., DAWID, I. B.: A size difference between mitochondrial DNA molecules of urodele and anuran amphibia. J. Cell Biol. **39**, 222 (1968).

WOLSTENHOLME, D. R., KOIKE, K., RENGER, H. C.: A. Cellular and molecular mechanisms of carcinogenesis. B. Regulation of gene expression. Proceedings Xth Intern. Cancer Congress. Chicago: Yearbook Medical publishers Inc. Oncology **1** 628 (1970).

WOODCOCK, C. L. F., MORAN, H.: Electron microscopy of DNA conformation in spinach chloroplasts. J. molec. Biol. **31**, 627 (1969).

WUNDERLICH, V., SCHUTT, M., CORAFFI, A.: Über Differenzen im DNS-Gehalt von Mitochondrien aus Tumor- und Normalgeweben. Acta biol. med. germ. **17**, K 27 (1966).

YAMAMOTO, G., ODA, T.: Studies on nucleic acids in Rous sarcoma virus induced mouse ascites sarcoma cells. Distribution and electron microscopy of nucleic acids in subcellular fractions and circular DNA in mitochondrial fraction. Acta med. Okayama **24**, 287 (1970).

YOTSUYANAGI, Y.: Un mode de differenciation de la membrane. mitochondriale évoquant le lysosome bactérien. C. R. Acad. Sci. (Paris) **262**, 1348 (1966)

The manuscript of this review was sent to the editor on July 31, 1971 and received as galley proofs on May 30, 1973. We had therefore not time enough to apply important modifications to our text and bibliography. However, the main conclusions of this work are still valid, despite a two years' delay. The following list of references — although not exhaustive, has been added on the proofs:

1. Replication of mt DNA

ARNBERG, A., VAN BRUGGEN, E. F. J., TER SCHEGGET, J., BORST, P.: The presence of DNA molecules with a displacement loop in standard mitochondrial DNA preparations. Biochim. biophys. Acta (Amst.) **246**, 353 (1971).

FLAVELL, R. A., BORST, P., TER SCHEGGET, J.: DNA synthesis by isolated mitochondria. IV. Isolation of an intermediate containing newly synthesized DNA in full length light strands. Biochim. biophys. Acta (Amst.) **272**, 341 (1972).

KASAMATSU, H., ROBBERSON, D. L., VINOGRAD, J.: A novel closed-circular mitochondrial DNA with properties of a replicating intermediate. Proc. nat. Acad. Sci. (Wash.) **68**, 2252 (1971).

Kasamatsu, H., Vinograd, J.: Unidirectionality of replication in mouse mitochondrial DNA. Nature (Lond.) **241**, 103 (1973).

Robberson, D. L., Clayton, D. A.: Replication of mitochondrial DNA in mouse L cells and their thymidine kinase derivatives; displacement replication on a covalently closed circular template. Proc. nat. Acad. Sci. (Wash.) **69**, 3810 (1972).

Robberson, D. L., Kasamatsu, H., Vinograd, J.: Replication of mitochondrial DNA. Circular replicative intermediates in mouse L cells. Proc. nat. Acad. Sci. (Wash.) **69**, 737 (1972).

Ter Shegget, J., Flavell, R. A., Borst, P.: DNA synthesis by isolated mitochondria. III. Characterization of D-loop DNA, a novel intermediate in mt DNA synthesis. Biochim. biophys. Acta (Amst.) **254**, 1 (1971).

2. mt DNA in cancer cells

Wolstenholme, D. R., Koike, K., Cochran-Fouts, P.: Single strand containing-replicating molecules of circular mitochondrial DNA. J. Cell Biol. **56**, 230 (1973).

Wolstenholme, D. R., McLaren, J. D., Koike, K., Elbine, L.: Catenated oligomeric circular DNA molecules from mitochondria of malignant and normal mouse and rat tissues. J. Cell Biol. **56** 247 (1973).

3. Circular DNA

Bendow, R. M., Eisenberg, M., Sinsheimer, R. L.: Multiple length DNA molecules of bacteriophages ϕ X 174. Nature (Lond.) New Biol. **237**, 141 (1972).

Cohen, S. N., Silver, P., McCoubrey, A. E.: Isolation of catenated forms of R factor DNA from minicells. Nature (Lond.) New Biol. **231**, 249 (1971).

Fuke, M., Inselburg, J.: Electron microscopic studies of replicating and catenated colicin factor El DNA isolated from minicells. Proc. nat. Acad. Sci. (Wash.) **69**, 89 (1972).

Inselburg, J., Fuke, M.: Isolation of catenated and replicating DNA molecules of colicin factor E1 from minicells. Proc. nat. Acad. Sci. (Wash.) **68**, 2839 (1971).

Jarnisch, R., Levine, A.: DNA replication in SV 40-infected cells V. circular and catenated oligomers of SV 40 DNA. Virology **44**, 480 (1971).

Subject Index

Molecular Biology, Biochemistry and Biophysics

Editors: A. Kleinzeller, G. F. Springer, H. G. Wittmann

Volume 1
J. H. van't Hoff:
Imagination in Science
Translated into English with notes and a general introduction by G. F. Springer
1 portrait. VI, 18 pages. 1967. DM 6,60; US $ 2.80
ISBN 3-540-03933-3

Volume 2
K. Freudenberg, A. C. Neish:
Constitution and Biosynthesis of Lignin
10 figs. IX, 129 pages. 1968
Cloth DM 28,—; $ 11.50
ISBN 3-540-04274-1

Volume 3
T. Robinson:
The Biochemistry of Alkaloids
37 figs. X, 149 pages. 1968
Cloth DM 39,—; US $ 16.00
ISBN 3-540-04275-X

Volume 4
A. S. Spirin, L. P. Gavrilova:
The Ribosome
Second edition in preparation

Volume 5
B. Jirgensons:
Optical Activities of Proteins and Other Macromolecules
Second revised and enlarged edition. 71 figs. Approx. 190 pages. 1973 in preparation
ISBN 3-540-06340-4

Volume 6
F. Egami, K. Nakamura
Microbial Ribonucleases
5 figs. IX, 90 pages. 1969
Cloth DM 28,—; US $ 11.50
ISBN 3-540-04657-7

Volume 7
F. Hawkes
Nucleic Acids and Cytology
34 figs. Approx. 288 pages in preparation
ISBN 3-540-05209-7

Volume 8
Protein Sequence Determination
A Sourcebook of Methods and Techniques
Editor: S. B. Needleman
77 figs. XXI, 345 pages. 1970
Cloth DM 84,—; US $ 34.50
ISBN 3-540-05210-0

Volume 9
R. Grubb
The Genetic Markers of Human Immunoglobulins
8 figs. XII, 152 pages. 1970
Cloth DM 42,—; US $ 17.30
ISBN 3-540-05211-9

Volume 10
R. J. Lukens
Chemistry of Fungicidal Action
8 figs. XIII, 136 pages. 1971
Cloth DM 42,—; US $ 17.30
ISBN 3-540-05405-7

Volume 11
P. Reeves
The Bacteriocins
9 figs. XI, 142 pages. 1972
Cloth DM 48,—; US $ 19.70
ISBN 3-540-05735-8

Volume 12
T. Ando, M. Yamasaki, K. Suzuki:
Protamines: Isolation, Characterization, Structure and Function
24 figs.
Approx. 148 pages. 1973
Cloth DM 48,—; US $ 19.70
ISBN 3-540-06221-1

Volume 13
P. Jollès, A. Paraf:
Chemical and Biological Basis of Adjuvants
24 figs. Approx. 156 pages 1973
Cloth DM 48,—; US $ 19.70
ISBN 3-540-06308-0

Volume 14
Micromethods in Molecular Biology
Editor: V. Neuhoff
275 figs. (2 in color)
XV. 428 pages. 1973
Cloth DM 98,—; US $ 40.20
ISBN 3-540- 06319-6

Distribution rights for U.K., Commonwealth, and Traditional British Market (excluding Canada): Chapman & Hall Ltd. London

Prices are subject to change without notice

Springer-Verlag Berlin Heidelberg New York
München · London · Paris
Sydney · Tokyo · Wien

U.C.W. ABERYSTWYTH
LIBRARY